Semiconductor Nanoscale Devices: Materials and Design Challenges

Edited by

Ashish Raman

Department of Electronics and Communication Engineering
Dr. B. R. Ambedkar National Institute of Technology
Jalandhar, Punjab, India

Prabhat Singh

Department of Electronics and Communication Engineering
Dr. B. R. Ambedkar National Institute of Technology
Jalandhar, Punjab, India

Naveen Kumar

Device Modelling Group, James Watt School of Engineering
University of Glasgow, Glasgow
United Kingdom

&

Ravi Ranjan

Tyndall National Institute, Lee Maltings Complex Dyke
Parade, Cork, Cork, Ireland

Semiconductor Nanoscale Devices: Materials and Design Challenges

Editors: Ashish Raman, Prabhat Singh, Naveen Kumar and Ravi Ranjan

ISBN (Online): 978-981-5313-20-8

ISBN (Print): 978-981-5313-21-5

ISBN (Paperback): 978-981-5313-22-2

need for a court order if at any point you breach any terms of this License Agreement. In no event will any delay or failure by Bentham Science Publishers in enforcing your compliance with this License Agreement constitute a waiver of any of its rights.

3. You acknowledge that you have read this License Agreement, and agree to be bound by its terms and conditions. To the extent that any other terms and conditions presented on any website of Bentham Science Publishers conflict with, or are inconsistent with, the terms and conditions set out in this License Agreement, you acknowledge that the terms and conditions set out in this License Agreement shall prevail.

Bentham Science Publishers Pte. Ltd.
80 Robinson Road #02-00
Singapore 068898
Singapore
Email: subscriptions@benthamscience.net

BENTHAM SCIENCE

CONTENTS

FOREWORD i

PREFACE ii

DEDICATION iii

LIST OF CONTRIBUTORS iv

CHAPTER 1 NANOSCALE TECHNOLOGIES: DESIGN CHALLENGES AND ADVANCEMENTS 1
Sumit Srivastava, Abhinav Jaiswal and *Arman Khan*
   INTRODUCTION 2
   FUNDAMENTAL ASPECTS OF NANOSCALE DEVICE DESIGN 4
      Quantum Effects 4
      Material Properties 5
      Fabrication Techniques 6
   CHALLENGES IN NANOSCALE TECHNOLOGIES 8
      Manufacturing Constraints 8
      Reliability Issues 9
      Thermal Management 9
      Interdisciplinary Collaboration 10
   ADVANCEMENTS IN DEVICE DESIGN 10
      Novel Materials 10
      Advanced Fabrication Techniques 10
      Hybrid Integration 10
      Quantum Technologies 11
      3D Printing 11
      Flexible Electronics 11
      Nanophotonics 11
      Neuromorphic Computing 11
      Biomedical Nanotechnology 11
      Energy Harvesting and Storage 12
      Nanoelectromechanical Systems (NEMS) 12
      Nanofluidics 12
      Nanorobotics 12
      Nanoscale Sensing and Imaging 12
      Environmental Applications 13
   INTEGRATION AND INTERCONNECT CHALLENGES 13
      Compatibility 13
      Scalability 14
      Reliability 14
      Interconnect Challenges 15
   DESIGN OPTIMIZATION TECHNIQUES 16
      Simulation Tools 16
      Finite Element Analysis (FEA) 17
      Computational Fluid Dynamics (CFD) 17
      Machine Learning Algorithms 17
         Supervised Learning 17
         Unsupervised Learning 17
         Reinforcement Learning 18
      Design Automation Techniques 18

    Parametric Modeling 18
    Design Space Exploration 18
    Optimization Algorithms 18
  Significance of Design Optimization in Nanoscale Devices 19
APPLICATION ACROSS DIFFERENT FIELDS 19
  Electronics 19
  Photonics 20
  Biotechnology 20
  Energy 21
CONCLUSION 21
KEY POINTS 22
ACKNOWLEDGEMENTS 23
AUTHORS' CONTRIBUTION 23
REFERENCES 24

CHAPTER 2 MATERIALS USED IN THE DESIGN OF SEMICONDUCTOR DEVICES 27
Trinath Talapaneni, Vatsala Chaturvedi and *Ankireddy Narendra*
INTRODUCTION 28
FUNDAMENTALS OF SEMICONDUCTOR MATERIALS 28
  Necessity of Semiconductor Materials 28
  History of Semiconductor Materials 29
  Applications of Semiconductor Materials 30
SEMICONDUCTOR MATERIALS FROM PERIODIC TABLE 32
  Silicon 32
  Germanium 32
  Gray Tin 32
TYPES OF SEMICONDUCTOR MATERIALS BASED ON DOPING 33
  Intrinsic Semiconductors 33
  Extrinsic Semiconductors 33
    N-type Semiconductor 33
    P-type Semiconductor 34
    Compensated Semiconductors 34
    Highly Doped Semiconductors 34
COMPOUND SEMICONDUCTOR MATERIALS 34
  Silicon Carbide 34
    Crystal Structure of SiC 35
    Physical Properties of SiC 36
  Boron Nitride 37
    Crystal Structure of Boron Nitride 37
    Physical Properties of Boron Nitride 38
    Various Methods for Synthesizing Boron Nitride 39
  Red Selenium 41
    Crystal Structure of Red Selenium 41
    Physical Properties of Red Selenium 42
    Various Methods of Synthesizing Red Selenium Semiconductors 42
  Boron Phosphide 43
    Various Methods of Synthesizing Boron Phosphide Semiconductors 45
  Boron Arsenide 46
    Physical Properties of Boron Arsenide 46
    Various Methods of Synthesizing Boron Arsenide Semiconductors 47
  Aluminium Nitride 48

   Physical Properties of Aluminium nitride 48
   Various Methods of Synthesizing Aluminium Nitride Semiconductors 50
  Aluminium Phosphide 51
   Physical Properties of Aluminium Phosphide 51
   Various Methods of Synthesizing Aluminium Phosphide Semiconductors 53
  Aluminium Arsenide 54
   Physical Properties of Aluminium Arsenide 54
   Various Methods for Synthesizing Aluminium Arsenide Semiconductors 55
  Gallium Nitride 56
   Physical Properties of Gallium Nitride 57
   Various Methods for Synthesizing Gallium Nitride Semiconductors 58
  Gallium Phosphide 59
   Physical Properties of Gallium Phosphide 59
   Various Methods for Synthesizing Gallium Phosphide Semiconductors 60
  Gallium Arsenide 61
   Physical Properties of Gallium Arsenide 62
   Various Methods for Synthesizing Gallium Arsenide Semiconductors 63
  Zinc Oxide 64
   Physical Properties of Zinc Oxide 65
   Various Methods for Synthesizing Zinc Oxide Semiconductors 65
  Cadmium Arsenide 67
   Physical Properties of Cadmium Arsenide 67
   Various Methods for Synthesizing Cadmium Arsenide Semiconductors 68
  Zinc Phosphide 69
   Physical Properties of Zinc Phosphide 69
   Various Methods for Synthesizing Zinc Phosphide Semiconductors 70
  Zinc Antimonide 72
   Physical Properties of Zinc Antimonide 72
   Various Methods for Synthesizing Zinc Antimonide Semiconductors 73
 CONCLUSION 75
 KEY POINTS 75
 ACKNOWLEDGEMENTS 75
 REFERENCES 76

**CHAPTER 3 A COMPREHENSIVE OVERVIEW OF THE FOUNDATIONS OF
SEMICONDUCTOR MATERIALS** 80
Agnibha Dasgupta, Soumya Sen, Prabhat Singh and Ashish Raman
 INTRODUCTION 81
  Types of Semiconductor Materials 81
 PROPERTIES OF SEMICONDUCTOR MATERIALS 86
  SILICON 86
  III-V Semiconductor (*e.g.*, GaAs) - Direct Band Gap Material 87
  Heterostructures and Quantum Wells 88
  Vegard's Law for Blending Alloys 89
 APPLICATION OF SEMICONDUCTOR MATERIALS 90
  Silicon Semiconductors 90
   Integrated Circuits 90
   Photovoltaic Solar Cell 91
   Microelectromechanical Systems (MEMS) 93
  III-V Semiconductors 95
   Light Emitting Diodes (LEDs) 95

High Electron Mobility Transistors (HEMTs) 96
SEMICONDUCTOR MATERIALS MARKET 98
CONCLUSION 101
KEY POINTS 102
ACKNOWLEDGMENT 102
REFERENCES 103

CHAPTER 4 INNOVATIVE MATERIALS SHAPING THE FUTURE: A DEEP DIVE INTO THE DESIGN OF SEMICONDUCTOR DEVICES 110
Peeyush Phogat, Shreya, Ranjana Jha and Sukhvir Singh
INTRODUCTION 110
SOLAR CELLS: ENHANCING EFFICIENCY AND DURABILITY 112
OVERVIEW OF SOLAR CELL TECHNOLOGIES 112
LATEST SEMICONDUCTOR MATERIALS IN SOLAR CELL DESIGN 113
INSIGHTS INTO THE EVOLVING LANDSCAPE OF PHOTOVOLTAIC TECHNOLOGIES 119
CAPACITORS AND SUPERCAPACITORS: UNLEASHING ENERGY STORAGE CAPABILITIES 120
    Role of Semiconductor Materials in Energy Storage 120
SCRUTINY OF CAPACITORS AND SUPERCAPACITORS 121
NOVEL SEMICONDUCTOR MATERIALS FOR HEIGHTENED PERFORMANCE AND LONGEVITY 123
THERMOELECTRIC DEVICES: CONVERTING WASTE HEAT TO ELECTRICAL ENERGY 125
    Thermoelectricity and its Applications 125
UNIQUE PROPERTIES OF SEMICONDUCTOR MATERIALS IN THERMOELECTRIC DESIGN 127
    P-type Thermoelectric Materials 127
    N-type Thermoelectric Materials 128
CONVERTING WASTE HEAT INTO VALUABLE ELECTRICAL ENERGY 130
SENSORS: AMPLIFYING SENSITIVITY, SELECTIVITY, AND RESPONSE TIMES 130
    Importance of Semiconductor Materials in Sensor Technologies 130
EXAMINATION OF SENSORS FOR VARIOUS APPLICATIONS 132
SEMICONDUCTOR MATERIALS DESIGNED FOR ENHANCED SENSING CAPABILITIES 133
ELECTROCATALYSIS: SEMICONDUCTOR MATERIALS IN HER AND OER 135
    Overview of Electrocatalysis 135
SEMICONDUCTOR MATERIALS IN WATER SPLITTING 135
CATALYZING REACTIONS FOR SUSTAINABLE ENERGY SOURCES 136
CONCLUSION 138
KEY POINTS 138
ACKNOWLEDGEMENTS 139
REFERENCES 139

CHAPTER 5 MEASUREMENT TECHNIQUES FOR DETERMINING THE THERMAL CONDUCTIVITY OF BULK SAMPLES AND THIN FILMS 155
Simrandeep Kour, Rikky Sharma, Sameena Sulthana and Rupam Mukherjee
INTRODUCTION 156
STEADY-STATE METHOD FOR BULK SAMPLES 157
    Absolute Technique 157
COMPARATIVE TECHNIQUE 160
PARALLEL THERMAL CONDUCTANCE TECHNIQUE 162

TRANSIENT METHOD FOR BULK SAMPLES 163
   The Pulse-power Method 163
   The Laser Flash Technique 165
   3ωTechnique for Thin Films 166
CONCLUSION 168
KEY POINTS 169
ACKNOWLEDGEMENTS 169
REFERENCES 170

CHAPTER 6 STRUCTURAL ANALYSIS OF FEEDBACK FIELD EFFECT TRANSISTOR AND ITS APPLICATIONS 173
Simranjit Singh, Ashish Raman, Ravi Ranjan and *Prabhat Singh*
INTRODUCTION 173
   Principle of Positive Feedback 175
STRUCTURE OF FBFET 176
   Recessed SOI FBFET 177
   Vertical FBFET 179
   Si-SiGe Heterostructure FBFET 181
   Characteristics of FBFET 183
FBFET APPLICATIONS 184
   Logic Device 184
   Memory Cells 184
CONCLUSION 185
KEY POINTS 186
ACKNOWLEDGEMENTS 186
REFERENCES 186

CHAPTER 7 GAN-BASED HIGH ELECTRON MOBILITY TRANSISTOR 193
Nipun Sharma, Ashish Raman and *Ravi Ranjan*
INTRODUCTION 193
WORKING PRINCIPLE 195
   Bound Charge Calculation at the Interface 195
ALGaN/GaN HEMT STRUCTURE 197
   Substrate 197
   Nucleation and Buffer Layer 198
   Channel Layer 198
   Barrier Layer 198
NORMALLY-ON HEMT 199
NORMALLY-OFF HEMT 200
APPLICATIONS 200
NORMALLY OFF TECHNIQUES 201
   Recessed Gate Technology 202
   Thin Barrier Layer 203
   Buried p-region 203
   P-GaN Gate 205
   Fluorine Implantation 207
CONCLUSION 208
KEY POINTS 208
ACKNOWLEDGEMENTS 209
REFERENCES 209

CHAPTER 8 ADVANCED SEMICONDUCTOR SENSING TECHNOLOGIES: MATERIALS AND DESIGN CHALLENGES AT THE NANOSCALE 213
Shreya, Peeyush Phogat, Ranjana Jha and Sukhvir Singh
   INTRODUCTION 214
   FUNDAMENTALS OF SEMICONDUCTOR SENSING 215
      Types of Semiconductor Sensing Technologies 216
         Resistive Sensing 216
         Capacitive Sensing 217
         Optical Sensing 218
         Piezoresistive Sensing 218
         Chemical Sensing 219
         Biomedical Sensing 220
      Quantum Sensitivity in Semiconductor Sensors 221
      Selective Binding Mechanisms 222
      Temporal Dynamics and Response Time 223
   NANOMATERIALS IN SEMICONDUCTOR SENSING 225
      Metal and Noble Metals 226
      Metal Oxide Nanoparticles 227
      Carbon-based Nanomaterials 228
      Transition Metal Dichalcogenides (TMDs) 229
      Polymer and Bio-nanomaterials 231
   FABRICATION AND TESTING OF NANOSCALE SEMICONDUCTOR SENSORS 231
      Architecture and Layout 232
      Fabrication Techniques for Nanoscale Semiconductor Sensing Devices 233
         Nanoscale Thin-film Deposition 233
         Photolithography for Nanoscale Patterning 235
         Self-assembly Methods for Nanostructured Materials 236
      Characterization Techniques 238
         Scanning Electron Microscopy (SEM) 238
         Transmission Electron Microscopy (TEM) 239
         Atomic Force Microscopy 240
         X-ray Photoelectron Spectroscopy (XPS) 241
         Electrical Characterization 242
      Power Consumption and Efficiency 244
      Reliability and Durability Assessment 245
   RECENT DEVELOPMENTS IN SEMICONDUCTOR SENSING TECHNOLOGIES 246
      Applications of Semiconductor Sensors 247
      Challenges and Opportunities 248
   CONCLUSION 251
   KEY POINTS 252
   REFERENCES 252

CHAPTER 9 ENGINEERING TFET BIOSENSORS: DESIGN OPTIMIZATION, ANALYTICAL MODELING, AND RADIATION CONSIDERATIONS 263
Priyanka Goma and Ashwani K. Rana
   INTRODUCTION 264
   DESIGN TECHNIQUES FOR TFET BIOSENSORS 265
      Material Selection 265
      Device Geometry Optimization 266
      Surface Functionalization 266
      Gate Dielectric Engineering 266

Bioreceptor Integration 266
Signal Amplification 266
Noise Reduction 267
Microfluidic Integration 267
On-chip Reference Electrode 267
Packaging and Encapsulation 267
ANALYTICAL MODELING OF THE SURFACE POTENTIAL OF TFET BIOSENSORS FOR DNA DETECTION 267
RADIATION-INDUCED EFFECTS ON TFET 274
IMPACT OF X-RAYS ON THE SENSITIVITY OF TFET BIOSENSOR 276
CONCLUSION 281
KEY POINTS 281
ACKNOWLEDGEMENTS 282
REFERENCES 282

CHAPTER 10 A NEW PARADIGM SHIFT IN THE SEMICONDUCTOR INDUSTRY FOR 6G TECHNOLOGY: A REVIEW 286
Karabi Baruah and *Prachi Gupta*
INTRODUCTION 286
EVOLUTION OF 1G TO 6G 288
First Generation (1G) 288
Second Generation (2G) 289
Third Generation (3G) 289
Fourth Generation (4G) 289
Fifth Generation (5G) 289
Sixth Generation (6G) 290
OUTLINE OF THE 6G MARKET AND PRESENT SITUATION 292
Massive Ultra-reliable Low Latency Communication (URLLC) 294
eMBB (Enhanced Mobile Broadband) 294
Massive eMBB 294
APPLICATIONS OF 6G IN VARIOUS ASPECTS 295
Security 295
Telecommunication 297
Manufacturing Industry 297
Environmental Performance 297
Healthcare 297
Smart Cities 297
Entertainment 298
Education and Transportation 298
Semiconductor Components in 6G 299
Millimeter-wave and Terahertz Transceivers 299
Advanced Semiconductor Components for Signal Processing and Modulation 299
Edge Processing Units and AI Accelerators 299
Power-efficient Processors and Memory 299
Antennas and Beamforming Components 299
Components of Quantum Communication 300
Sensors and Sensor Fusion Chips 300
Security Hardware Modules 300
Photonic Components for Optical Communication 300
Personalized System-on-Chip (SoC) Solutions 300
Impact of 6G Technology on Semiconductor Industry 301

   Increased Demand for Advanced Chips 301
   Innovation in Semiconductor Design 301
   Advanced Materials Development 301
   Higher Frequency Components 301
   Integration of AI and Machine Learning 301
   Security and Privacy Enhancements 302
   Global Economic Impact 302
   Supply Chain Adjustments 302
   Research and Development Investments 302
  Required Properties of Advanced Semiconductor Materials in 6G 302
   High Electron Mobility 302
   Heterogeneous Integration 303
   Wide Bandgap 303
   Low Noise 303
   Quantum Properties 303
   High Thermal Conductivity 303
   Low Power Consumption 303
   Flexibility and Stretchability 304
   Photonics Integration 304
   Compatibility with Fabrication Processes 304
  Emerging Semiconductor Materials in 6G Technology 304
   Gallium Nitride (GaN) 304
   Silicon Carbide (SiC) 304
   2D Materials 305
   Graphene 305
   III-V Compound Semiconductors [e.g., Gallium Arsenide (GaAs), Indium Phosphide (InP)] 305
   Silicon Germanium (SiGe) 305
   Nanowires and Nanotubes 305
   Quantum Dots 305
   Diamond Semiconductors 306
   Organic Semiconductors 306
   Perovskite 306
 CONCLUSION 306
 KEY POINTS 307
 ACKNOWLEDGEMENTS 308
 REFERENCES 308

CHAPTER 11 EXPLORING THE DEPTHS OF SIGMA-DELTA ANALOG-TO DIGITAL CONVERTERS: A COMPREHENSIVE REVIEW 311
Ravita, Ashish Raman and Ramesh K Sunkaria
INTRODUCTION 311
OVERVIEW OF ANALOG-TO-DIGITAL CONVERTERS (ADCs) 312
 Successive Approximation ADCs 313
 Flash ADCs 313
 Sigma-Delta ADCs 313
 Pipeline ADCs 313
NYQUIST RATE ANALOG-TO-DIGITAL CONVERTERS 314
ADC CHARACTERISTICS 314
IMPORTANCE OF SIGMA-DELTA ADCs 316
SIGMA-DELTA MODULATION TECHNIQUE 317

BASIC PRINCIPLE 318
QUANTIZATION 318
OVERSAMPLING AND NOISE SHAPING 320
ARCHITECTURES OF SIGMA-DELTA ADCs 321
    First-order Sigma-delta Converters 322
    Higher-order Sigma-delta Converters 322
    Multi-bit Sigma-delta Converters 324
DESIGN CONSIDERATIONS 324
    Stability Analysis 325
    Non-idealities and Error Sources 325
    Circuit Implementation Techniques 325
    Linearity and Distortion 325
    Power Consumption 325
    Clocking and Timing 325
    Layout and Parasitic Aspects 326
    Technology Scaling 326
OVERSAMPLING TECHNIQUES 326
DIGITAL DECIMATION FILTERS 327
APPLICATIONS OF SIGMA-DELTA ADCs 328
    Communications 328
    Wireless Transceivers 328
    Software-defined Radio (SDR) 328
    Cellular Networks 329
    Broadband Communication Systems 329
    Sensor Interfaces 329
    Medical Devices 329
    Audio Processing 330
    Professional Audio Equipment 330
    Consumer Audio Devices 330
    Audio Interfaces and Sound Cards 330
    High-resolution Audio 330
    Medical Instrumentation 331
    Medical Imaging 331
    Patient Monitoring 331
    Diagnostic Equipment 332
    Research Instruments 332
CONCLUSION 332
KEY POINTS 333
ACKNOWLEDGEMENTS 334
REFERENCES 334

CHAPTER 12 PHOTOVOLTAIC PERFORMANCE ESTIMATION OF THIN FILM
LATERAL PN-JUNCTION SOLAR DEVICES AND COMPREHENSIVE CONSIDERATION
OF PERFORMANCES OF VARIOUS HOMO- AND HETERO-JUNCTION STRUCTURES ... 337
    Yasuhisa Omura
INTRODUCTION 338
DEVICE STRUCTURES ASSUMED 339
THEORETICAL BASE 341
    Preparing the Theoretical Procedure 341
    Additional Model Option to Consider the Hetero-junction 345
    Temperature Dependences of Physical Parameters 347

CALCULATION RESULTS AND DISCUSSION 348

    Calculation Results of Carrier Diffusion from the Metallurgical Junction in Homo-junction

    Devices 348

    Calculated Performance Parameters as a Function of Temperature in Homo-junction

    Devices 349

    Calculation Results of Carrier Diffusion from the Metallurgical Junction in Hetero-junction

    Devices 350

    Performance Comparison of Homo-junction Devices and Hetero-junction Devices 352

    Performance Comparison of Various Lateral Pn-junction Solar Devices 353

    Design Issue of Lateral Pn Junction Film Solar Battery 354

CONCLUSION 355

KEY POINTS 355

REFERENCES 355

APPENDIX 360

    APPENDIX A: SOLVING EQ. (15) 362

    APPENDIX B: POSSIBLE SOLUTIONS FOR EQ. (A10) 364

    APPENDIX C: DERIVATION OF EQUATION (23) 365

    APPENDIX D: FERMI-LEVEL DEFINITIONS OF MATERIALS 365

SUBJECT INDEX 367

FOREWORD

"Semiconductor Nanoscale Devices: Materials and Design Challenges" provides a timely and comprehensive examination of the advancements in this rapidly evolving field. This book addresses the intricate balance between theoretical understanding and application, offering valuable insights into the material properties and design principles that define the behavior of nanoscale devices. The pursuit of smaller, more efficient transistors has led to the exploration of novel materials and innovative design structures. Nano-FET devices, with their potential to operate at low power and high frequencies, exemplify the kind of breakthroughs that are possible. However, the transition from conventional technologies to nanotechnologies brings with it a host of challenges that must be understood and overcome.

The editors and authors have meticulously compiled a body of work that not only charts the current landscape of semiconductor nanoscale devices, but also points towards future directions and possibilities. Their collective expertise and dedication to advancing knowledge in this domain are evident throughout the chapters. "Semiconductor Nanoscale Devices: Materials and Design Challenges" is more than just a textbook; it is a guide to the future of semiconductor technology. As we stand on the brink of new technological horizons, this book will undoubtedly serve as a critical reference for those striving to push the boundaries of what is possible in VLSI design.

This book is an essential resource for engineers, researchers, and students who are navigating the complexities of nanoscale device technology. It bridges the gap between foundational concepts and advanced research, making it accessible to those new to the field, while also providing depth for experienced practitioners. The detailed exploration of quantum effects, scaling issues, and material properties offers a robust framework for understanding and innovating in nanoscale device design.

I am confident that readers will find this book an invaluable addition to their professional libraries, providing both the inspiration and the knowledge necessary to drive forward the next generation of semiconductor devices.

Dharmendra Singh Yadav
Department of Electronics and Communication Engineering
National Institute of Technology
Kurukshetra, Haryana
India

PREFACE

The relentless miniaturization in semiconductor technology has paved the way for nanoscale devices to become pivotal components in modern electronic systems. These advancements have brought about unprecedented opportunities and challenges, especially in materials selection and device design. "Semiconductor Nanoscale Devices: Materials and Design Challenges" aims to provide a comprehensive exploration of these cutting-edge technologies, offering insights into both the theoretical foundations and practical implementations.

As the VLSI industry continues to evolve, the reduction in transistor size has been instrumental in integrating more functionality onto silicon wafers and minimizing power consumption. This progress has led to the realization of nano-FET devices using various innovative materials and structures, demonstrating significant potential for low-power and high-frequency applications. The continuous pursuit of enhancing performance while addressing the complexities of nanoscale phenomena underscores the importance of a comprehensive guide to these advancements.

"Semiconductor Nanoscale Devices: Materials and Design Challenges" serves as a concise benchmark for beginners and experienced practitioners alike. It is tailored for those who are just getting started with nanoscale device technology and for those looking to design integrated circuits using novel FET devices. This book aims to be a valuable resource, inspiring new discoveries, innovations, and advancements at the forefront of electronic engineering. We hope that this book will serve as a guide and inspiration for researchers, engineers, and students, unlocking the potential of nanoscale semiconductor devices and contributing to the continuous evolution of electronic technology.

Ashish Raman
Department of Electronics and Communication Engineering
Dr. B. R. Ambedkar National Institute of Technology
Jalandhar, Punjab, India

Prabhat Singh
Department of Electronics and Communication Engineering
Dr. B. R. Ambedkar National Institute of Technology
Jalandhar, Punjab, India

Naveen Kumar
Device Modelling Group
James Watt School of Engineering
University of Glasgow, Glasgow
United Kingdom

&

Ravi Ranjan
Tyndall National Institute
Lee Maltings Complex Dyke Parade, Cork
Cork, Ireland

DEDICATION

I dedicate this book to my beloved mother, Veena Saxena, my father, R. R. Saxena, my wife, Deepti Saxena, and my daughter, Arshika Saxena. Their unwavering love and support have been the cornerstone of my success. I am deeply grateful for everything they have done and continue to do for me. This book is a testament to their belief in me and a symbol of my heartfelt appreciation.

— Dr. Ashish Raman

I dedicate this book to my loving mother Shakuntala Singh, father Dinesh Singh, and brother Prasoon Singh, as a token of my appreciation for everything they have done and continue to do for me. Their love and support are the foundation of my success, and I am blessed to have them as a part of my life. Their love and belief in me mean everything. This book is dedicated to them as a symbol of my gratitude for all they have done.

— Dr. Prabhat Singh

This book is dedicated to my beloved wife, Nisha Chaudhary, and our wonderful son, Zishaan Kumar, whose love and encouragement have been my greatest motivation. Special thanks to my parents, Smt. Indu Devi, and Shri. Rajpal Singh, for their unwavering support and sacrifices, and to my brother, Nitish Kumar, for always being there for me. I am deeply grateful to my mentors, Dr. Ashish Raman and Prof. Vihar Georgiev, for their invaluable guidance and wisdom. I also extend my heartfelt appreciation to my supervisors, colleagues, and friends, whose support and encouragement have been instrumental in the completion of this work.

— Dr. Naveen Kumar

I want to begin by thanking my entire family for their unwavering support and encouragement throughout this journey. My mother Kusum Kumari and father Raj Nandan, their constant faith in me gave me the courage. Their wisdom and guidance have been invaluable, and I am forever grateful. I dedicate this book to my spouse, Chandni Kumari, son Reyansh Raj, and daughter Reeva Raj as their patience and understanding have been my bedrock. I thank them for believing in me and giving me the time.

— Dr. Ravi Ranjan

List of Contributors

Abhinav Jaiswal	Department of Electronics and Communication, M.J.P Rohilkhand University, Bareilly, Uttar Pradesh, India
Arman Khan	Department of Electronics and Communication, M.J.P Rohilkhand University, Bareilly, Uttar Pradesh, India
Ankireddy Narendra	Department of Electrical Engineering, OP Jindal University, Raigarh, 496109, Chhattisgarh, India
Agnibha Dasgupta	GE Vernova T&D India Limited, GE VERNOVA; Services Specialist - DIG Grid Support, New Delhi, India
Ashish Raman	Department of Electronics and Communication Engineering, Dr. B. R. Ambedkar National Institute of Technology, Jalandhar, Punjab, India
Ashwani K. Rana	Department of Electronics and Communication Engineering, National Institute of Technology, Hamirpur, Himachal Pradesh, India
Karabi Baruah	SOET, CMR University, Lakeside Campus, Bangalore, India
Nipun Sharma	Dr. B. R. Ambedkar National Institute of Technology, Jalandhar, Punjab, India
Prabhat Singh	Department of Electronics and Communication Engineering, Dr. B. R. Ambedkar National Institute of Technology, Jalandhar, Punjab, India
Peeyush Phogat	Research Lab for Energy Systems, Department of Physics, Netaji Subhas University of Technology, New Delhi, India
Priyanka Goma	Department of Electronics and Communication Engineering, National Institute of Technology, Hamirpur, Himachal Pradesh, India
Prachi Gupta	SOET, CMR University, Lakeside Campus, Bangalore, India
Ranjana Jha	Research Lab for Energy Systems, Department of Physics, Netaji Subhas University of Technology, New Delhi, India
Rupam Mukherjee	Department of Physics, Presidency University, Bangalore, Karnataka, 560064, India
Ravi Ranjan	Tyndall National Institute, Lee Maltings Complex Dyke Parade, Cork, Cork, Ireland
Rikky Sharma	Department of Physics, Lovely Professional University, Phagwara, Punjab, 144001, India
Ramesh K Sunkaria	Dr. B. R. Ambedkar National Institute of Technology, Jalandhar, Punjab, India
Ravita	Dr. B. R. Ambedkar National Institute of Technology, Jalandhar, Punjab, India
Sumit Srivastava	Department of Electronics and Communication, M.J.P Rohilkhand University, Bareilly, Uttar Pradesh, India
Soumya Sen	University of Engineering and Management, Jaipur, Rajasthan, India
Shreya	Research Lab for Energy Systems, Department of Physics, Netaji Subhas University of Technology, New Delhi, India
Sukhvir Singh	Research Lab for Energy Systems, Department of Physics, Netaji Subhas University of Technology, New Delhi, India

Simrandeep Kour	Department of Physics, Lovely Professional University, Phagwara, Punjab, 144001, India
Sameena Sulthana	Department of Physics, Presidency University, Bangalore, Karnataka, 560064, India
Simranjit Singh	Dr. B. R. Ambedkar National Institute of Technology, Jalandhar, Punjab, India
Trinath Talapaneni	Department of Metallurgical Engineering, OP Jindal University, Raigarh, 496109, Chhattisgarh, India
Vatsala Chaturved	Department of Metallurgical Engineering, OP Jindal University, Raigarh, 496109, Chhattisgarh, India
Yasuhisa Omura	ORDIST, Kansai University, Suita, Osaka 564-8680, Japan

Semiconductor Nanoscale Devices, 2025, 1-26

CHAPTER 1

Nanoscale Technologies: Design Challenges and Advancements

Sumit Srivastava¹, Abhinav Jaiswal[1,*] and Arman Khan[1,*]

¹ Department of Electronics and Communication, M.J.P Rohilkhand University, Bareilly, Uttar Pradesh, India

Abstract: This chapter delves into nanoscale technologies within semiconductor devices, covering design principles, challenges, and recent advancements. It examines the fundamental aspects of nanoscale device design, addressing key challenges and highlighting the latest developments in the field. The chapter navigates integration and interconnect challenges, design optimization techniques, and diverse applications across various fields. Nanoscale technologies, fundamental to semiconductor innovation, offer a spectrum of opportunities and hurdles. By addressing design intricacies and technological barriers, researchers aim to unlock the full potential of nanoscale devices. Additionally, the chapter discusses optimization strategies to enhance device performance and functionality. It sheds light on the intricate interplay between nanoscale technologies and their applications in electronics, photonics, and biotechnology. By comprehensively examining design methodologies and real-world applications, this chapter provides valuable insights into the evolving landscape of nanoscale technologies within the semiconductor domain. Focusing on recent advancements, the chapter explores how these technologies are integrated into current semiconductor devices and the challenges associated with their implementation. It also highlights the importance of continuous research and development to overcome existing technological barriers. The discussion extends to various design optimization techniques aimed at improving device efficiency, reliability, and overall performance. Overall, this chapter serves as a comprehensive guide to understanding the complexities and innovations of nanoscale technologies in semiconductor devices, offering readers an in-depth look at the design principles, challenges, and advancements shaping the future of this critical field.

Keywords: Applications, Advancements, Challenges, Design principles, Design optimization, Integration, Interconnect, Nanoscale technologies.

---

* **Corresponding authors Abhinav Jaiswal and Arman Khan:** Department of Electronics and Communication, M.J.P Rohilkhand University, Bareilly, Uttar Pradesh, India; E-mails: mrjaisabhi.14@gmail.com and armankhan032002@gmail.com

Ashish Raman, Prabhat Singh, Naveen Kumar & Ravi Ranjan (Eds.)

INTRODUCTION

In the realm of modern engineering, nanoscale technologies have emerged as a transformative force, ushering in a new era of innovation and exploration. These technologies operate at the molecular and atomic levels, offering unparalleled opportunities to manipulate matter with unprecedented precision [1 - 4].

The chapter at hand embarks on a comprehensive journey through the landscape of nanoscale technologies, aiming to provide a thorough understanding of their design principles, challenges, and recent advancements [5].

At the core of nanoscale technologies lies a profound understanding of the fundamental principles governing matter at the nanoscale. Designing devices at such minute scales requires a deep appreciation of quantum mechanics, material properties, and advanced fabrication techniques. Quantum effects, such as tunneling and confinement, become increasingly pronounced at the nanoscale, necessitating a departure from classical design methodologies. Moreover, the choice of materials and fabrication processes plays a pivotal role in determining the performance and functionality of nanoscale devices [6 - 9].

Despite their immense potential, nanoscale technologies are not without their challenges. Manufacturing constraints, reliability issues, and thermal management pose significant hurdles in the development and deployment of nanoscale devices. The intricacies of nanoscale phenomena demand innovative solutions and interdisciplinary collaboration to overcome these challenges effectively. Furthermore, the scaling laws that govern traditional engineering principles often break down at the nanoscale, necessitating novel approaches to device design and optimization [10].

Recent years have witnessed remarkable advancements in nanoscale device design, driven by breakthroughs in materials science, fabrication techniques, and device architectures. These advancements have unlocked new possibilities for nanoscale devices, enabling improvements in performance, efficiency, and functionality. From novel materials with tailored properties to innovative device architectures with enhanced performance characteristics, the field of nanoscale technologies is evolving at an unprecedented pace [11 - 14].

Integration and interconnect challenges pose additional complexities in the realization of nanoscale devices. As devices shrink to ever-smaller dimensions, the compatibility, scalability, and reliability of integration processes become increasingly critical [15]. Interconnect challenges, including signal propagation delay and cross-talk, further compound the integration process, requiring sophisticated solutions to ensure seamless operation of nanoscale systems [16].

In the quest for optimal device performance, design optimization techniques play a crucial role. Simulation tools, machine learning algorithms, and design automation techniques enable engineers to explore vast design spaces and identify optimal solutions efficiently [17]. These techniques empower designers to achieve optimal device performance while minimizing development time and cost, accelerating the pace of innovation in nanoscale technologies [18].

Beyond the realm of fundamental research and development, nanoscale technologies have found applications across diverse fields, including electronics, photonics, biotechnology, and energy. From ultra-efficient electronics and high-performance sensors to advanced drug delivery systems and renewable energy technologies, the potential applications of nanoscale technologies are vast and varied [19 - 22]. By harnessing the unique properties of nanomaterials and devices, researchers and engineers are pushing the boundaries of what is possible in fields ranging from healthcare to environmental sustainability [23].

Fig. (1) shows the nanoscale MOSFET (Metal-oxide-semiconductor Field-effect Transistor), a key semiconductor device for signal amplification and switching at nanometer scales. Its design involves miniaturization to enhance performance, but faces challenges, like quantum effects, leakage currents, and manufacturing precision, requiring innovative solutions for efficient operation and reliability.

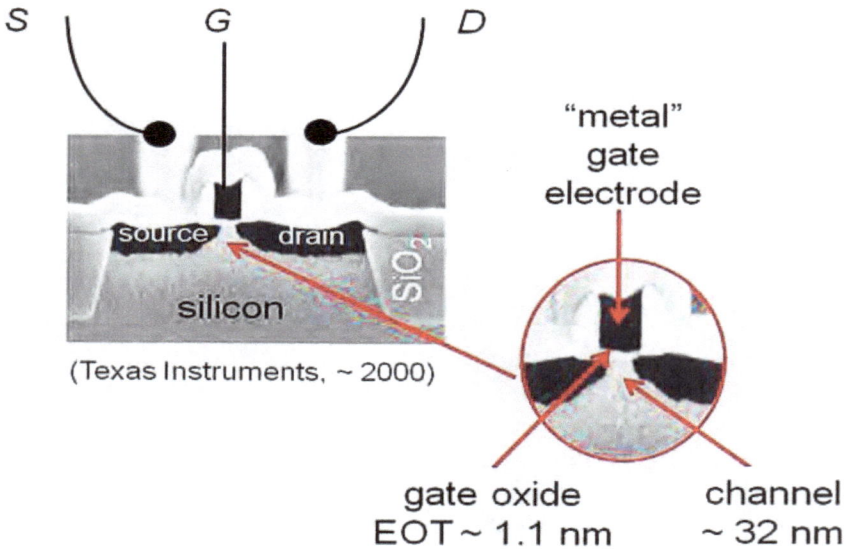

Fig. (1). Nanoscale MOSFET.

In conclusion, nanoscale technologies represent a paradigm shift in modern engineering, offering unprecedented opportunities for innovation and advancement. By delving into their design principles, addressing key challenges, and highlighting recent advancements, this chapter aims to provide a comprehensive overview of nanoscale technologies.

FUNDAMENTAL ASPECTS OF NANOSCALE DEVICE DESIGN

Nanoscale technologies have revolutionized the field of engineering by enabling the manipulation of matter at the atomic and molecular levels. At the heart of these technologies lies the intricate design of devices, where quantum effects, material properties, and fabrication techniques converge to shape their functionality and performance. In this comprehensive exploration, we delve into the fundamental aspects of nanoscale device design, elucidating the principles that underpin their creation and the critical considerations in their development [24 - 27].

Quantum Effects

Quantum mechanics governs the behavior of particles at the nanoscale, leading to phenomena that defy classical intuition. Tunneling and confinement are two key quantum effects that play a pivotal role in nanoscale device design. Tunneling refers to the quantum mechanical phenomenon where particles traverse energy barriers that would be insurmountable in classical physics. In nanoscale electronic devices, such as tunneling transistors, tunneling enables the flow of charge carriers through thin barriers, facilitating low-power operation and high-speed switching. However, excessive tunneling can lead to leakage currents, compromising device performance and reliability. Understanding and controlling tunneling processes are crucial for optimizing device performance, while minimizing undesirable effects [28 - 32].

In Fig. (2), it can be seen that the quantum interference in transistors arises due to the electron wave nature at nanoscales, impacting device performance. As transistors shrink, electron paths become comparable to device dimensions, leading to interference effects. This challenges precise control of electron flow, causing fluctuations in device behavior and compromising reliability, demanding novel design strategies for stable operation.

Confinement arises when particles are restricted to nanoscale dimensions, leading to discrete energy levels and size-dependent properties. Semiconductor quantum dots exemplify confinement, with electrons and holes confined within a three-dimensional potential well. By tuning the size of quantum dots, engineers can precisely control their electronic and optical properties, making them valuable components in optoelectronic devices and quantum computing applications [33].

Material Properties

The choice of materials is fundamental to nanoscale device design, as nanomaterials exhibit unique properties that differ from their bulk counterparts. Quantum confinement, surface effects, and enhanced surface-to-volume ratios contribute to the distinctive behavior of nanomaterials, offering opportunities for enhancing device functionality and performance [34].

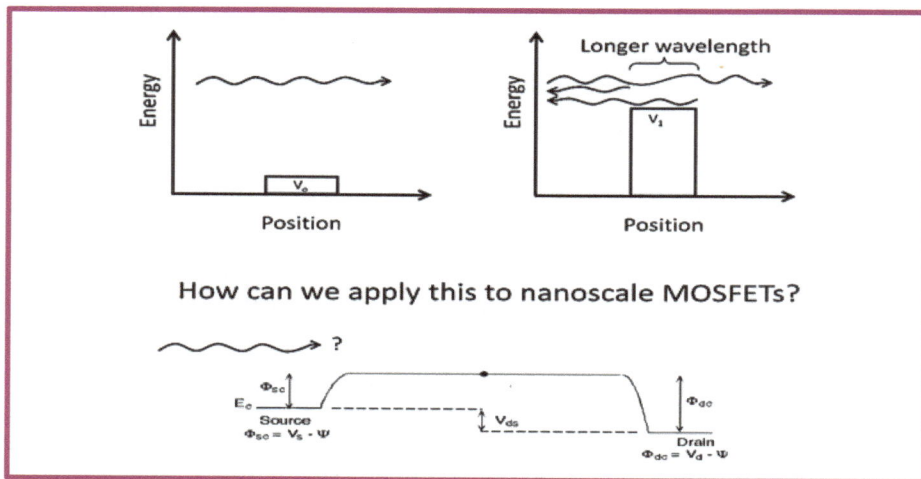

Fig. (2). Quantum interference faced in transistors.

Carbon-based nanomaterials, such as Carbon Nanotubes (CNTs) and graphene, exemplify the remarkable properties of nanomaterials. CNTs, with their exceptional mechanical strength, electrical conductivity, and thermal properties, hold promise for applications in flexible electronics, high-performance composites, and Nanoelectromechanical Systems (NEMS). Graphene, a single layer of carbon atoms arranged in a two-dimensional honeycomb lattice, exhibits extraordinary electrical conductivity, optical transparency, and mechanical flexibility, making it an ideal candidate for next-generation electronics, sensors, and energy storage devices [35, 36].

Fig. (3) provides the schematic representation of nanoscale devices, like h-BN (hexagonal Boron Nitride), TMDC (Transition Metal Dichalcogenides), black phosphorus, and graphene, showcasing diverse materials for novel electronic applications. Designing nanoscale devices with these materials involves harnessing unique properties while overcoming challenges, such as fabrication precision, interface engineering, and scalability, to enable practical implementation.

In addition to carbon-based materials, semiconductor nanomaterials, such as quantum dots, nanowires, and two-dimensional Transition Metal Dichalcogenides (TMDs), offer unique opportunities for device design and integration. Quantum dots, with their size-tunable bandgap and high photoluminescence quantum yield, find applications in Light-emitting Diodes (LEDs), displays, and biological imaging. Nanowires provide a versatile platform for the realization of nanoscale transistors, sensors, and photodetectors, owing to their high surface-to-volume ratio and unique electronic properties. TMDs, such as molybdenum disulfide ($MoS2$) and tungsten diselenide ($WSe2$), exhibit layer-dependent electronic properties, making them attractive candidates for Field-effect Transistors (FETs), photodetectors, and optoelectronic devices [37 - 39].

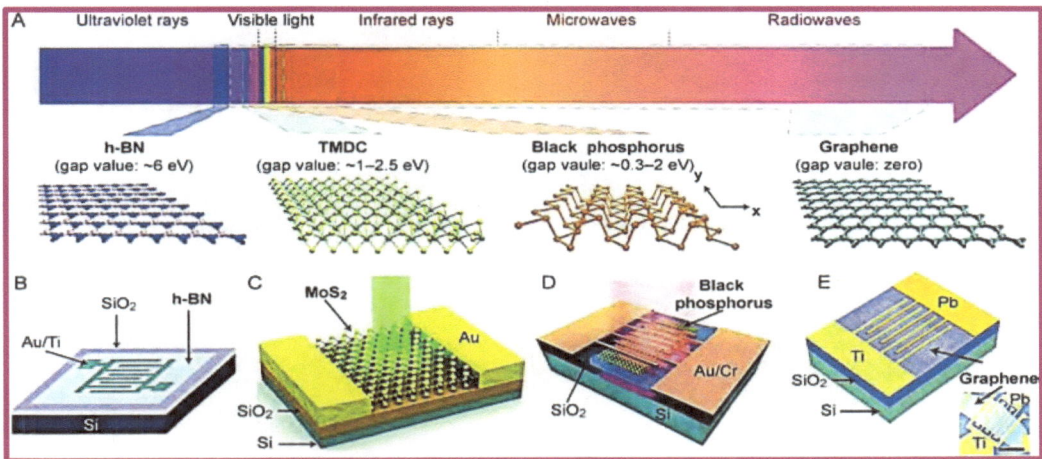

Fig. (3). Schematic representation of various nanoscale devices.

Fabrication Techniques

Fabricating nanoscale devices with precision and reproducibility requires advanced fabrication techniques capable of achieving sub-nanometer resolution and high throughput. Traditional lithographic techniques, such as photolithography and electron beam lithography, are limited by the diffraction of light and the resolution of patterning tools, making them unsuitable for nanoscale device fabrication [40].

To overcome these limitations, researchers have developed alternative fabrication techniques, such as nanoimprint lithography, self-assembly, and Molecular Beam Epitaxy (MBE) [40]. Fig. (**4**) elaborates that Chemical Vapor Deposition (CVD) for Carbon Nanotubes (CNTs) involves synthesizing CNTs by decomposing carbon-containing gases on a substrate. In nanoscale device design, CVD offers precise control over CNT growth for tailored properties. Challenges include achieving uniformity, controlling chirality, and minimizing defects, crucial for reliable integration into electronic devices with desired performance.

Nanoimprint lithography involves the mechanical deformation of a resist layer to create nanoscale patterns on a substrate, offering high resolution, scalability, and cost-effectiveness. Self-assembly techniques leverage molecular interactions or external fields to spontaneously organize molecules or nanoparticles into ordered structures, enabling the creation of nanoscale patterns and devices with nanometer-scale precision. Molecular beam epitaxy enables the precise deposition of atomic layers onto substrates, allowing for the growth of epitaxial thin films with atomic-scale control, making it suitable for semiconductor heterostructures and quantum well devices [41 - 45].

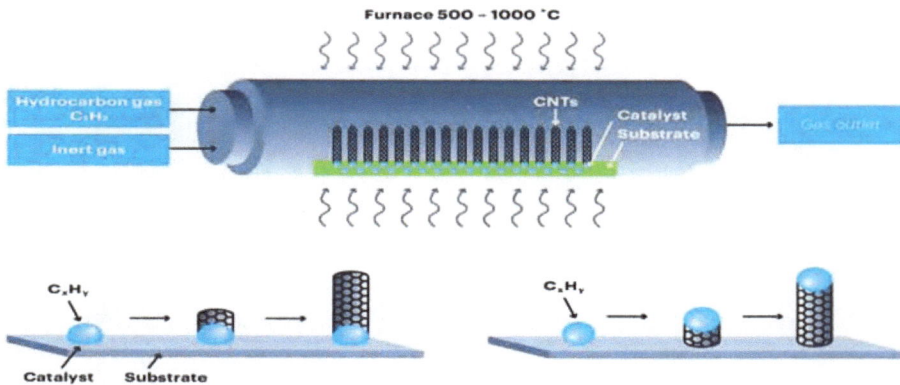

Fig. (4). Chemical Vapour Deposition (CVD) for Carbon Nanotubes (CNTs).

Fig. (**5**) provides the schematic representation of CNT-NEMS (Carbon Nanotube Nano-electro-mechanical Systems) devices fabrication, illustrating the process of integrating carbon nanotubes into nanoscale electromechanical systems. Designing such devices involves precise manipulation and assembly of CNTs to construct functional components, like sensors or actuators. Challenges include achieving controlled alignment, ensuring uniformity, and minimizing defects for optimal device performance. By leveraging these advanced fabrication techniques, engineers can realize complex nanoscale devices with unprecedented precision and functionality. However, the choice of fabrication method must be carefully tailored to the specific requirements of the device, considering factors, such as scalability, cost, and material compatibility [42].

CHALLENGES IN NANOSCALE TECHNOLOGIES

Nanoscale technologies represent a revolutionary approach to engineering, offering unparalleled opportunities for innovation and advancement. However, despite their promise, nanoscale technologies encounter significant challenges that impede their widespread adoption and commercialization. Nanoscale technologies hold immense potential for transforming various industries, but their widespread adoption is hindered by significant challenges. Manufacturing constraints, reliability issues, and thermal management pose obstacles to the development and deployment of nanoscale devices [45]. However, through innovative solutions and interdisciplinary collaboration, researchers can overcome these challenges and unlock the full potential of nanotechnology. By addressing these key challenges, nanoscale technologies can revolutionize fields, such as electronics, healthcare, energy, and environmental monitoring, paving the way for a more sustainable and technologically advanced future [46]. In this section, we delve into the key challenges facing nanoscale technologies, ranging from manufacturing constraints to reliability issues and thermal management, and explore the innovative solutions and interdisciplinary collaboration needed to overcome these obstacles.

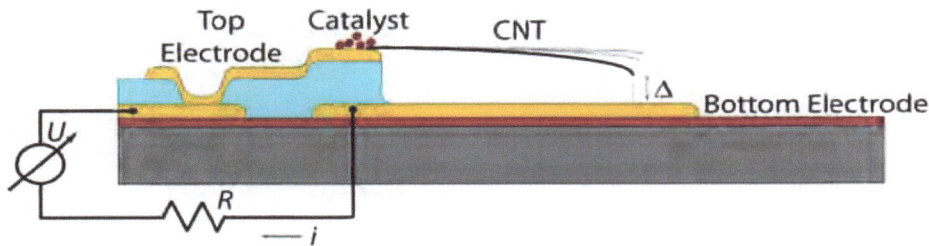

Fig. (5). Schematic representation of CNT-NEMS devices fabrication.

Manufacturing Constraints

One of the primary challenges in nanoscale technologies is the scalability and reproducibility of manufacturing processes. While laboratory-scale fabrication techniques may yield promising results, scaling up production to industrial levels presents numerous hurdles [47]. Traditional lithographic methods, such as photolithography and electron beam lithography, encounter limitations in achieving sub-nanometer resolution and high throughput. Additionally, the cost-effectiveness of manufacturing nanoscale devices in large quantities remains a significant concern [41].

To address these challenges, researchers are exploring alternative fabrication techniques that offer scalability, cost-effectiveness, and precise control over nanoscale features. Nanoimprint lithography, for example, enables the mass production of nanoscale patterns by mechanically deforming a resist layer on a substrate. Self-assembly techniques, such as block copolymer lithography and DNA origami, leverage molecular interactions to organize nanoparticles into ordered structures with nanometer-scale precision. By developing innovative manufacturing processes tailored to nanoscale devices, researchers can overcome scalability challenges and accelerate the commercialization of nanotechnology [38].

Reliability Issues

Another critical challenge in nanoscale technologies is ensuring the reliability and longevity of nanoscale devices under various operating conditions. As devices shrink to nanoscale dimensions, they become more susceptible to defects, environmental influences, and reliability degradation mechanisms. For example, in nanoscale electronic devices, such as transistors and memory cells, defects in the semiconductor material or at interfaces can lead to performance degradation and device failure over time [34].

To mitigate reliability issues, researchers are developing robust design methodologies and novel materials that exhibit enhanced stability and durability at the nanoscale. Advanced characterization techniques, such as Scanning Transmission Electron Microscopy (STEM) and Atomic Force Microscopy (AFM), enable the visualization and analysis of nanoscale defects and degradation mechanisms [34]. Furthermore, the integration of self-healing materials and redundant pathways can enhance the resilience of nanoscale devices against environmental stressors and wear-out mechanisms [37].

Thermal Management

Thermal management presents a significant challenge in nanoscale technologies, as miniaturization leads to increased power densities and localized heating effects. In nanoscale electronic devices, such as integrated circuits and microprocessors, heat dissipation becomes more challenging due to limited space for thermal dissipation and reduced thermal conductivity of nanomaterials [23, 38].

To address thermal management challenges, researchers are exploring innovative cooling solutions and materials with enhanced thermal properties. Microfluidic cooling systems, employing liquid coolant channels integrated into the device substrate, offer efficient heat removal and temperature regulation at the nanoscale. Additionally, nanomaterials, such as carbon nanotubes and graphene, exhibit high thermal conductivity, making them promising candidates for thermal interface materials and heat spreaders in nanoscale devices [41].

Interdisciplinary Collaboration

Addressing the key challenges in nanoscale technologies requires interdisciplinary collaboration between researchers from diverse fields, including materials science, electrical engineering, physics, and chemistry. By leveraging expertise from multiple disciplines, researchers can develop holistic solutions to complex problems and accelerate technological advancements in nanotechnology [27, 42].

ADVANCEMENTS IN DEVICE DESIGN

Recent advancements in nanoscale device design have propelled the field forward, offering new opportunities for innovation and discovery. Breakthroughs in materials science, fabrication techniques, and device architectures have enabled the development of next-generation nanoscale devices with enhanced performance, functionality, and versatility. From novel materials and advanced fabrication techniques to hybrid integration and quantum technologies, these advancements pave the way for transformative applications in electronics, photonics, biomedicine, and beyond. As researchers continue to push the boundaries of nanoscale device design, we can expect further breakthroughs that will shape the future of technology and society [23, 45].

Novel Materials

Advancements in materials science have led to the discovery and development of new nanomaterials with unique properties and functionalities. Examples include two-dimensional materials, like graphene, Transition Metal Dichalcogenides (TMDs), and perovskite nanocrystals, which offer enhanced electrical, optical, and mechanical properties for various applications in nanoscale devices [42].

Advanced Fabrication Techniques

Emerging fabrication techniques, such as nanoimprint lithography, directed self-assembly, and atomic layer deposition, enable precise control over nanoscale features and patterns, facilitating the mass production of nanoscale devices with sub-nanometer resolution and high throughput [37].

Hybrid Integration

Hybrid integration techniques combine disparate materials and components, such as organic and inorganic materials, to create multifunctional nanoscale devices with enhanced performance and versatility. This approach enables the development of flexible electronics, optoelectronic devices, and biomedical sensors [37].

Quantum Technologies

Advancements in quantum technologies, including quantum dots, superconducting qubits, and topological materials, have unlocked new possibilities for nanoscale device design. These technologies enable the realization of quantum computers, quantum communication networks, and quantum sensors with unprecedented capabilities [23, 34, 40].

3D Printing

Additive manufacturing techniques, such as 3D printing, allow for the rapid prototyping and customization of nanoscale devices. By layering materials with nanometer precision, 3D printing enables the creation of complex nanoscale structures and devices with tailored functionalities [30].

Flexible Electronics

Flexible and stretchable electronics based on organic semiconductors, carbon nanotubes, and other nanomaterials, offer new opportunities for wearable devices, smart textiles, and biomedical implants. These advancements in flexible electronics enable conformal integration onto curved surfaces and flexible substrates [32].

Nanophotonics

Nanophotonic devices, such as plasmonic nanoparticles, photonic crystals, and metasurfaces, manipulate light at the nanoscale for applications in sensing, imaging, and telecommunications. These devices offer enhanced light-matter interactions, enabling miniaturized optical components and high-speed data transmission [29, 39].

Neuromorphic Computing

Nanoscale devices inspired by the structure and function of the human brain, known as neuromorphic devices, hold promise for energy-efficient computing and artificial intelligence applications. Memristive devices, phase-change materials, and spintronics enable the realization of neuromorphic computing architectures with synaptic plasticity and parallel processing capabilities [12, 21].

Biomedical Nanotechnology

Advancements in biomedical nanotechnology, including nanomedicine, drug delivery systems, and biosensors, offer new approaches to disease diagnosis, treatment, and monitoring. Nanoparticle-based drug carriers, biofunctionalized nanoparticles, and nanoscale imaging probes enable targeted delivery of therapeutics and real-time monitoring of biological processes [14, 19].

Energy Harvesting and Storage

Nanoscale devices for energy harvesting and storage, such as nanostructured photovoltaics, thermoelectric generators, and supercapacitors, enable efficient conversion and storage of renewable energy sources. These advancements contribute to the development of sustainable energy solutions and portable power sources for electronics and IoT devices [32].

Nanoelectromechanical Systems (NEMS)

Advancements in Nanoelectromechanical Systems (NEMS) have enabled the development of highly sensitive sensors, resonators, and actuators at the nanoscale. These devices leverage the mechanical properties of nanomaterials, such as carbon nanotubes and nanowires, to detect and manipulate signals with exceptional precision, paving the way for applications in healthcare, environmental monitoring, and telecommunications [23].

Nanofluidics

Nanofluidic devices manipulate fluids at the nanoscale, offering precise control over fluid flow, mixing, and confinement. These devices enable applications in drug delivery, DNA sequencing, and chemical analysis, where small sample volumes and high-throughput processing are critical. Advancements in nanofluidic fabrication techniques and surface modification strategies enhance the performance and functionality of nanofluidic devices for various applications [12, 38].

Nanorobotics

Nanorobotics involves the design and manipulation of nanoscale robots or machines capable of performing tasks with nanometer precision. These nanorobots hold promise for applications in targeted drug delivery, tissue engineering, and nanoscale assembly. Recent advancements in nanorobotics include the development of DNA nanorobots, self-propelled nanomotors, and nanoparticle-based drug carriers, enabling precise control over nanoscale processes and interactions [39, 40].

Nanoscale Sensing and Imaging

Advancements in nanoscale sensing and imaging techniques enable the visualization and characterization of nanomaterials and nanodevices with unprecedented resolution and sensitivity. Scanning probe microscopy, super-resolution microscopy, and single-molecule imaging techniques offer insights into the behavior and properties of nanoscale structures, facilitating advancements in materials science, biology, and nanotechnology [10, 17].

Environmental Applications

Nanoscale technologies hold promise for addressing environmental challenges, including pollution remediation, water purification, and renewable energy generation. Nanomaterials, such as carbon nanotubes, graphene, and nanocatalysts, enable efficient removal of contaminants from air and water, while nanostructured photovoltaics and energy storage devices contribute to the development of sustainable energy solutions. Additionally, nanoscale sensors and monitoring devices facilitate real-time environmental monitoring and detection of pollutants, supporting efforts towards environmental sustainability and conservation [12, 45].

INTEGRATION AND INTERCONNECT CHALLENGES

Integrating nanoscale devices into larger systems presents numerous challenges related to compatibility, scalability, reliability, and interconnectivity. Addressing these challenges requires interdisciplinary collaboration and innovative solutions spanning materials science, fabrication techniques, design methodologies, and testing protocols. By overcoming these obstacles, researchers and engineers can unlock the full potential of nanoscale technologies and realize their transformative impact across various fields, from electronics and healthcare to energy and environmental monitoring [14, 41].

In this section, we delve into the integration and interconnect challenges facing nanoscale technologies and discuss strategies for overcoming them [21].

Compatibility

One of the primary challenges in integrating nanoscale devices is ensuring compatibility with existing materials, processes, and systems. Nanomaterials often exhibit unique properties that may differ from conventional materials, requiring careful consideration during integration. Moreover, compatibility issues may arise when interfacing nanoscale devices with macroscopic components or systems, such as integrated circuits and sensors [23].

To address compatibility challenges, researchers are exploring materials and fabrication techniques that facilitate seamless integration with existing systems. Hybrid integration approaches, combining nanomaterials with conventional materials through bonding, layering, or deposition techniques, enable the realization of hybrid devices with enhanced functionalities. Additionally, surface modification techniques and interface engineering strategies can enhance compatibility and adhesion between nanoscale and macroscopic components, ensuring reliable integration without compromising performance [25, 34].

Scalability

Scalability is another significant challenge in integrating nanoscale devices into larger systems, particularly for mass production and commercialization. While laboratory-scale fabrication techniques may yield promising results, scaling up production to industrial levels requires overcoming numerous hurdles, including cost-effectiveness, reproducibility, and yield optimization [10, 18].

To address scalability challenges, researchers are developing scalable fabrication processes and manufacturing platforms tailored to nanoscale devices. Roll-to-roll printing, for example, enables continuous and high-throughput production of flexible electronics and sensors, leveraging techniques such as gravure printing and screen printing. Additionally, advancements in nanolithography and self-assembly techniques enable the creation of nanoscale patterns and structures over large areas with high precision and throughput, facilitating the mass production of nanodevices [15, 19].

Reliability

Ensuring the reliability and long-term performance of nanoscale devices is essential for their integration into practical applications. Nanomaterials and nanodevices may exhibit unique failure mechanisms, reliability degradation pathways, and susceptibility to environmental factors, such as temperature variations, moisture, and mechanical stress [10, 34].

To address reliability challenges, researchers are developing robust design methodologies and testing protocols to evaluate the reliability of nanoscale devices under various operating conditions. Accelerated aging tests, reliability modeling, and failure analysis techniques enable the identification and mitigation of potential failure modes and degradation mechanisms. Furthermore, the integration of self-diagnostic and self-healing functionalities into nanoscale devices enhances their resilience and reliability, ensuring uninterrupted operation in real-world environments [5, 9].

Interconnect Challenges

Interconnect challenges, including signal propagation delay, cross-talk, and power dissipation, pose additional obstacles to the integration of nanoscale devices into larger systems. As device dimensions shrink and operating frequencies increase, interconnect performance becomes increasingly critical for overall system performance and functionality [3, 6].

To address interconnect challenges, researchers are exploring novel interconnect materials, architectures, and design techniques optimized for nanoscale devices. Low-loss dielectric materials, such as low-k dielectrics and airgaps, reduce signal propagation delay and power dissipation in nanoscale interconnects, enabling high-speed communication with minimal energy consumption [38, 47].

Fig. (**6**) shows the interconnect delay variation with different widths in nanometer-scale devices, elucidating how signal propagation is influenced by width reduction. As widths decrease, resistance and capacitance increase, elevating delay. Designing efficient interconnects at this scale demands addressing these effects *via* innovative materials, precise fabrication, and optimized architectures to ensure signal integrity and minimize performance degradation.

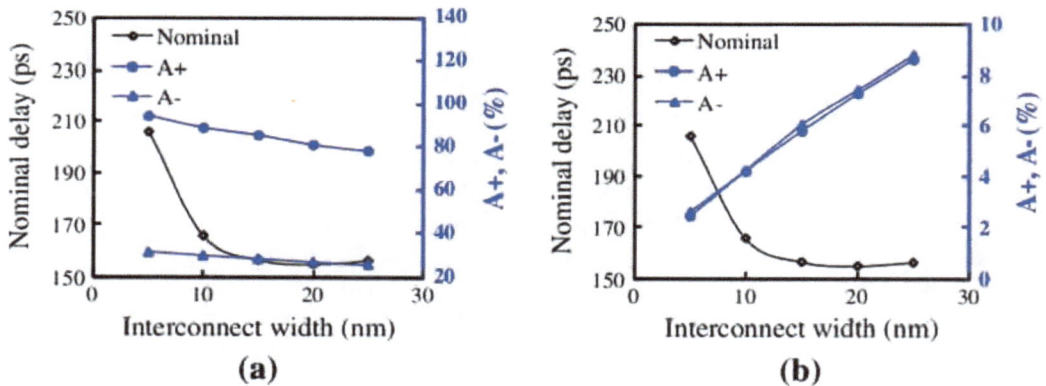

Fig. (6). Interconnect delay for different widths.

Fig. (**7**) shows the propagation delay due to interconnect occurrences in nanoscale devices. It refers to the time taken for a signal to travel through interconnecting wires. As dimensions shrink, resistance and capacitance rise, increasing delay. Design challenges involve minimizing these effects through advanced materials, layout optimization, and signal routing strategies to ensure high-speed and reliable operation. Additionally, advanced packaging technologies, such as Through Silicon Vias (TSVs) and wafer-level packaging, enable the integration of nanoscale devices into three-dimensional (3D) stacked configurations, reducing interconnect lengths and mitigating signal integrity issues.

DESIGN OPTIMIZATION TECHNIQUES

Design optimization techniques are essential for advancing nanoscale device design, enabling engineers to maximize device performance, efficiency, and reliability while minimizing design iterations and development time. Simulation tools, machine learning algorithms, and design automation techniques provide powerful tools for exploring design spaces, identifying optimal configurations, and accelerating the design optimization process. By leveraging these techniques, researchers and engineers can overcome design challenges, push the boundaries of nanotechnology, and realize the full potential of nanoscale devices in various applications, from electronics and healthcare to energy and environmental monitoring. In this section, we explore the various design optimization techniques employed in nanoscale device design and their significance in advancing nanotechnology [43].

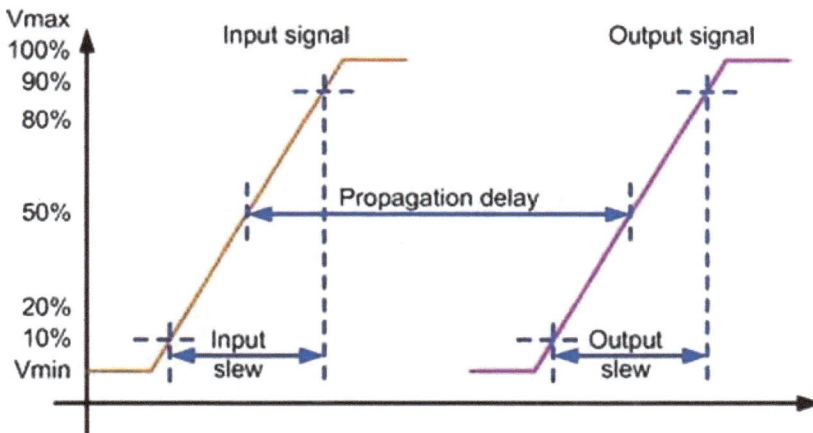

Fig. (7). Propagation delay due to interconnect occurrences..

Simulation Tools

Simulation tools are indispensable for the analysis and optimization of nanoscale devices before fabrication. These tools leverage mathematical models and numerical methods to simulate device behavior under varying operating conditions. By providing insights into device performance and identifying potential issues, simulation tools aid engineers in optimizing design parameters. There are several types of simulation tools commonly used in nanotechnology, listed as follows:

Finite Element Analysis (FEA)

FEA is a prevalent simulation technique used for analyzing the mechanical behavior of nanoscale structures, such as nanowires, nanotubes, and nanocomposites. FEA allows engineers to evaluate stress distribution, deformation, and failure modes in nanoscale devices. By optimizing device geometry, material selection, and manufacturing processes based on FEA results, engineers can enhance mechanical performance and reliability [12, 17].

Computational Fluid Dynamics (CFD)

CFD is employed to simulate fluid flow and heat transfer in nanofluidic devices, microfluidic channels, and nanoscale sensors. By optimizing device geometry, flow patterns, and thermal management strategies, CFD simulations enable engineering devices, such as transistors, diodes, and sensors. These tools enable engineers to analyze device characteristics, like current-voltage (I-V) curves, charge distribution, and signal propagation. By optimizing device performance, power consumption, and reliability through EDA simulations, engineers can ensure the functionality and efficiency of nanoscale electronic devices [19, 37, 41].

Machine Learning Algorithms

Machine learning algorithms are increasingly employed to optimize the design of nanoscale devices by leveraging large datasets, complex models, and iterative optimization processes. These algorithms facilitate the exploration of design spaces, identification of optimal configurations, and acceleration of the design optimization process. Several types of machine learning algorithms commonly used in nanoscale device design are provided below:

Supervised Learning

Supervised learning algorithms, including neural networks and support vector machines, are trained on labeled datasets of device performance metrics and design parameters. These algorithms learn patterns and relationships between input and output variables, enabling them to predict device performance for new design configurations and recommend optimal design parameters to achieve desired performance targets [40, 42].

Unsupervised Learning

Unsupervised learning algorithms, such as clustering and dimensionality reduction techniques, are employed for exploring high-dimensional design spaces and identifying patterns or clusters of optimal design configurations. By gaining insights into the underlying structure of design spaces, engineers can identify promising regions for further exploration and optimization [41, 46].

Reinforcement Learning

Reinforcement learning algorithms, like Q-learning and deep reinforcement learning, are used for sequential decision-making tasks in nanoscale device design. These algorithms learn an optimal policy for selecting design actions that maximize a long-term reward signal. By dynamically adjusting design parameters based on feedback from simulation results or experimental data, reinforcement learning algorithms guide iterative design improvements, leading to optimized device performance [23, 32].

Design Automation Techniques

Design automation techniques streamline the design process by automating repetitive tasks, optimizing design workflows, and facilitating collaboration between engineers. These techniques, integrated within Computer-aided Design (CAD) environments, enable engineers to efficiently explore, analyze, and optimize nanoscale device designs. Several design automation techniques commonly employed in nanoscale device design are mentioned as follows:

Parametric Modeling

Parametric modeling techniques allow engineers to define and manipulate design parameters within CAD software. By linking design parameters to simulation models, engineers can perform sensitivity analyses and optimization studies to identify optimal design configurations that meet specified performance criteria [22, 30].

Design Space Exploration

Design space exploration techniques involve systematically exploring a multidimensional space of design parameters to identify regions of interest, trade-offs, and optimal solutions. These techniques leverage optimization algorithms, sampling methods, and surrogate models to efficiently search large design spaces and identify promising design configurations [12, 32].

Optimization Algorithms

Optimization algorithms, such as genetic algorithms, particle swarm optimization, and simulated annealing, are used for finding optimal solutions to complex design optimization problems. These algorithms iteratively evaluate candidate designs, adjust design parameters based on objective functions and constraints, and converge toward optimal solutions through successive generations or iterations.

Significance of Design Optimization in Nanoscale Devices

Design optimization plays a crucial role in maximizing the performance and efficiency of nanoscale devices, enabling engineers to achieve superior device performance, reduced power consumption, and enhanced reliability. By leveraging simulation tools, machine learning algorithms, and design automation techniques, engineers can explore design spaces, identify optimal configurations, and accelerate the design optimization process. These techniques enable engineers to address design challenges, such as manufacturing variability, reliability issues, and performance trade-offs, leading to the development of next-generation nanoscale devices with unprecedented capabilities and applications [37 - 45].

APPLICATION ACROSS DIFFERENT FIELDS

Nanoscale technologies have emerged as a transformative force across diverse fields, revolutionizing industries and driving innovation in areas, such as electronics, photonics, biotechnology, and energy. From enhancing the performance of electronic devices to enabling breakthroughs in medical diagnostics and renewable energy, the applications of nanoscale technologies are vast and varied. In this section, we explore the myriad ways in which nanoscale technologies are shaping the future across different fields.

Electronics

Nanoscale technologies have revolutionized the field of electronics, enabling the development of smaller, faster, and more energy-efficient devices. Nanoelectronic components, such as nanotransistors, nanowires, and quantum dots, offer superior performance compared to their macroscopic counterparts. These nanoscale devices have paved the way for advances in computing, communication, and sensing technologies [20, 32].

Quantum computing represents one of the most promising applications of nanoscale technologies in electronics. By harnessing the principles of quantum mechanics, quantum computers leverage the unique properties of quantum bits (qubits) to perform complex calculations at unprecedented speeds. Nanoscale fabrication techniques enable the precise control and manipulation of individual qubits, leading to the development of scalable quantum computing architectures [23].

In addition to quantum computing, nanoscale technologies are driving advancements in wearable electronics, flexible displays, and high-performance sensors. Nanomaterials, such as graphene, carbon nanotubes, and quantum dots, exhibit unique electrical, optical, and mechanical properties that make them ideal candidates for next-generation electronic devices. By integrating nanoscale components into electronic circuits, engineers can create ultra-thin, lightweight, and flexible devices with enhanced functionality and performance [29, 30].

Photonics

Photonics is another field that has benefited greatly from nanoscale technologies, with applications ranging from telecommunications and data storage to imaging and sensing. Nanophotonic devices, such as photonic crystals, plasmonic nanostructures, and nanoscale waveguides, enable the manipulation and control of light at the nanoscale [44].

One of the key applications of nanophotonics is in the development of high-speed optical communication systems. Nanoscale waveguides and optical resonators allow for the confinement and manipulation of light on a chip scale, enabling the realization of ultra-compact and energy-efficient Photonic Integrated Circuits (PICs). These PICs are crucial for meeting the increasing bandwidth demands of modern communication networks [23].

Nanoscale technologies are also driving innovations in imaging and sensing applications. Nanoplasmonic sensors, for example, utilize the interaction between light and metallic nanostructures to detect and analyze biological and chemical molecules with high sensitivity and specificity. By engineering the properties of plasmonic nanostructures at the nanoscale, researchers can design sensors capable of detecting single molecules or nanoparticles, opening up new possibilities for medical diagnostics, environmental monitoring, and homeland security [12, 23].

Biotechnology

In the field of biotechnology, nanoscale technologies are revolutionizing drug delivery, medical imaging, and diagnostics. Nanomaterials, such as liposomes, polymeric nanoparticles, and quantum dots, offer unique properties that make them ideal candidates for targeted drug delivery and imaging applications.

One of the most promising applications of nanoscale technologies in biotechnology is in the field of nanomedicine. Nanoparticle-based drug delivery systems can improve the efficacy and safety of therapeutic agents by enhancing their solubility, stability, and targeted delivery to diseased tissues. Additionally, nanoscale imaging agents enable high-resolution imaging of biological structures and processes, facilitating early disease detection and personalized medicine [34 - 42].

Nanoscale technologies are also driving advancements in regenerative medicine and tissue engineering. Nanofibrous scaffolds and biomaterials mimic the structure and function of the extracellular matrix, providing a conducive environment for cell growth, differentiation, and tissue regeneration. By precisely controlling the architecture and properties of nanoscale scaffolds, researchers can engineer tissues and organs for transplantation and regenerative therapies [45].

Energy

Energy is another field where nanoscale technologies hold immense promise for addressing global challenges, such as renewable energy generation, energy storage, and energy efficiency. Nanomaterials, such as quantum dots, nanowires, and perovskite nanocrystals, offer unique properties that make them ideal candidates for enhancing the performance of energy conversion and storage devices [47].

One of the key applications of nanoscale technologies in energy is in the development of next-generation solar cells. Nanostructured photovoltaic materials, such as quantum dot solar cells and perovskite solar cells, exhibit enhanced light absorption and charge transport properties, leading to higher efficiency and lower production costs. Additionally, nanomaterial-based coatings and surface treatments can improve the durability and stability of solar panels, prolonging their lifespan and reducing maintenance requirements [34].

Nanoscale technologies are also driving innovations in energy storage devices, such as batteries and supercapacitors. Nanostructured electrode materials, such as graphene, carbon nanotubes, and transition metal oxides, offer higher surface area, faster charge-discharge rates, and longer cycle life compared to conventional materials. By leveraging nanoscale engineering techniques, researchers can develop high-performance energy storage devices with improved energy density, power density, and cycling stability [46].

CONCLUSION

Nanoscale technologies stand at the forefront of scientific and technological advancement, offering unparalleled opportunities for innovation and societal impact. Throughout this chapter, we have explored the intricate design principles, encountered the challenges, and witnessed the remarkable advancements in nanoscale device technologies. From fundamental aspects to real-world applications, the journey through nanoscale technologies has been both enlightening and inspiring. In this concluding section, we reflect on the significance of nanoscale technologies, their potential for driving innovation, and their role in addressing societal challenges. We emphasize the importance of continued research, collaboration, and innovation to unlock the full potential of nanoscale technologies and pave the way for a brighter future. Nanoscale technologies hold immense promise for driving innovation across a wide range of fields, from electronics and photonics to biotechnology and energy. By leveraging the unique properties of nanomaterials and nanoscale devices, researchers and engineers are developing groundbreaking solutions to address pressing global challenges and improve the quality of life for people around the world.

One of the key strengths of nanoscale technologies lies in their ability to address design challenges and optimize device performance. Through meticulous design principles and innovative fabrication techniques, engineers can create nanoscale devices with enhanced functionality, reliability, and efficiency. By pushing the boundaries of what is possible at the nanoscale, researchers are unlocking new opportunities for technological innovation and scientific discovery. Furthermore, nanoscale technologies play a vital role in exploring diverse applications across different fields. From nanoelectronics and quantum computing to nanomedicine and renewable energy, the potential applications of nanoscale technologies are vast and varied. By harnessing the power of nanotechnology, researchers are developing novel solutions to some of the most pressing challenges facing humanity, from disease diagnosis and treatment to environmental sustainability and energy security. However, despite the remarkable progress made in nanoscale technologies, challenges remain. Manufacturing constraints, reliability issues, and integration challenges continue to pose obstacles to the widespread adoption of nanoscale devices. Addressing these challenges requires interdisciplinary collaboration, innovative solutions, and a commitment to pushing the boundaries of what is possible. In conclusion, nanoscale technologies hold immense promise for driving innovation and addressing societal challenges. By addressing design challenges, optimizing device performance, and exploring diverse applications, nanoscale technologies would continue to shape the future of technology and propel toward a more sustainable and interconnected world. As we embark on this journey of exploration and discovery, let us remain committed to pushing the boundaries of nanoscale technologies and unlocking their full potential for the benefit of humanity.

KEY POINTS

• Quantum effects: Understanding and harnessing quantum phenomena, such as quantum confinement, tunneling, and coherence, are essential for designing nanoscale devices with tailored electronic and optical properties.

- Material properties: Nanomaterials exhibit unique properties due to their small size, high surface area, and quantum effects, influencing device behavior. Consideration of surface effects, mechanical properties, and material behavior is critical in device design.
- Fabrication techniques: Various fabrication methods, including top-down, bottom-up, and hybrid approaches, enable the precise creation of nanoscale structures with controlled dimensions and functionalities, driving advancements in nanotechnology.
- Design considerations: Designing reliable nanoscale devices requires careful consideration of performance metrics, reliability, and scalability. Balancing trade-offs between performance parameters while ensuring long-term stability is crucial for successful device implementation.
- Applications across industries: Nanoscale devices find applications across diverse fields, such as electronics, biomedicine, energy, and the environment. Their use spans from nanoelectronics and quantum computing to nanomedicine and energy storage, offering innovative solutions to various challenges.
- Interdisciplinary research: Nanoscale device design involves collaboration across disciplines, such as physics, chemistry, materials science, engineering, and biology. Interdisciplinary approaches drive innovation and address complex technological and societal problems.
- Future perspectives: Anticipated advancements in nanoscale device design include improved performance, functionality, and scalability. Overcoming challenges, such as maintaining quantum coherence and addressing reliability concerns, could shape the future of nanotechnology.
- Ethical and societal implications: Alongside technological advancements, nanoscale device design raises ethical and societal considerations, including privacy, security, environmental impact, and equitable access to technology. Addressing these concerns is essential for the responsible development and deployment of nanotechnology.

ACKNOWLEDGEMENTS

The authors thank the Department of Electronics and Communication, M.J.P Rohilkhand University, Bareilly, Uttar Pradesh, India, for providing the resources and environment necessary for their research.

AUTHORS' CONTRIBUTION

Sumit Srivastava, Abhinav Jaiswal, and Arman Khan contributed to the study conception and design and the analysis and interpretation of results; Abhinav Jaiswal and Arman Khan collected the data and drafted the manuscript. All authors have reviewed the results and approved the final version of the chapter.

REFERENCES

[1]   G. Cao, D. Wang, and S.C. Heilshorn, "Nanoscale materials in protein and peptide delivery", In: *Nanoscale Materials in Targeted Drug Delivery, Theragnosis and Tissue Regeneration* Elsevier, 2019. https://www.sciencedirect.com/science/article/pii/B978012814031400011X

[2]   C. Chen, T. Xia, and G. Nie, "Nanoscale materials in biomedicine: delivery, imaging, and sensing", *Advanced Materials,* vol. 31, no. 13, p. 1806723, 2019.

[3]   J.H. Choi, and C.D. Montemagno, "Review of recent advances in nanoscale information technology and potential applications to defense", In: *Technical Report, Defense Technical Information Center (DTIC).*, 2005. https://apps.dtic.mil/dtic/tr/fulltext/u2/a435816.pdf

[4]   L. Dorogin, and J. Linnros, "Nanoscale materials in photonics: synthesis, characterization, and applications", In: *Nanomaterials* vol. 8. , 2018, no. 9, p. 701. https://www.mdpi.com/2079-4991/8/9/701

[5]   J. Goldberger, R. He, Y. Zhang, S. Lee, H. Yan, H. J. Choi, and P. Yang, "Single-crystal gallium nitride nanotubes", *Nature,* vol. 422, no. 6932, pp. 599-602, 2003. [http://dx.doi.org/10.1038/nature01418]

[6]   G. W. Guglietta, and J. H. Thomas, "Nanotechnology: Assessing the risks", *The Journal of Law, Medicine & Ethics,* vol. 31, no. 4, pp. 511-524, 2003. [http://dx.doi.org/10.1111/j.1748-720X.2003.tb00142.x]

[7]   Y. Huang, X. Duan, Q. Wei, and C. M. Lieber, "Directed assembly of one-dimensional nanostructures into functional networks", *Science,* vol. 291, no. 5504, pp. 630-633, 2001. [http://dx.doi.org/10.1126/science.291.5504.630]

[8]   N. Kumar, A. Dixit, A. Rezaei, T. Dutta, C.P. García, and V. Georgiev, "Insights into the ultra-steep subthreshold slope gate-all-around feedback-FET for memory and sensing applications", *2023 IEEE Nanotechnology Materials and Devices Conference (NMDC),* 2023pp. 617-620 [http://dx.doi.org/10.1109/NMDC57951.2023.10343913]

[9]   S. Y. Park, D. N. Heo, H. N. Lim, G. C. Schatz, and D. E. Kim, "Hollow gold nanoparticles for nanomedicine: Tunable cytotoxicity", *ACS Nano,* vol. 9, no. 10, pp. 9986-9996, 2015.

[10]   K. Sanderson, "Nanotechnology: The Big Picture", *Nature,* vol. 531, no. 7592, pp. S96-S97, 2016.

[11]   C. Wang, M. Han, S. Liu, and J. Li, "Recent advances in nanoscale device fabrication techniques", *Journal of Nanomaterials,* p. 2020, 2020.

[12]   J. Lee, J. Kim, and S. Lee, *Nanoscale heat transfer: Fundamentals and engineering applications.* CRC Press, 2018. https://www.taylorfrancis.com/books/e/9781315189416

[13]   Smith, J. K., & Jones, L. M. *Nanoscale Materials Handbook.* Springer, 2019. https://www.springer.com/gp/book/9783319657223

[14]   F. Yang, and G.S. Shekhawat, *Nanoscale sensors: A guide for the design and fabrication of nanoscale sensors.* CRC Press, 2017. https://www.taylorfrancis.com/books/e/9781315363981

[15]   H. Yin, and S. Ren, *Nanoscale materials in environmental engineering: Development, application, and management.* CRC Press., 2020. https://www.taylorfrancis.com/books/e/9781351066045

[16]   L. Zhang, and Z. Li, *Nanoscale materials in energy applications.* CRC Press., 2018. https://www.taylorfrancis.com/books/e/9781315162105

[17]   H. Kim, and J. Park, *Nanoscale materials in biology and medicine.* CRC Press., 2017. https://www.taylorfrancis.com/books/e/9781315365558

[18]   R.A. Jones, and M.T. Smith, *Nanoscale materials in electronics.* CRC Press., 2019. https://www.taylorfrancis.com/books/e/9781351076006

[19]   Y. Chen, and J. Li, *Nanoscale materials in catalysis.* CRC Press., 2016. https://www.taylorfrancis.com/books/e/9781315363479

[20]   X. Li, and J. Zhang, *Nanoscale materials in drug delivery.* CRC Press., 2019. https://www.taylorfrancis.com/books/e/9780429157623

[21]   Z. Wang, and J. Liu, *Nanoscale materials in aerospace engineering.* CRC Press., 2020. https://www.taylorfrancis.com/books/e/9781351092075

[22]   P. Singh, and D.S. Yadav, "Impactful study of f-shaped tunnel fet", *Silicon,* vol. 14, no. 10, pp. 5359-5365, 2022.
       [http://dx.doi.org/10.1007/s12633-021-01319-6]

[23]   X. Lu, and C. Wang, *Nanoscale materials in agriculture and food science.* CRC Press., 2017. https://www.taylorfrancis.com/books/e/9781315366647

[24]   X. Li, and Y. Ma, *Nanoscale materials in automotive technology.* CRC Press., 2018. https://www.taylorfrancis.com/books/e/9781315377353

[25]   Y. Wang, and L. Zhou, *Nanoscale materials in construction engineering.* CRC Press., 2019. https://www.taylorfrancis.com/books/e/9781351189899

[26]   H. Chen, and Q. Guo, *Nanoscale materials in environmental pollution control and remediation.* CRC Press, 2016. https://www.taylorfrancis.com/books/e/9781315374185

[27]   Y. Liu, and W. Hu, *Nanoscale materials in textiles.* CRC Press., 2017. https://www.taylorfrancis.com/books/e/9781315363196

[28]   L. Zhang, and X. Li, *Nanoscale materials in water desalination.* CRC Press., 2019. https://www.taylorfrancis.com/books/e/9781351075979

[29]   J. Wang, and H. Zhang, *Nanoscale materials in petroleum engineering.* CRC Press., 2018. https://www.taylorfrancis.com/books/e/9781315387505

[30]   Smith, G., & Johnson, R. *Nanoscale materials in space exploration* CRC press, 2020. https://www.taylorfrancis.com/books/e/9781351187758

[31]   Y. Xu, and Q. Wang, *Nanoscale materials in renewable energy.* CRC Press., 2019. https://www.taylorfrancis.com/books/e/9781351187796

[32]   N. Kumar, C.P. García, A. Dixit, A. Rezaei, and V. Georgiev, "Charge dynamics of amino acids fingerprints and the effect of density on FinFET-based Electrolyte-gated sensor", *Solid-State Electron.,* vol. 210, p. 108789, 2023.
       [http://dx.doi.org/10.1016/j.sse.2023.108789]

[33]   K. Johnson, and S. Brown, *Nanoscale materials in defense technologies.* CRC Press., 2017. https://www.taylorfrancis.com/books/e/9781315371122

[34]   H. Bhaskaran, S. Rohr, Ed., *Nanoscale devices: Fabrication, functionalization, and accessibility from the macroscopic world.* Wiley-VCH: Weinheim, Germany, 2019.

[35]   A.N. Cleland, and M.L. Roukes, "A nanometre-scale mechanical electrometer", *Nature,* vol. 392, no. 6672, pp. 160-162, 1998.
       [http://dx.doi.org/10.1038/32373]

[36]   Y.C. Cheng, "High mobility WSe2 p- and n-type field-effect transistors contacted by highly doped graphene for low-resistance contacts", *Nano Lett.,* vol. 16, no. 5, pp. 3448-3455, 2016.

[37]   P. Singh, A. Raman, D.S. Yadav, N. Kumar, A. Dixit, and M.D.H.R. Ansari, "Ultra thin finger-like source region-based TFET: Temperature sensor", *IEEE Sens. Lett.,* vol. 8, no. 5, pp. 1-4, 2024.
       [http://dx.doi.org/10.1109/LSENS.2024.3390689]

[38]   D.J. Srolovitz, "Materials science: Nanoindents test the metal", *Nat. Mater.,* vol. 2, no. 10, pp. 621-622, 2003.

[39]   E. Fort, and S. Grésillon, "Surface enhanced fluorescence", *J. Phys. D Appl. Phys.,* vol. 41, no. 1, p. 013001, 2008.

[http://dx.doi.org/10.1088/0022-3727/41/1/013001]

[40]   N. Kumar, R. Dhar, C. Pascual Garcia, and V.P. Georgiev, "A novel computational framework for simulations of bio-field effect transistors", *ECS Trans.,* vol. 111, no. 1, pp. 249-260, 2023.
[http://dx.doi.org/10.1149/11101.0249ecst]

[41]   C.T. Campbell, "Ultrathin metal films and particles on oxide surfaces: structural, electronic and chemisorptive properties", *Surf. Sci. Rep.,* vol. 27, no. 1-3, pp. 1-111, 1997.
[http://dx.doi.org/10.1016/S0167-5729(96)00011-8]

[42]   R.E. Russo, "Laser ablation and nanoparticle production in a background gas: Implications for direct writing of electronic and photonic devices", *J. Appl. Phys.,* vol. 95, no. 6, pp. 2896-2906, 2004.

[43]   D.W. Pohl, "Scanning near-field optical microscopy (SNOM)", *J. Vac. Sci. Technol. B,* vol. 9, no. 3, pp. 1269-1273, 1991.

[44]   P. Singh, and D.S. Yadav, "Assessment of temperature and ITCs on single gate L-shaped tunnel FET for low power high frequency application", *Eng. Res. Express,* vol. 6, no. 1, p. 015319, 2024.
[http://dx.doi.org/10.1088/2631-8695/ad32b0]

[45]   M.B. Cortie, and X. Xu, "Size-Dependent Melting of Gold Nanoparticles", *Nanotechnology,* vol. 16, no. 6, pp. 841-844, 2005.

[46]   P. Singh, and D.S. Yadav, "Performance analysis of ITCs on analog/RF, linearity and reliability performance metrics of tunnel FET with ultra-thin source region", *Appl. Phys., A Mater. Sci. Process.,* vol. 128, no. 7, p. 612, 2022.
[http://dx.doi.org/10.1007/s00339-022-05741-4]

[47]   R.R. Schlittler, "Interaction of hydrogen with a Si(100)-2x1 surface studied with the scanning tunneling microscope", *Phys. Rev. Lett.,* vol. 62, no. 6, pp. 649-652, 1989.
[PMID: 10040293]

CHAPTER 2

Materials Used in the Design of Semiconductor Devices

Trinath Talapaneni[1,*], Vatsala Chaturvedi[1] and Ankireddy Narendra[2]

[1] *Department of Metallurgical Engineering, OP Jindal University, Raigarh, 496109, Chhattisgarh, India*

[2] *Department of Electrical Engineering, OP Jindal University, Raigarh, 496109, Chhattisgarh, India*

Abstract: The design and advancement of semiconductor devices are fundamentally rooted in the diverse range of materials utilized, each selected for its unique properties and contributions to device performance. This chapter explores the necessity and history of semiconductor materials, tracing their evolution and wide-ranging applications. Central to this discussion are the elemental semiconductors derived from the periodic table, focusing on silicon, germanium, and gray tin, which have historically underpinned the semiconductor industry. Also, this chapter differentiates between intrinsic and extrinsic semiconductors, highlighting their respective roles and characteristics in device functionality. Intrinsic semiconductors, with their pure form, contrast with extrinsic semiconductors, which are doped to enhance specific electrical properties, catering to various application needs. Furthermore, the study delves into compound semiconductor materials, showcasing their importance in modern technology. Compounds, like silicon carbide, boron nitrate, red selenium, boron phosphide, and boron arsenide, are examined for their exceptional electrical and thermal properties. The chapter also discusses aluminum-based compounds, including aluminum nitride, phosphide, and arsenide, and their applications in high-power and high-frequency devices. The study extends to gallium-based compounds, like gallium nitride, phosphide, and arsenide, known for their high electron mobility and applications in optoelectronics. Additionally, zinc and cadmium compounds, such as zinc oxide, cadmium arsenide, zinc phosphide, and zinc antimonide, are analyzed to enhance device performance and efficiency. This comprehensive study underscores the critical role of diverse semiconductor materials in the ongoing innovation and optimization of electronic, optoelectronic, and power devices, meeting the escalating demands of modern technology.

Keywords: Compound semiconductors, Extrinsic, Intrinsic, Materials, Semiconductor devices.

---

* **Corresponding author Trinath Talapaneni:** Department of Metallurgical Engineering, OP Jindal University, Raigarh, 496109, Chhattisgarh, India; E-mail: trinath.talapaneni@opju.ac.in

Ashish Raman, Prabhat Singh, Naveen Kumar & Ravi Ranjan (Eds.)

INTRODUCTION

Semiconductors are widely used in various applications, such as power semiconductor devices, integrated circuits, amplifiers, power control, telecommunications, optoelectronics, medical devices, *etc* [1–3]. Thus, they are in demand in most applications; hence, they require more research and development. The primary focus of research on semiconductor devices is on the material used in them. So this chapter provides fundamental aspects of semiconductor materials and how semiconductor materials are classified from the periodic table; moreover, to achieve better performance, there is a need for compound or hybrid semiconductor devices. Thus, this chapter provides a clear understanding of all the above aspects.

FUNDAMENTALS OF SEMICONDUCTOR MATERIALS

Based on contemporary technology, semiconductor materials power everything from solar cells to computers and cell phones.

Necessity of Semiconductor Materials

For several reasons, semiconductor materials are essential to contemporary technology [4].

 i. Electronic components, including Integrated Circuits (ICs), diodes, and transistors, are built on semiconductors. Almost all electronic systems, including those found in computers, cellphones, televisions, and automobile electronics, depend on these gadgets.
 ii. Semiconductors facilitate the digital revolution because they provide the foundation for binary logic processes. Binary digits (bits) of information are processed using digital circuits made of transistors, which can also function as switches. Computers and digital communication systems operate on this foundation.
 iii. Electrical power can be managed and transformed through power transistors, thyristors, and other semiconductor devices. This technology finds application in power supplies, motor drives, renewable energy systems, electric vehicles, and other electronic devices.
 iv. Semiconductors make communication systems' signal processing, reception, and transmission possible. Semiconductors are crucial in Digital Signal Processing (DSP) processors and Radio Frequency (RF) amplifiers, essential components of wireless devices, internet, and telecommunication networks.
 v. Medical diagnostics, imaging technologies (such as MRIs and CT scans), and biotechnology applications all involve semiconductor-based sensors and

imaging equipment. They make improvements in illness identification, monitoring, and treatment possible.

vi. Materials for semiconductors are necessary for many applications involving the environment and energy. These comprise photovoltaic cells for solar energy conversion, semiconductor-based catalysts for pollution control, and sensors for environmental monitoring.

Semiconductor materials are essential to modern society's operation because they fuel an extensive array of technologies that affect almost every facet of human existence, from communication and entertainment to healthcare and environmental sustainability.

History of Semiconductor Materials

The history of semiconductor materials is a fascinating journey spanning decades of scientific discovery and technological growth. Here is a provided a quick rundown:

The late 19th and early 20th century saw the beginning of our understanding of semiconductors and the development of various semiconductor switches, as shown in Fig. (**1**). Michael Faraday noted in 1833 that a tiny covering of Ag2S had the "extraordinary effect" of lowering silver's electrical resistance. Karl Ferdinand Braun laid the foundation for semiconductor diodes in 1874 when he discovered the rectifying characteristic of metal-semiconductor junctions.

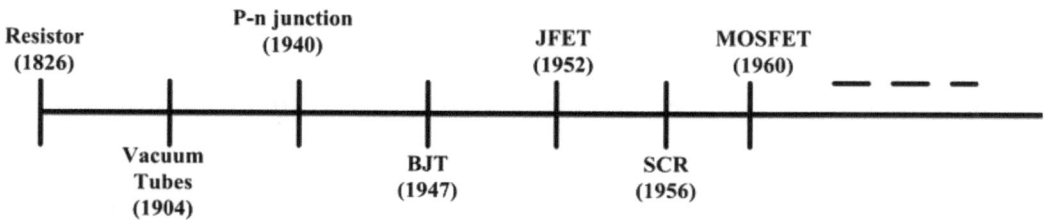

Fig. (1). History of semiconductor switches.

In the early 20th century, scientists, like Lee De Forest, Jagadish Chandra Bose, and Ferdinand Braun, advanced the understanding of semiconductor materials and developed primary semiconductor devices, like the crystal detector for radio receivers.

Due to germanium's (Ge) availability and semiconductor qualities, researchers concentrated on studying semiconductor materials in the 1940s. John Bardeen, Walter Brattain, and William Shockley of Bell Laboratories created the first point-contact transistor in 1947 with the help of germanium, which was employed in early semiconductor diodes and transistors.

During the late 1950s to the early 1960s, silicon (Si) became the most widely used semiconductor material because of its stability, quantity, and ease of manufacturing. The mass fabrication of Integrated Circuits (ICs) was made possible by Jean Hoerni's 1959 discovery of the silicon planar process at Fairchild Semiconductor, revolutionizing semiconductor manufacturing.

A turning point in the history of semiconductors was the Integrated Circuit (IC), created in the early 1960s by Robert Noyce at Fairchild Semiconductor and Jack Kilby at Texas Instruments. By combining several electronic components onto a single semiconductor substrate, ICs have revolutionized electronics and enabled the downsizing and widespread use of electronic devices.

Semiconductor technology has developed quickly since the 1960s because of Moore's law and unrelenting innovation. Complementary Metal Oxide Semiconductor (CMOS) technology has seen remarkable advancements alongside the introduction of specialized semiconductor materials, like gallium arsenide (GaAs), for targeted applications. Moreover, there is an ongoing trend of scaling down semiconductor devices, pushing the boundaries of miniaturization and performance.

In the twenty-first century, semiconductor materials are still being developed to suit the needs of new technologies, including artificial intelligence, quantum computing, renewable energy, and the Internet of Things (IoT). Graphene, carbon nanotubes, and perovskites are examples of novel materials that are the subject of research activities to push the limits of semiconductor technology.

The development of semiconductor materials over time is a monument to human inventiveness and the unwavering search for scientific knowledge, producing revolutionary technologies that have shaped the contemporary world.

Applications of Semiconductor Materials

Semiconductor materials' distinct electrical, optical, and mechanical characteristics make them useful in various industries [5]. The following are some critical applications, as shown in Fig. (2).

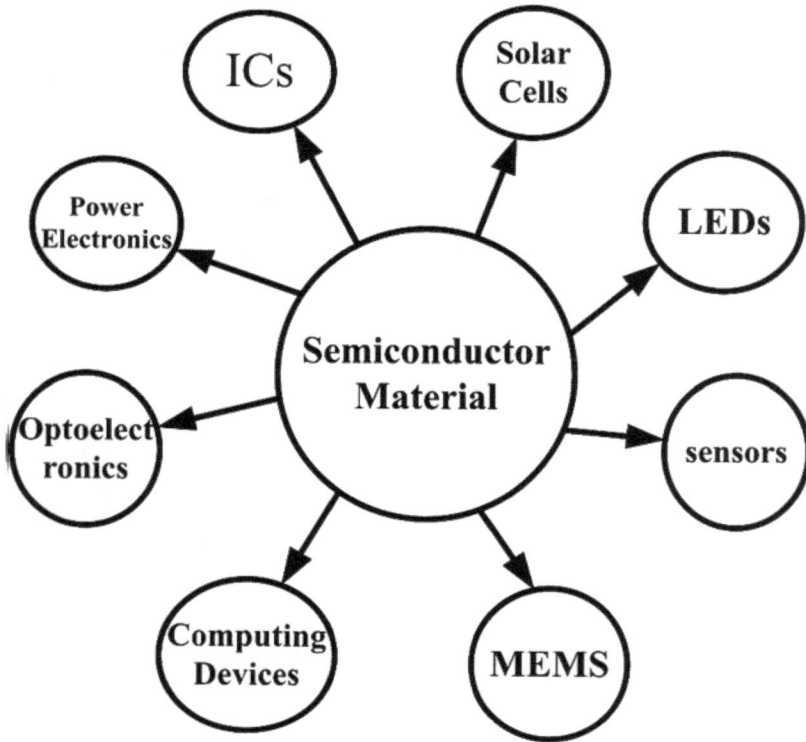

Fig. (2). Applications of semiconductor materials.

- Modern electronics are built on semiconductors. They are employed in producing Integrated Circuits (ICs), transistors, and diodes, which are the building blocks of electronic devices, including PCs, tablets, smartphones, and televisions.
- Solar cells employ semiconductor materials to turn light into power. Although silicon-based solar cells are the most widely used type, photovoltaic systems can also use Copper Indium Gallium Selenide (CIGS) and cadmium telluride (CdTe) semiconductors.
- Light-emitting Diodes (LEDs) and Organic Light-emitting Diodes (OLEDs) are used as semiconductor-based light sources in lighting applications as they are durable and energy-efficient; LEDs are perfect for various lighting applications, such as commercial, automotive, and residential lighting.
- Optoelectronic devices that interact with light require semiconductors. Telecommunications, optical communication networks, sensing applications, and optical storage devices comprise photodetectors, photodiodes, laser diodes, and optical sensors.
- Power electronics uses semiconductors, such as transistors, thyristors, and diodes, to convert and regulate electrical power. Power supply, motor drives,

renewable energy systems, electric vehicles, and grid infrastructure are just a few of the uses.

- Many physical, chemical, and biological factors can be detected and measured with semiconductor-based sensors. Examples include sensors used in consumer electronics, automotive, industrial, medical, and environmental applications; these include temperature, pressure, gas, humidity, and biosensors.

- Microelectromechanical System (MEMS) devices, which combine electrical and mechanical components on a semiconductor substrate, employ semiconductors. Accelerometers, gyroscopes, pressure sensors, microphones, inkjet print heads, and biomedical devices are among the gadgets that employ MEMS technology.

- Semiconductor materials are used in non-volatile memory devices, crucial for data storage in electronic devices, like memory cards, Universal Serial Bus (USB), and Solid-state Drives (SSDs). Examples of these devices are flash memory and phase-change memory.

- The fundamental units of quantum computing, semiconductor-based qubits, are the subject of ongoing study and development. Researchers are looking into semiconductor materials, like silicon and superconductors, to see if they may be used to create scalable quantum computing systems.

These are only a few instances of the diverse range of industries where semiconductor materials are used, underscoring their vital role in developing technology and creating the contemporary world.

SEMICONDUCTOR MATERIALS FROM PERIODIC TABLE

Typically, periodic table group IV elements are used to make semiconductor materials.

Silicon

The silicon atom has 14 electrons, as shown in Fig. (**3**). Four electrons are available in the valance band. So, this configuration provides the silicon to be used as a semiconductor material.

Germanium

Germanium is also one of the semiconductors, like silicon, but it has a wide band gap, which requires more potential to transform it from a non-conducting to a conduction state.

Gray Tin

It is also one of the semiconductors having higher carrier mobility than silicon and germanium.

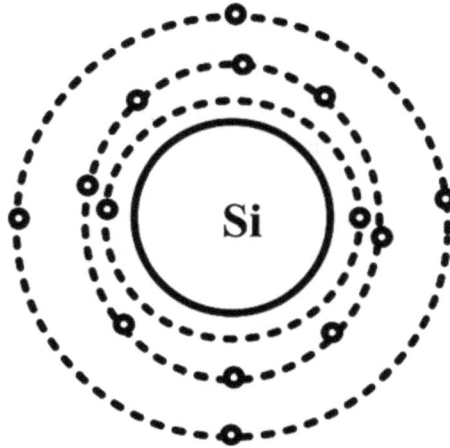

Fig. (3). Electron configuration of Si atom.

TYPES OF SEMICONDUCTOR MATERIALS BASED ON DOPING

The kind of doping semiconductor materials receive establishes their electrical conductivity, which allows for classification. The primary categories are as follows:

Intrinsic Semiconductors

Pure semiconductor materials devoid of deliberate dopants are known as intrinsic semiconductors. Since their electron and hole counts are equivalent, heat excitation is the only factor determining their electrical conductivity. Germanium (Ge) and pure silicon (Si) are two examples.

Extrinsic Semiconductors

Extrinsic semiconductors contain carefully measured dopants applied to alter their electrical characteristics. Depending on the kind of dopant added, they fall into one of the following two categories:

N-type Semiconductor

Compared to intrinsic semiconductors, n-type semiconductors have excess negative charge carriers (electrons) due to doping with materials that supply extra electrons. For n-type semiconductors, common dopants include phosphorus (P), arsenic (As), and antimony (Sb).

P-type Semiconductor

When elements that produce electron deficits, or "holes", are doped into p-type semiconductors, the outcome is an excess of positive charge carriers (holes) relative to the intrinsic semiconductor. For p-type semiconductors, common dopants include boron (B), gallium (Ga), and indium (In).

Compensated Semiconductors

Compensated semiconductors are materials engineered with a harmonized electrical conductivity, achieved by doping them with both n-type and p-type dopants to counterbalance each other's influence. This balancing is frequently employed in semiconductor devices to obtain particular electrical properties.

Highly Doped Semiconductors

Dopant concentrations in highly doped semiconductors are substantially higher than in extrinsic semiconductors. These materials are employed in specialized applications, such as power devices, sensors, and high-speed electronics, because of their distinctive electrical properties.

Based on its particular conductivity type and doping concentration, every doped semiconductor has distinct electrical properties and is employed in various electronic and optoelectronic devices. Comprehending the doping process is essential for customizing semiconductor materials to fulfill the demands of specific applications in contemporary semiconductor technologies, photonics, and electronics.

COMPOUND SEMICONDUCTOR MATERIALS

Silicon Carbide

Silicon carbide, known scientifically as moissanite (SiC), occurs naturally and was first discovered in 1893. In industrial processes, it is produced by subjecting quartz sand to reduction with surplus coke or anthracite at a temperature of 2000 to 2500°C in an electric furnace. This process is represented by the following chemical equations:

$$SiO_2 + 2\ C \rightarrow Si + 2CO \tag{1}$$

and

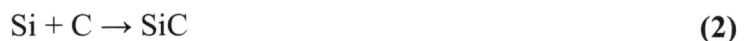

$$Si + C \rightarrow SiC \tag{2}$$

With a density of 3.22 g/cm^3, silicon carbide possesses a decomposition temperature of 2880 K. Additionally, its heat of formation ($\Delta H \circ$ 298) is measured at 66.16 kJ/mol [6].

Crystal Structure of SiC

The crystal structure of silicon carbide (SiC) [7, 8] is a crucial aspect of its material properties and plays a significant role in determining its suitability for various applications. SiC can exist in numerous crystal structures or polytypes, commonly 3C, 4H, and 6H polytypes. Here is provided an overview of the crystal structures of silicon carbide:

1. β-SiC, or 3C-SiC, embodies a cubic polytype of silicon carbide reminiscent of diamond and silicon in its crystal structure. Each silicon atom forms a tetrahedral bond with four carbon atoms within this framework. In comparison, each carbon atom reciprocally bonds with four silicon atoms, creating a highly ordered lattice akin to the arrangement found in diamond and silicon. 3C-SiC is metastable at ambient conditions, but can be grown heteroepitaxially on silicon substrates.
2. Hexagonal (4H) Polytype: The hexagonal polytype of silicon carbide, denoted as 4H-SiC, features a hexagonal crystal structure. It comprises stacked silicon and carbon atom layers organized in a hexagonal lattice. In 4H-SiC, there are four layers of silicon atoms, followed by one layer of carbon atoms in the stacking sequence, resulting in a 1:1 stacking ratio. This polytype exhibits superior electronic properties and is frequently employed in semiconductor devices requiring high power and operating at high frequencies.
3. Rhombohedral (6H) Polytype: The rhombohedral polytype of silicon carbide, known as 6H-SiC, has a similar hexagonal crystal structure to 4H-SiC, but with a different stacking sequence. In 6H-SiC, there are six layers of silicon atoms, followed by one layer of carbon atoms in the stacking sequence, resulting in a 1:1 stacking ratio, like 4H-SiC. This polytype also exhibits excellent electronic properties and is utilized in various semiconductor applications.

These are the three most common silicon carbide polytypes. Their structure is shown in Fig. (**4**), but other polytypes with different stacking sequences and crystal symmetries also exist. The selection of polytypes is subject to the application's precise needs, including considerations, such as electronic characteristics, crystal integrity, and growth feasibility.

In addition to the polytypes mentioned above, silicon carbide can also form different crystal faces or terminations, such as the Si (0001) and C (000-1) faces in hexagonal polytypes, which can influence surface properties and device performance in epitaxial growth processes and semiconductor device fabrication.

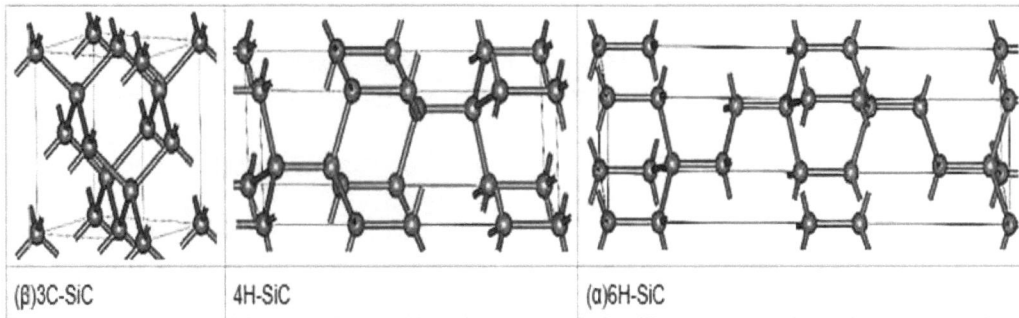

Fig. (4). Crystal structure of major SiC polytypes.

Physical Properties of SiC

When silicon carbide (SiC) is utilized in semiconductor devices, its physical properties are crucial in determining device performance and reliability. Here are listed the critical physical properties of silicon carbide relevant to semiconductor applications:

1. Wide Bandgap: Silicon carbide has a wide bandgap ranging from approximately 2.2 to 3.3 electron volts (eV), depending on the crystal structure and doping. This wide bandgap enables SiC-based semiconductor devices to work at higher temperatures and handle higher voltages than silicon-based devices. It also lowers leakage currents and breakdown voltages, enhancing performance and efficiency.
2. High Thermal Conductivity: Silicon carbide exhibits high thermal conductivity, typically around 3 to 5 times higher than silicon. This property allows SiC-based semiconductor devices to efficiently dissipate heat generated during operation, reducing thermal stress and improving device reliability, particularly in high-power applications.
3. High Electron Mobility: SiC showcases impressive electron mobility, indicating the swiftness with which electrons move through the material under the influence of an electric field. The elevated electron mobility found in SiC facilitates swift switching speeds and increased operational frequencies in semiconductor devices, rendering them well-suited for demanding applications, like RF amplifiers and power converters that necessitate high-frequency operation and robust power handling capabilities.
4. Chemical Inertness: Silicon carbide exhibits exceptional chemical inertness and resilience against corrosion from various acids and alkalis. This chemical

stability ensures the long-term reliability of SiC-based semiconductor devices, even in harsh operating environments with exposure to corrosive substances or high temperatures.

5. High Breakdown Voltage: SiC exhibits a high critical electric field strength, leading to higher breakdown voltages than silicon-based devices. This characteristic enables semiconductor devices based on SiC to function effectively at elevated voltage levels, mitigating the risk of electrical breakdowns and rendering them well-suited for demanding high-voltage power electronics applications.

6. Thermal Expansion Coefficient: The Coefficient of Thermal Expansion (CTE) of SiC is relatively low compared to silicon. This property helps to reduce thermal stress in SiC-based semiconductor devices during temperature cycling, improving device reliability and longevity.

Silicon carbide single crystals exhibit distinctive properties when doped with different impurities. Group V elements, including N, P, As, Sb, Bi, Li, and O, introduce n-type conductivity to the crystals, denoted by green coloration. Conversely, group III elements, such as B, Al, Ga, and In, and group II elements, like Be, Mg, and Ca, act as acceptors. Consequently, SiC crystals doped with these elements exhibit p-type conductivity and typically a blue or black coloration. Additionally, deviations from stoichiometric composition towards increased silicon content result in crystals with n-type electrical conductivity. Conversely, excess carbon yields crystals with p-type conductivity [9].

Boron Nitride

Boron, a metalloid with a relatively low weight, tends to establish stable covalent bonds within crystalline structures, predominantly exhibiting an oxidation state of III. Recent findings have revealed Boron Nitrides (BNs) in natural minerals and various forms, showcasing their versatility. These variations comprise hexagonal Boron Nitride (h-BN), which resembles graphite, and cubic Boron Nitride (c-BN). The hexagonal Boron Nitride (h-BN) is the quintessential crystal structure, representing the diverse range of BN compounds [10].

Crystal Structure of Boron Nitride

Boron Nitride (BN) exists in several different structural forms, similar to carbon (which can form diamond, graphite, graphene, *etc.*). The prevalent crystalline structures of boron nitride encompass hexagonal Boron Nitride (h-BN) and cubic Boron Nitride (c-BN).

Hexagonal Boron Nitride (h-BN)

Within the h-BN configuration, each boron atom is encircled by three nitrogen atoms, resulting in a planar arrangement. The hexagonal arrangement gives rise to its name. With a layered structure akin to graphite and characterized by weak van der Waals forces between its layers, h-BN is often compared to "white graphite" due to its resemblance to graphite despite its white color. Functioning as an insulator, it possesses a broad band gap.

Cubic Boron Nitride (c-BN)

The structure of c-BN resembles a diamond, as each boron atom forms bonds with four nitrogen atoms arranged in a tetrahedral configuration. It is challenging and is often used as an abrasive in cutting tools and wear-resistant applications. It demonstrates characteristics akin to diamond, including elevated thermal conductivity and robust chemical stability. c-BN is sometimes referred to as "borazon". Fig. (**5**) shows the crystal structure of various polymorphs of boron nitride.

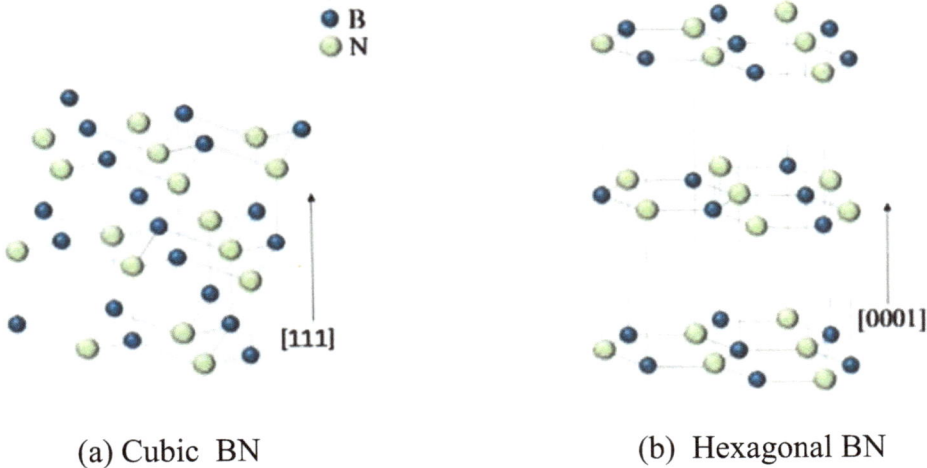

(a) Cubic BN   (b) Hexagonal BN

Fig. (5). Crystal structure of (**a**) cubic and (**b**) hexagonal BN [11].

Physical Properties of Boron Nitride

Boron Nitride (BN) is utilized in semiconductor devices because of its exclusive amalgamation of physical properties that make it suitable for various applications in the semiconductor industry. The comparison of h-BN and c-BN structural and physical characteristics are listed out in Table **1**. Here are mentioned some fundamental physical properties of boron nitride relevant to semiconductor devices:

1. **Thermal Conductivity:** Boron nitride has excellent thermal conductivity, especially its cubic form (c-BN). This property is crucial for semiconductor devices as it helps in efficient heat dissipation, prevents overheating, and ensures the stability and reliability of the devices.

2. **Electrical Insulation:** Boron nitride is an excellent electrical insulator with a wide band gap, particularly in its hexagonal form (h-BN). This property is essential for isolating different components within semiconductor devices, preventing unwanted electrical conduction between them, and ensuring proper functionality.

3. **Chemical Stability:** BN is chemically inert and resistant to many chemicals, acids, and alkalis. This stability is advantageous in semiconductor device fabrication processes where the material may be exposed to various chemicals and environments, ensuring the durability and longevity of the devices.

4. **Dielectric Properties:** Boron nitride exhibits high dielectric strength, making it suitable for insulating semiconductor devices, such as gate dielectrics in Field-effect Transistors (FETs) and insulating capacitor layers. Its high dielectric constant allows for efficient charge storage and transfer in these devices.

5. **High Temperature Stability:** Boron nitride can withstand high temperatures without significant degradation. This property is crucial in semiconductor device fabrication processes, often involving high-temperature steps, such as annealing, deposition, and etching.

6. **Mechanical Properties:** While not as relevant in all semiconductor applications, the mechanical properties of boron nitride, particularly its hardness and wear resistance, can be advantageous in specific contexts, such as protective coatings and encapsulation layers.

7. **Optical Transparency:** In its hexagonal form (h-BN), boron nitride is transparent in the visible and infrared spectra. This property can be helpful in optoelectronic applications and devices requiring transparency, such as photonic integrated circuits.

8. **Low Dielectric Constant:** Besides its high dielectric strength, boron nitride has a relatively low dielectric constant. This property is desirable in semiconductor devices as it helps minimize parasitic capacitance, enabling faster operation and improved performance.

Various Methods for Synthesizing Boron Nitride

There are several methods for synthesizing Boron Nitride (BN) semiconductors, each with advantages and limitations. Here are some of the standard methods used:

Table 1. The diverse structural and physical characteristics of different Boron Nitride (BN) polymorphs.

Material	c-BN	h-BN
Structure	Zinc blende	Hexagonal
Space group	F 43m	P63/mmc
Lattice parameter	a=3.6169 [12]	a=2.5043; c = 6.6562 [13]
Density	3.4847	2.2 [11]
Bandgap	6.4 [14]	5.971 [15]
Thermal conductivity	13 [16]	2.35 in plane
Thermal expansion at room temperature	1.15 [17]	-2.72 in plane

Chemical Vapor Deposition (CVD)

CVD is a highly employed technique for producing thin boron nitride films. In CVD, precursors containing boron and nitrogen are introduced into a reactor chamber at high temperatures, where they react to form BN films on a substrate. CVD allows precise film thickness, morphology, and composition control and is suitable for large-scale production.

Hydrothermal Synthesis

Hydrothermal synthesis involves placing boron-containing precursors and nitrogen sources into an aqueous solution and subjecting them to a reaction under high temperature and pressure conditions. The method often produces bulk BN materials with controlled crystallinity, particle size, and morphology. Scaling up hydrothermal synthesis for industrial production is feasible, and it offers a relatively straightforward process compared to alternative methods.

Physical Vapor Deposition (PVD)

PVD methods, like sputtering and Pulsed Laser Deposition (PLD), can apply thin films of BN onto substrates. During sputtering, boron nitride targets undergo bombardment by high-energy ions, leading to the ejection of atoms, which then deposit onto the substrate. PLD involves ablating a BN target with a laser to produce a plume of material deposited onto a substrate. PVD techniques offer suitable film thickness and uniformity control, but may require high vacuum conditions.

Sol-gel Synthesis

Sol-gel synthesis comprises the hydrolysis and condensation of boron and nitrogen-containing precursors to form a colloidal suspension or gel, which can then be processed into various forms, including thin films and nanoparticles. This approach provides versatility in regulating the composition and structure of the synthesized BN materials and can be integrated with other techniques for further processing.

Template-assisted Synthesis

Template-assisted synthesis entails the utilization of templates or scaffolds to regulate the morphology and structure of synthesized BN materials [18]. BN nanotubes can be synthesized using carbon nanotubes as templates, where boron and nitrogen precursors are introduced into the carbon nanotube channels and react to form BN. Utilizing this technique enables meticulous manipulation of both the dimensions and characteristics of the produced boron nitride materials.

Mechanical Exfoliation

Similar to the exfoliation of graphene from graphite, BN layers can be mechanically exfoliated from bulk BN crystals using techniques, such as scotch tape exfoliation [19]. This technique produces high-quality BN flakes with controlled thickness, but is limited in scalability and efficiency.

Red Selenium

Selenium (Se), a member of group VI alongside tellurium (Te), exhibits remarkable semiconductor properties derived from its unique electron configuration with six outer electrons in the $5s^2 5p^4$ setup. These properties include high photoconductivity, anisotropic thermal conductivity, and piezoelectric solid and thermoelectric responses. Selenium exists in three stable allotropes: an amorphous form comprised of Se8 rings (red), a polymeric form containing Se8 chains (black), and a crystalline structure featuring helical Se chains arranged in hexagonal patterns (gray), commonly referred to as trigonal selenium [20–22].

Crystal Structure of Red Selenium

The crystal structure of red selenium is composed of rings of Se8 molecules. These rings are interconnected to form a disordered network without a regular repeating pattern, characteristic of an amorphous structure. This lack of long-range order distinguishes it from crystalline forms of selenium, such as gray selenium, where atoms are arranged in a regular and repeating pattern [22].

Physical Properties of Red Selenium

Red selenium, specifically in its amorphous or red allotrope form, possesses several physical properties that make it suitable for semiconductor device applications [23–25].

1. **Photoconductivity:** Red selenium exhibits high photoconductivity, meaning its electrical conductivity increases when exposed to light. This property is essential in photovoltaic cells and light sensors used in semiconductor devices.
2. **Semiconducting Behavior:** Red selenium acts as a semiconductor, allowing for the control of electrical conductivity. This property is crucial for fabricating diodes, transistors, and other semiconductor devices.
3. **Thermoelectric Properties:** Red selenium demonstrates thermoelectric behavior, converting heat energy into electrical energy and *vice versa*. This property is utilized in thermoelectric generators and temperature sensors within semiconductor devices.
4. **Piezoelectricity:** Red selenium exhibits piezoelectric properties, generating an electric charge when subjected to mechanical stress. This characteristic is valuable in fabricating pressure sensors, accelerometers, and acoustic devices.
5. **Amorphous Structure:** The amorphous structure of red selenium provides flexibility in device fabrication processes, allowing for thin-film deposition techniques, like sputtering or evaporation. This property is advantageous for the production of integrated circuits and thin-film transistors.
6. **Optical Transparency:** Red selenium appears opaque in bulk form, so it can achieve transparency when fabricated into thin films. This unique attribute proves advantageous for various optoelectronic applications, including Light-emitting Diodes (LEDs) and photodetectors.
7. **Stability:** Red selenium is stable under typical operating conditions, ensuring the reliability and longevity of semiconductor devices incorporating this material.

Various Methods of Synthesizing Red Selenium Semiconductors

Several methods for synthesizing red selenium semiconductors offer different advantages regarding scalability, purity, and control over the material's properties. Some of the standard techniques include the following:

1. **Chemical Vapour Deposition (CVD):** In this method, selenium precursor compounds are vaporized and deposited onto a substrate, where they react to form red selenium. CVD provides precise film thickness and composition management, making it well-suited for large-scale manufacturing of thin films.

2. **Sputtering:** Sputtering involves bombarding a selenium target with energetic ions in a vacuum chamber, causing selenium atoms to dislodge and deposit onto a substrate. This technique is widely used for depositing uniform layers of red selenium onto substrates and is compatible with various substrate materials.

3. Solution-based Methods:

• **Hydrothermal Synthesis:** Red selenium nanoparticles can be synthesized by heating a selenium precursor solution under high pressure and temperature conditions in an autoclave.

• **Chemical Precipitation:** Selenium ions can be precipitated from a solution containing selenium salts by adding a reducing agent. This method allows for synthesizing red selenium nanoparticles with controlled size and morphology.

4. Physical Vapor Deposition (PVD): PVD techniques, such as thermal or electron beam evaporation, can deposit red selenium thin films by evaporating selenium atoms from a heated source and condensing them onto a substrate.

5. Sol-gel Method: In this approach, selenium precursor compounds are hydrolyzed and polymerized to form a sol. This is then converted into a gel and annealed to produce red selenium nanoparticles or thin films.

6. Hydrogen Selenide (H_2Se) Decomposition: Red selenium can be synthesized by thermally decomposing hydrogen selenide gas under controlled conditions. This method is suitable for producing high-purity red selenium with precise control over its crystallinity and morphology.

7. Electrodeposition: Electrochemical methods can deposit red selenium films onto conductive substrates by applying a voltage across the substrate and a selenium-containing electrolyte solution.

Boron Phosphide

Boron Phosphide (BP), classified as an III-V compound semiconductor, possesses a zinc-blende structure, presenting distinctive properties as both n- and p-type materials. Within this structure, each boron atom is tetrahedrally linked to four phosphorus atoms and reciprocally [26]. This arrangement creates a lattice structure with alternating boron and phosphorus atoms. The zinc-blende structure is characterized by a face-centered cubic (FCC) lattice, with boron and phosphorus atoms occupying alternating positions within the lattice. The boron and phosphorus atoms are of similar sizes, allowing them to form strong covalent bonds with each other. The atomic structure of the Boron Phosphide monolayer is shown in Fig. (**6**). This crystal structure gives BP its characteristic properties, including its mechanical strength, thermal stability, and electronic properties, such as its band gap and electron mobility.

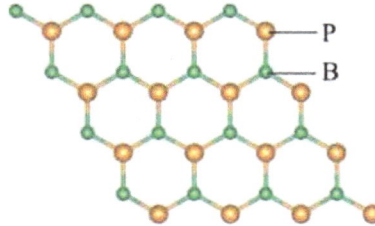

Fig. (6). Atomic structure of Boron Phosphide [27].

1. Strong Covalent Bonding: BP adopts a tetragonal bonded configuration with a small lattice constant, demonstrating sturdy covalent bonds with minimal ionic attributes.

2. Mechanical Strength: With an elastic constant comparable to β-SiC, BP exhibits remarkable mechanical resilience.

3. High Thermal and Chemical Stability: Featuring a high melting point, elevated Debye temperature, and exceptional thermal conductivity, BP demonstrates resilience to thermal and chemical stresses.

4. Wide Band Gap: BP possesses a broad band gap featuring an indirect transition, augmenting its applicability across diverse electronic applications.

5. Distinct Lattice Properties: Unlike conventional semiconductors, BP's lattice exhibits significant unharmonicity, offering unique structural characteristics.

Despite these strengths, BP faces challenges compared to other III-V compound semiconductors:

1. Crystal Growth Complexity: The remarkable melting points of BP, exceeding 3000°C, and its substantial decomposition pressures, around 105 atm at 2500°C, pose significant challenges in preparing thoroughly characterized single crystals.

2. Material Handling Difficulty: BP's refractory hardness and brittleness contribute to the complexity of its handling process, presenting challenges in its practical use.

3. Lower Electron Mobility: BP exhibits lower electron mobility than other III-V compound semiconductors, limiting its performance in specific applications.

Various Methods of Synthesizing Boron Phosphide Semiconductors

Several methods are employed to synthesize boron phosphide (BP) semiconductors, each offering unique advantages regarding scalability, purity, and control over material properties. Some standard techniques include:

1. Chemical Vapor Deposition (CVD): CVD entails the thermal decomposition of precursor gases containing boron and phosphorus onto a substrate under elevated temperatures. This method allows for precise control over film thickness, composition, and crystallinity, making it suitable for large-scale production and thin-film deposition.

2. Physical Vapor Deposition (PVD): PVD methods like sputtering or evaporation enable the deposition of thin films of BP onto substrates. These methods offer high purity and control over film thickness but may have limitations regarding scalability and uniformity.

3. Hydrothermal Synthesis: BP is synthesized through hydrothermal synthesis, which involves subjecting boron and phosphorus precursors to elevated temperatures and pressures in an aqueous solution. This method can produce high-quality crystalline BP nanoparticles or bulk crystals with precise control over size and morphology.

4. Solid-State Reaction: Solid-state reaction methods involve heating elemental boron and phosphorus precursors at high temperatures to form BP. This approach is straightforward and adaptable for scale-up, although it may necessitate high temperatures and extended reaction durations to attain optimal purity and crystalline quality.

5. Solution-Based Synthesis: Solution-based methods, such as sol-gel or solvothermal synthesis, involve dissolving boron and phosphorus precursors in a solvent and then precipitating BP through chemical reactions or thermal decomposition. These methods offer control over particle size, morphology, and composition and can be easily scaled up for large-scale production.

6. Mechanical Alloying: Mechanical alloying involves high-energy ball milling of elemental boron and phosphorus powders to form BP nanoparticles through solid-state reactions. This method offers simplicity, scalability, and control over particle size but may require prolonged milling times and post-processing steps to achieve desired properties.

Boron Arsenide

Boron arsenide (BAs) is a compound composed of boron and arsenic, denoted by the chemical formula BAs. It falls within the category of III-V semiconductors, aligning it with necessary materials like gallium arsenide (GaAs) and indium phosphide (InP) on the periodic table, all crucial to electronic applications. BAs adopts a face-centered zinc-blende crystal structure with an F43m space group, where boron and arsenic atoms interlace to form covalent bonds, showcasing a lattice constant of 4.78 A, as shown in Fig. (7) [28].

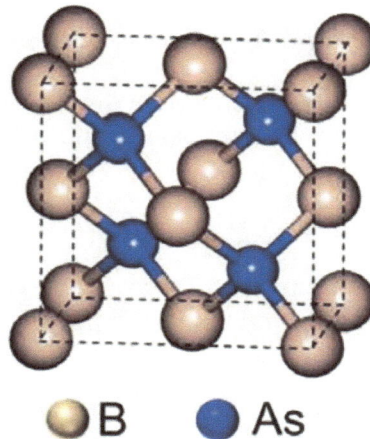

Fig. (7). Crystal structure of Boron Arsenide.

Physical Properties of Boron Arsenide

Boron arsenide (BAs) has numerous physical properties that make it attractive for semiconductor applications:

1. Thermal Conductivity: One of the most notable properties of BAs is its exceptionally high thermal conductivity, which can be around 1300 W/mK at room temperature [29]. This high thermal conductivity makes it desirable for applications where efficient heat dissipation is crucial, such as in high-power electronic devices.

2. Wide Bandgap: BAs typically have a wide band gap of 1.5 – 2.0 eV [30], making them suitable for high-temperature and high-power electronic applications. A broadband gap facilitates superior performance in high-temperature conditions and enables higher breakdown voltages.

3. Crystal Structure: BAs exhibit a zinc-blende face-centered crystal structure, similar to other III-V semiconductors. This crystal structure provides mechanical stability and facilitates the growth of high-quality crystalline materials, essential for semiconductor device fabrication.

4. Electrical Properties: BAs can be doped to control their electrical conductivity and tailor their properties for specific applications. By introducing impurities into the crystal lattice, the electrical properties of BAs can be modified to meet the requirements of different semiconductor devices.

5. Chemical Stability: Boron arsenide is chemically stable, essential for its reliability and longevity in semiconductor devices.

Various Methods of Synthesizing Boron Arsenide Semiconductors

Synthesizing boron arsenide (BA) semiconductors can be achieved through several methods, each with advantages and limitations. Here are some standard techniques:

Chemical Vapour Deposition (CVD)

Metal-Organic Chemical Vapour Deposition (MOCVD): In MOCVD, organometallic precursors of boron and arsenic are introduced into a reactor chamber along with a carrier gas. Under specific temperature and pressure conditions, these precursors decompose and deposit onto a substrate, forming a thin film of BA [31].

Hydride Vapour Phase Epitaxy (HVPE): HVPE involves reacting gaseous hydrides of boron and arsenic in a reactor chamber at high temperatures. BA thin films grow epitaxial on a substrate due to the reaction. HVPE is advantageous for large-area and high-throughput production.

Molecular Beam Epitaxy (MBE)

In MBE, boron and arsenic are evaporated from separate sources in ultra-high vacuum conditions onto a heated substrate. The evaporated atoms then condense and form a crystalline BA layer on the substrate surface with precise control over thickness and doping.

Physical Vapour Deposition (PVD)

Sputtering: In sputtering, the boron arsenide target is bombarded with high-energy ions in a vacuum chamber. This dislodges atoms from the target surface, depositing them onto a substrate to form a thin film of BAs [32].

Evaporation: BAs can also be synthesized by evaporating boron and arsenic sources simultaneously in a vacuum chamber. The evaporated species then condense onto a substrate to form a thin film.

Solution-based Methods

Sol-Gel Process: This method involves the hydrolysis and condensation of boron and arsenic precursors in a solvent to form a gel. The gel is dried and annealed to form BA nanoparticles or thin films.

Chemical Precipitation: BA nanoparticles can be synthesized by precipitating boron and arsenic salts under controlled conditions.

Hydrothermal Synthesis

BAs nanoparticles can be synthesized by subjecting an aqueous solution containing boron and arsenic precursors to high temperatures and pressures in a hydrothermal reactor.

Aluminium Nitride

Aluminium Nitride (AlN) possesses a wurtzite crystal structure [33], illustrated in Fig. (8) [34], and stands out as one of the premier band gap semiconductor materials. At room temperature, its band gap registers an impressive 6.2 eV [35]. Notably, its thermal conductivity dwarfs that of other semiconductors such as Silicon (Si), boasting a remarkable 285 W/(m•K) [36] compared to Si's 145 W/(m•K) [37]. This heightened thermal conductivity makes AlN particularly suitable for various applications. AlN's significance extends to photodetectors [38], where its low thermo-optic coefficient (4.26×10^{-5} /K at a wavelength of 1000 nm [39] enhances its power handling capabilities. As a result, AlN emerges as an ideal candidate for metasurfaces, especially in the UV wavelength spectrum, where conventional semiconductor materials tend to suffer from absorption issues.

Physical Properties of Aluminium nitride

Aluminium Nitride (AlN) boasts several advantageous physical properties that make it highly desirable for semiconductor devices:

1. Large Bandgap: AlN possesses a relatively large bandgap of around 6.2 eV at room temperature. This characteristic is crucial for applications requiring high breakdown voltage and insulation, such as power electronics.

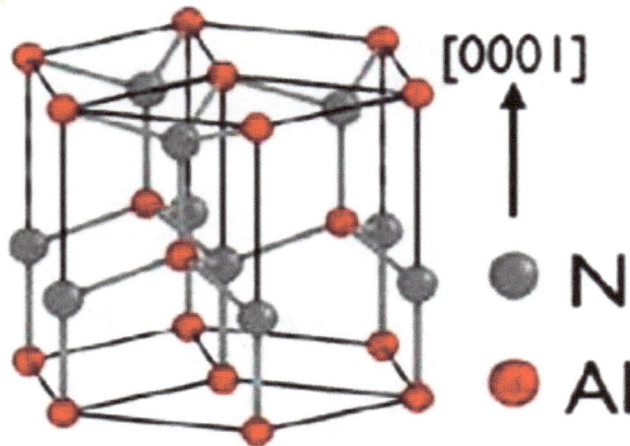

Fig. (8). Crystal structure of Aluminium Nitride.

2. High Thermal Conductivity: AlN exhibits an exceptional thermal conductivity of approximately 285 W/(m•K), surpassing many other semiconductor materials and facilitating efficient heat dissipation in semiconductor devices, improving their reliability and performance.

3. Wurtzite Crystal Structure: The wurtzite crystal structure of AlN provides inherent mechanical stability and high thermal conductivity along the crystallographic c-axis. This structural feature enables the fabrication of high-quality epitaxial layers for semiconductor devices.

4. Low Thermo-Optic Coefficient: AlN exhibits a low thermo-optic coefficient, indicating minimal changes in its refractive index with fluctuations in temperature. This property is advantageous for optical device applications, as it helps maintain performance over a wide temperature range.

5. Chemical Stability: AlN exhibits excellent chemical stability, particularly in harsh environments and at elevated temperatures. This stability enhances the durability and reliability of semiconductor devices, especially those intended for use in demanding conditions.

6. Piezoelectric Properties: AlN is piezoelectric, which generates an electric charge when subjected to mechanical stress. This property finds application in various sensor and actuator devices, contributing to the versatility of AlN-based semiconductor technologies.

Various Methods of Synthesizing Aluminium Nitride Semiconductors

Aluminium Nitride (AlN) semiconductors can be synthesized using several methods, each offering unique advantages in terms of scalability, purity, and control over material properties. Some standard methods include:

Physical Vapor Deposition (PVD)

Sputtering: AlN thin films can be deposited onto substrates using sputtering techniques. During this procedure, high-energy ions are directed towards an AlN target, resulting in the ejection of atoms which are then deposited onto a substrate, thereby forming a thin film.

Molecular Beam Epitaxy (MBE): MBE allows precise control over the growth of crystalline structures by evaporating Al and N atoms in a high vacuum environment. This method is particularly suitable for growing high-quality epitaxial layers of AlN with precise thickness and composition control.

Chemical Vapor Deposition (CVD)

Metalorganic Chemical Vapor Deposition (MOCVD): In MOCVD, volatile metalorganic precursors react with nitrogen-containing gases at elevated temperatures to deposit AlN films onto substrates. This method enables the growth of high-quality

AlN films with excellent uniformity over large areas.

Hydride Vapor Phase Epitaxy (HVPE): HVPE involves the reaction of aluminum chloride ($AlCl_3$) with ammonia (NH_3) gas at high temperatures to deposit AlN layers onto substrates. HVPE is known for its scalability and cost-effectiveness in producing large-area AlN substrates.

Sol-Gel Method

Hydrothermal Synthesis: In this method, AlN nanoparticles are synthesized by hydrothermal treatment of precursor solutions containing aluminum salts and nitrogen sources under high temperature and pressure conditions. This approach offers control over nanoparticle size, morphology, and surface properties.

Carbothermal Reduction

Direct Nitridation of Aluminum: Aluminium oxide (Al_2O_3) is mixed with carbon (C) or a nitrogen source and heated to high temperatures (>1500°C) in a nitrogen atmosphere. This route leads to the formation of AlN powder through the carbothermal reduction reaction.

<u>*Template-Assisted Growth*</u>

Template-Assisted Chemical Vapor Deposition (TACVD): In TACVD, AlN nanostructures are synthesized by depositing AlN onto template substrates with predefined patterns or nanopores. This method allows for the fabrication of AlN nanostructures with controlled dimensions and morphologies.

<u>*Combustion Synthesis*</u>

Self-Propagating High-Temperature Synthesis (SHS): SHS involves the exothermic reaction between aluminum powder and a nitrogen-containing compound (*e.g.*, urea) to rapidly produce AlN powder at high temperatures. This method offers simplicity and rapid synthesis but may require careful control to achieve the desired purity and phase composition.

Aluminium Phosphide

The aluminum phosphide (AlP) crystal structure is typically described as a zinc blende or sphalerite structure. The crystalline arrangement described is a common feature shared by numerous III-V compound semiconductors, such as GaAs and InP. In the zinc blende structure, aluminum and phosphorus atoms occupy positions within a Face-centered Cubic (FCC) lattice. Each aluminum atom is tetrahedrally coordinated with four phosphorus atoms, and each phosphorus atom is likewise tetrahedrally coordinated with four aluminum atoms. This arrangement results in a 1:1 stoichiometry between aluminum and phosphorus atoms, as shown in Fig. (**9**) [40]. The zinc blende structure is characterized by its cubic symmetry and consists of alternating layers of aluminum and phosphorus atoms. This highly stable crystal structure exhibits important semiconductor properties, making aluminium phosphide suitable for various semiconductor applications.

Physical Properties of Aluminium Phosphide

Aluminium phosphide (AlP) has some semiconductor properties that make it of interest in specific niche applications. Here are the physical properties of aluminium phosphide relevant to its potential use in semiconductors:

1. Crystal Structure: Aluminium phosphide typically adopts the zinc blend crystal structure, common among III-V compound semiconductors. This crystal structure provides important semiconductor properties, such as direct bandgap and efficient charge transport.

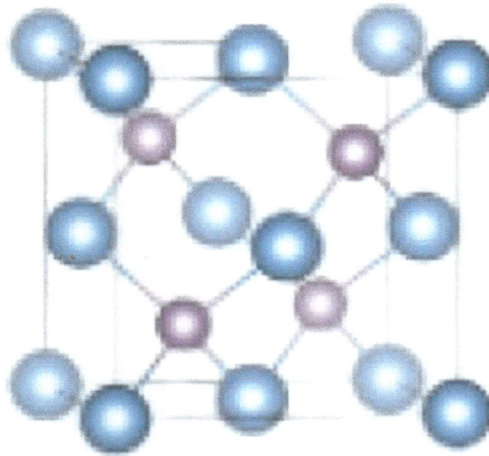

Fig. (9). Crystal structure of aluminium phosphide [40].

2. Bandgap: The band gap of aluminium phosphide depends on factors, such as doping and crystal quality. It is typically around 2.4 to 2.5 electron volts (eV) for undoped material, making it suitable for optoelectronic applications in the visible and near-infrared regions of the electromagnetic spectrum.

3. Carrier Mobility: The mobility of charge carriers (electrons and holes) in aluminium phosphide is essential for semiconductor device performance. The carrier mobility in AlP can vary depending on temperature, doping level, and crystal quality.

4. Thermal Conductivity: The thermal conductivity of aluminium phosphide is relevant for its use in semiconductor devices, particularly for heat dissipation considerations. The thermal conductivity of AlP is typically lower than common semiconductor materials, like silicon, which may affect its performance in high-power devices.

5. Dielectric Constant: The dielectric constant or relative permittivity of aluminium phosphide influences its behavior in electronic devices, mainly capacitive devices. The dielectric constant of AlP is moderate compared to other semiconductors, affecting its suitability for specific applications.

6. Optical Properties: Aluminium phosphide exhibits optical properties relevant to its use in optoelectronic devices. These properties include its refractive index, absorption spectrum, and photoluminescence characteristics, which can be tailored through material engineering and doping.

7. Mechanical Properties: While not as extensively studied as its electronic and optical properties, the mechanical properties of aluminum phosphide, such as hardness and Young's modulus, may also play a role in its performance in semiconductor devices, especially those subjected to mechanical stress.

Various Methods of Synthesizing Aluminium Phosphide Semiconductors

Synthesizing aluminum phosphide (AlP) semiconductors typically involves methods to produce high-quality crystalline material with controlled properties. Here are listed some standard methods used for synthesizing aluminum phosphide semiconductors:

Chemical Vapor Deposition (CVD)

1.1 Metal-organic Chemical Vapor Deposition (MOCVD): In MOCVD, organometallic precursors containing aluminum and phosphorus are thermally decomposed in hydrogen or other carrier gases at elevated temperatures. This method allows for precise control over deposition parameters, enabling the growth of high-quality epitaxial layers suitable for semiconductor device fabrication.

1.2 Hydride Vapor Phase Epitaxy (HVPE): HVPE involves the reaction of aluminum chloride ($AlCl_3$) and phosphorus precursors, such as phosphine (PH_3), in a high-temperature gas phase environment. This method can produce large-area AlP substrates and epitaxial layers with excellent crystalline quality.

Melt Growth Techniques

2.1 Liquid Encapsulated Czochralski (LEC) Growth: LEC involves melting high-purity aluminum and phosphorus feedstock in a crucible under controlled conditions. A seed crystal is then dipped into the melt and slowly withdrawn while rotating, allowing for the growth of single crystals of aluminum phosphide. LEC is suitable for producing large single crystals with low defect densities.

2.2 Zone Melting: Zone melting involves passing a molten zone through a polycrystalline AlP rod, resulting in the purification and recrystallization of the material. This method can refine the material and reduce impurity concentrations, improving semiconductor properties.

Solid-state Synthesis

3.1 Direct Reaction: Aluminium and phosphorus precursors are mixed in stoichiometric ratios and subjected to high temperatures (> 1000°C) in a sealed ampoule or furnace. This method results in the direct formation of aluminium phosphide through a solid-state reaction.

3.2 Mechanical Alloying: Mechanical alloying involves high-energy ball milling of aluminium and phosphorus powders in the presence of a grinding medium. This method can produce nanocrystalline or amorphous AlP powder, which can be annealed to obtain crystalline material.

Hydrothermal Synthesis

4.1 Hydrothermal Reaction: Aluminium and phosphorus precursors are dissolved in a suitable solvent, and the solution is subjected to high temperature and pressure conditions in a hydrothermal reactor. This method allows for the controlled synthesis of AlP nanoparticles or thin films with tailored morphologies and properties.

Chemical Synthesis

5.1 Solution-based Methods: Solution-based methods, such as sol-gel, precipitation, or hydrolysis, can synthesize aluminium phosphide nanoparticles or thin films from precursor solutions. These methods offer simplicity and scalability, but may require additional processing steps for semiconductor applications.

Aluminium Arsenide

Aluminium arsenide (AlAs) is a compound of aluminium and arsenic. It belongs to the III-V semiconductors group, comprising elements from group III (aluminum) and group V (arsenic) of the periodic table. Aluminum arsenide has a zinc blende crystal structure, common among III-V compound semiconductors. This structure is characterized by alternating aluminum and arsenic atom layers in a 1:1 ratio. Its bandgap energy is around 2.16 eV at room temperature, making it suitable for various electronic and optoelectronic applications [41].

Physical Properties of Aluminium Arsenide

The physical properties of aluminium arsenide (AlAs) are crucial in understanding its behaviour and applications in semiconductor devices. Here are mentioned some fundamental physical properties:

1. Lattice Constant: The lattice constant of aluminium arsenide is approximately 5.66 Angstroms (Å) at room temperature.

2. Density: The density of aluminum arsenide is around 3.7 g/cm^3.

3. Bandgap Energy: At room temperature, aluminum arsenide exhibits a direct bandgap semiconductor property, with a bandgap energy of around 2.16 electron volts (eV). This critical bandgap energy defines the wavelength of light that the material can either absorb or emit, rendering it indispensable for optoelectronic functionalities, like lasers and photo-detectors.

4. Thermal Conductivity: Aluminium arsenide exhibits relatively high thermal conductivity compared to other semiconductors. This property is advantageous in applications requiring efficient heat dissipation to maintain device performance and reliability.

5. Electron Mobility: Aluminium arsenide typically has high electron mobility, which is desirable for High Electron Mobility Transistor (HEMT) applications. High electron mobility enables faster electron transport, leading to better device performance in speed and efficiency.

6. Dielectric Constant: Aluminum arsenide's dielectric constant (relative permittivity) is around 11.1, influencing its electrical properties in capacitors and insulators.

7. Melting Point: Aluminium arsenide has a melting point of approximately 1570°C (2858°F). This high melting point is advantageous for high-temperature growth processes and fabrication techniques.

Various Methods for Synthesizing Aluminium Arsenide Semiconductors

Aluminium arsenide (AlAs) can be synthesized using various methods, each offering different levels of control over the material's properties and suitability for specific applications. Here are listed some standard methods for synthesizing aluminum arsenide semiconductors:

Molecular Beam Epitaxy (MBE)

MBE is a prevalent method for meticulously cultivating thin semiconductor films with exceptional precision. Within this process, aluminum and arsenic atoms are vaporized from distinct elemental reservoirs within a vacuum chamber. These atoms settle onto a heated substrate, crafting a crystalline thin film with meticulous thickness, composition, and crystal alignment management. MBE stands out as a favored technique in the semiconductor industry because it is renowned for generating top-tier epitaxial layers and intricate hetero-structures essential for cutting-edge semiconductor devices.

Metalorganic Chemical Vapour Deposition (MOCVD)

MOCVD is another popular method for depositing epitaxial layers of compound semiconductors. In MOCVD, volatile organometallic precursors containing aluminum and arsenic are introduced into a reactor chamber, and a carrier gas, typically hydrogen or nitrogen. The precursors decompose on a heated substrate at high temperatures, leading to aluminum arsenide thin film deposition. MOCVD offers good scalability and can be used for large-scale production of semiconductor wafers.

Hydride Vapour Phase Epitaxy (HVPE)

HVPE is a growth technique involving gaseous aluminum chloride ($AlCl_3$) and arsenic trichloride ($AsCl_3$) with hydrogen gas at high temperatures. Aluminum arsenide crystals are grown on a substrate placed in the reaction chamber, with the growth rate controlled by the precursor flow rates and reaction conditions. HVPE is known for its high growth rates and low cost, making it suitable for large-area deposition of aluminium arsenide layers.

Liquid Phase Epitaxy (LPE)

LPE involves the growth of epitaxial layers from a supersaturated solution of the semiconductor material. In LPE, a substrate is dipped into a molten solution containing aluminium and arsenic species, which crystallizes onto the substrate as it is slowly withdrawn. LPE offers simplicity and low cost compared to other methods, but may need to be improved in achieving high crystal quality and precise control over layer thickness.

Chemical Vapor Transport (CVT)

CVT is a technique used for growing bulk single crystals of aluminium arsenide. In CVT, a source material containing aluminium and arsenic is heated in the presence of a transporting agent, typically iodine. Aluminium arsenide vaporizes from the source material and deposits onto a cooler substrate, forming single crystals. CVT is suitable for producing large single crystals for research purposes, but may be less commonly used for device fabrication due to limitations in scalability and control over crystal quality.

Gallium Nitride

GaN is a binary compound of gallium (Ga) and nitrogen (N). Its chemical formula is GaN. Gallium nitride typically crystallizes in the wurtzite structure, which is a hexagonal crystal structure [42]. This structure provides GaN with unique electronic and optical properties. Depending on its crystal structure and composition, it has a wide bandgap energy, typically ranging from about 3.4 eV to 3.6 eV at room temperature. This wide bandgap allows GaN-based devices to operate at high temperatures and handle high power densities [43].

Physical Properties of Gallium Nitride

The physical properties of GaN play a crucial role in its applications as a semiconductor material. Here are listed some fundamental physical properties of GaN used in semiconductors:

1. Crystal Structure: Gallium nitride typically crystallizes in the wurtzite crystal structure, although zincblende (cubic) and rocksalt structures can also be found under certain conditions. The crystal structure arrangement of wurtzite is a hexagonal lattice of gallium and nitrogen atoms.

2. Bandgap Energy: GaN has a vast bandgap energy, typically ranging from about 3.4 eV to 3.6 eV at room temperature. This extensive bandgap permits GaN-based devices to work at elevated temperatures and handle high power densities. The energy of the bandgap determines the wavelength of light that GaN can emit or absorb, rendering it appropriate for utilization in applications, like Light-emitting Diodes (LEDs) and photodetectors.

3. Electron Mobility: GaN showcases remarkable electron mobility, indicating the speed at which electrons traverse the material when subjected to an electric field. This high electron mobility proves advantageous for high-frequency and high-power applications, facilitating faster switching speeds and reduced power losses in electronic devices.

4. Thermal Conductivity: Gallium nitride has relatively high thermal conductivity compared to other semiconductors, such as silicon and silicon carbide. This characteristic enables GaN-based devices to effectively dissipate the heat generated during operation, thereby enhancing device reliability and performance, particularly in high-power applications.

5. Dielectric Constant: The dielectric constant (also known as the relative permittivity) of GaN is relatively high, typically around 9-10. This property influences GaN-based devices' capacitance and performance in high-frequency applications.

6. Melting Point: The melting point of gallium nitride is approximately 1700°C. This high melting point allows for high-temperature device fabrication processes and ensures the stability of GaN-based devices in high-temperature environments.

7. Optical Properties: GaN exhibits interesting optical properties, including an extensive range of light emission wavelengths from Ultraviolet (UV) to visible and even into the near-infrared region. This makes GaN suitable for applications, such as LEDs, laser diodes, and photodetectors.

Various Methods for Synthesizing Gallium Nitride Semiconductors

Synthesizing gallium nitride (GaN) semiconductors typically involve processes to grow high-quality GaN crystals or epitaxial layers. Here are listed several methods commonly used for synthesizing GaN semiconductors:

Hydride Vapor Phase Epitaxy (HVPE)

HVPE is a method for growing bulk GaN crystals. It contains the reaction of gaseous gallium chloride ($GaCl_3$) and ammonia (NH_3) with elemental gallium and nitrogen sources at high temperatures (typically above 1000°C). Gallium nitride vaporizes from the source materials and deposits onto a seed crystal or substrate, which crystallizes to form a bulk GaN crystal. HVPE is known for its scalability and can produce large-area GaN wafers suitable for various applications [44].

Metalorganic Vapor Phase Epitaxy (MOVPE or MOCVD)

MOVPE, also called Metal-organic Chemical Vapor Deposition (MOCVD), is a widely used method for growing GaN epitaxial layers with high crystal quality. In MOVPE, precursors containing gallium and nitrogen, typically trimethylgallium (TMGa) and ammonia (NH_3), are introduced into a reactor chamber with a carrier gas. The precursors decompose on a heated substrate at high temperatures, producing GaN thin films with precise control over thickness, composition, and doping [45].

Hydrothermal Synthesis

Hydrothermal synthesis includes the growth of GaN crystals from an aqueous solution at high temperatures and pressures. This method dissolves gallium and nitrogen precursors in a solvent, and the solution is heated in a sealed vessel under high pressure. GaN crystals nucleate and grow from the solution onto a seed crystal or substrate. Hydrothermal synthesis can provide high-quality GaN crystals with controlled defect densities.

Ammonothermal Growth

Ammonothermal growth is a method for growing bulk GaN crystals using supercritical ammonia as the solvent and nitrogen source. Gallium, nitrogen precursors, and a seed crystal are placed in a high-pressure autoclave containing supercritical ammonia. Under high pressure and temperature conditions, GaN crystals grow from the seed crystal. Ammonothermal growth can produce high-quality GaN crystals with low defect densities and is particularly promising for applications requiring large, defect-free substrates [46].

Physical Vapor Transport (PVT)

PVT involves the sublimation of gallium nitride from a solid source material, followed by its deposition onto a cooler substrate. The source material is typically high-purity GaN powder or crystals. PVT can produce bulk GaN crystals, but it may be limited by challenges in controlling crystal quality and growth rate.

Gallium Phosphide

Gallium phosphide (GaP) is a semiconductor compound composed of gallium and phosphorus. It belongs to the III-V group of semiconductors, which means it is composed of elements from groups III and V of the periodic table.

Gallium phosphide has a zincblende crystal structure similar to that of diamond. Each gallium atom is tetrahedrally coordinated in this structure with four phosphorus atoms, and *vice versa*, as shown in Fig. (**10**). GaP has a direct bandgap energy of approximately 2.26 electron volts (eV) at room temperature. This bandgap suits optoelectronic applications, including LEDs and photovoltaic cells.

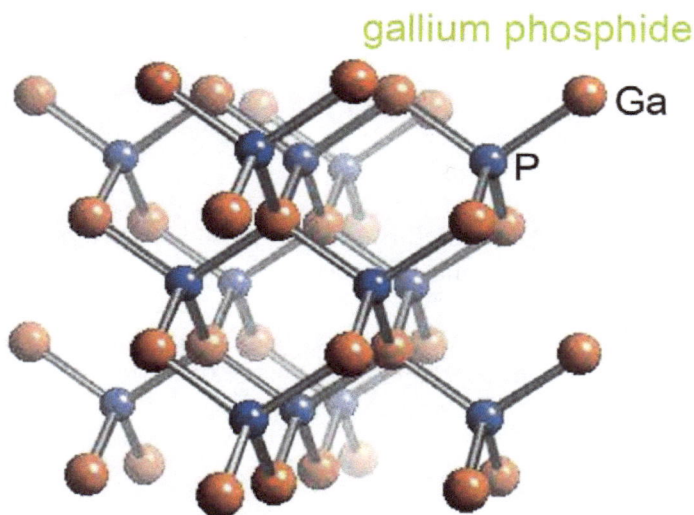

Fig. (10). Crystal structure of gallium phosphide.

Physical Properties of Gallium Phosphide

Gallium phosphide (GaP) boasts multiple physical properties, rendering it highly valuable in semiconductor applications. Here are some of the key physical properties of GaP relevant to its use in semiconductors:

1. Crystal Structure: GaP has a stable zinc blende crystal structure, allowing uniform and predictable electronic properties. This crystal structure contributes to GaP's optical and electrical characteristics.

2. Thermal Properties: GaP demonstrates robust thermal stability, rendering it well-suited for demanding high-temperature applications. Its capacity to endure elevated temperatures without notable deterioration proves beneficial for numerous semiconductor device manufacturing procedures.

3. Dielectric Properties: GaP has a relatively high dielectric constant, which is beneficial for electronic devices requiring insulating layers or capacitors. This property is essential in designing and fabricating integrated circuits and other semiconductor devices.

4. Electrical Conductivity: GaP is a semiconductor with an intrinsic electrical conductivity that can be modified through doping with specific impurities. GaP can be tailored to different semiconductor devices, including diodes, transistors, and solar cells by controlling the doping concentration.

5. Optical Properties: GaP exhibits optical transparency in the visible and near-infrared spectral regions. This property, combined with its direct bandgap, makes GaP suitable for optoelectronic devices, such as LEDs, photodetectors, and lasers.

6. Mechanical Properties: GaP is a complex and brittle material with a Mohs hardness of around 4.5. Its mechanical properties influence its suitability for various semiconductor device fabrication processes, including cutting, polishing, and bonding.

Various Methods for Synthesizing Gallium Phosphide Semiconductors

Gallium phosphide (GaP) can be synthesized using various methods, each with its own advantages and limitations. Here are some of the common methods for synthesizing GaP semiconductors:

Metal-organic Vapor Phase Epitaxy (MOVPE/MOCVD)

MOVPE, also known as MOCVD (Metal-organic Chemical Vapor Deposition), is a widely used method for synthesizing GaP thin films and heterostructures. In MOVPE, precursor gases containing gallium and phosphorus are reacted at high temperatures on a substrate, where GaP layers are deposited layer by layer. This method allows specific control over the thickness, composition, and doping of the GaP layers, making it apt for producing semiconductor devices on the industrial scale.

Hydride Vapor Phase Epitaxy (HVPE)

HVPE is another vapor-phase epitaxy technique used to grow GaP layers, primarily for large-area and bulk GaP crystal growth. In HVPE, gaseous precursors containing gallium and phosphorus, typically gallium chloride (GaCl) and phosphine (PH_3), are reacted at high temperatures in the presence of hydrogen. This method is particularly suitable for growing thick GaP layers and bulk single crystals with high crystalline quality.

Molecular Beam Epitaxy (MBE)

MBE is a deposition technique used to grow thin epitaxial layers of GaP with precise control over thickness and composition. In MBE, streams of gallium and phosphorus molecules are precisely directed onto a heated substrate within ultra-high vacuum environments, facilitating the meticulous deposition of GaP layers atom by atom. MBE offers excellent control over the growth process and is often used for fabricating high-quality semiconductor devices.

Liquid Phase Epitaxy (LPE)

LPE involves the growth of GaP layers from a supersaturated solution of gallium and phosphorus in a suitable solvent, typically gallium and phosphorus-containing metals, such as gallium or indium. Substrates are dipped into the solution, and GaP layers are deposited as the solution cools and GaP precipitates out. LPE is the most cost-effective method for growing thin GaP layers, but it is less commonly used than vapor-phase epitaxy techniques.

Solution Growth Techniques

Solution-based techniques, like Chemical Vapor Transport (CVT) and flux growth, can also synthesize GaP crystals. These methods involve the reaction of gallium and phosphorus-containing precursors in a solvent or flux material at high temperatures to form GaP crystals. Solution growth techniques are often used for producing large single crystals of GaP for research or specialized applications.

Gallium Arsenide

Gallium arsenide (GaAs) is a semiconductor compound composed of gallium and arsenic. It belongs to the III-V semiconductors group, comprising elements from groups III and V of the periodic table. GaAs have a zincblende crystal structure similar to that of diamond. In this arrangement, each Ga atom is tetrahedrally surrounded by four arsenic atoms, and *vice versa*, as depicted in Fig. (**11**). GaAs exhibits a direct bandgap energy of around 1.42 electron volts (eV) at room temperature. This bandgap allows for efficient absorption and emission of light in the near-infrared spectrum, making GaAs suitable for optoelectronic devices, such as lasers, LEDs, and photo-detectors.

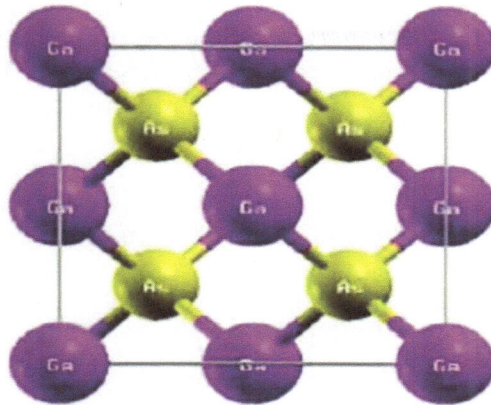

Fig. (11). Crystal structure of gallium arsenide [47].

Physical Properties of Gallium Arsenide

Gallium arsenide (GaAs) exhibits several physical properties, making it highly valuable in semiconductor applications. Here are some key physical properties of GaAs relevant to its use in semiconductors:

1. High Electron Mobility: GaAs exhibits high electron mobility compared to silicon, making it attractive for high-frequency and high-speed electronic devices. The high electron mobility of GaAs enables the fabrication of high-performance transistors and integrated circuits, particularly in applications requiring fast switching speeds and high-frequency operation.

2. Thermal Conductivity: GaAs has higher thermal conductivity than silicon. This property allows GaAs devices to dissipate heat efficiently, making them suitable for high-power applications where thermal management is critical.

3. Dielectric Constant: GaAs has a relatively high dielectric constant, which is beneficial for electronic devices requiring insulating layers or capacitors. The high dielectric constant of GaAs enables the fabrication of high-performance integrated circuits with enhanced electrical properties.

4. Optical Properties: GaAs exhibits excellent optical properties, including high absorption coefficients and low optical losses in the near-infrared spectrum. These properties make GaAs apt for various optoelectronic applications, including solar cells, infrared photodetectors, and optical communication systems.

5. Piezoelectric Effects: GaAs exhibits piezoelectric effects, generating an electric potential in response to mechanical stress. This property can be utilized in various sensor applications, such as pressure sensors and accelerometers.

6. Wide Bandgap Engineering: GaAs can be alloyed with other compound semiconductors, such as aluminum gallium arsenide (AlGaAs), to engineer materials with tailored electronic and optical properties. This capability allows for the fabrication of heterostructures and devices with specific performance characteristics.

Various Methods for Synthesizing Gallium Arsenide Semiconductors

Synthesizing gallium arsenide (GaAs) semiconductors involves several methods, each with its advantages and limitations. Here are some of the common methods used for synthesizing GaAs:

Metalorganic Vapor Phase Epitaxy (MOVPE/MOCVD)

MOVPE, also known as MOCVD (Metalorganic Chemical Vapor Deposition), is a widely used method for growing epitaxial layers of GaAs. In MOVPE, precursors containing gallium and arsenic are introduced into a reaction chamber, where they react at high temperatures to deposit GaAs layers onto a substrate. MOVPE allows precise control over the growth conditions, including temperature, pressure, and precursor flow rates, enabling the fabrication of high-quality GaAs films with customized properties.

Hydride Vapor Phase Epitaxy (HVPE)

HVPE is another vapor-phase epitaxy technique for growing GaAs layers, particularly for bulk crystal growth and thick epitaxial layers. In HVPE, gaseous precursors containing gallium and arsenic, such as gallium chloride (GaCl) and arsine (AsH_3), are reacted at high temperatures in a carrier gas. HVPE is suitable for producing large-area GaAs substrates and bulk crystals with high crystalline quality, making it valuable for industrial-scale production.

Molecular Beam Epitaxy (MBE)

MBE is a deposition technique used for growing thin epitaxial layers of GaAs with atomic precision. In MBE, gallium and arsenic molecular beams are directed onto a heated substrate in ultra-high vacuum conditions, where GaAs layers are formed atom by atom. MBE offers excellent control over the growth process and is particularly useful for fabricating heterostructures and advanced semiconductor devices with precise layer thicknesses and compositions.

Liquid Phase Epitaxy (LPE)

LPE involves the growth of GaAs layers from a supersaturated solution of gallium and arsenic in a suitable solvent. Substrates are dipped into the solution, and GaAs layers are deposited as the solution cools and GaAs precipitates out. LPE is a relatively simple and cost-effective method suitable for producing thin GaAs layers and is often used for research and small-scale production.

Flux Growth Techniques

Flux growth involves the growth of GaAs crystals from a molten flux containing gallium and arsenic. Gallium and arsenic sources, such as gallium metal and arsenic powder, are mixed with a flux material and heated to high temperatures to form GaAs crystals. Flux growth techniques can produce large single crystals of GaAs with high purity and crystalline quality for specialized applications.

Zinc Oxide

Zinc Oxide (ZnO) holds great promise for applications in semiconductor devices due to its unique properties [48–50]. Notably, it boasts a direct and wide bandgap situated in the near Ultraviolet (UV) spectral range. Additionally, it exhibits a substantial free-exciton binding energy, ensuring the persistence of excitonic emission processes even at or above room temperature. Structurally, ZnO crystallizes in the wurtzite configuration, as shown in Fig. (**12**). In this arrangement, each zinc ion (Zn^{2+}) is tetrahedrally surrounded by four oxygen ions (O^{2-}), conversely forming a hexagonal close-packed (hcp) lattice for oxygen ions. Notably, zinc ions occupy half of the tetrahedral sites within this lattice.

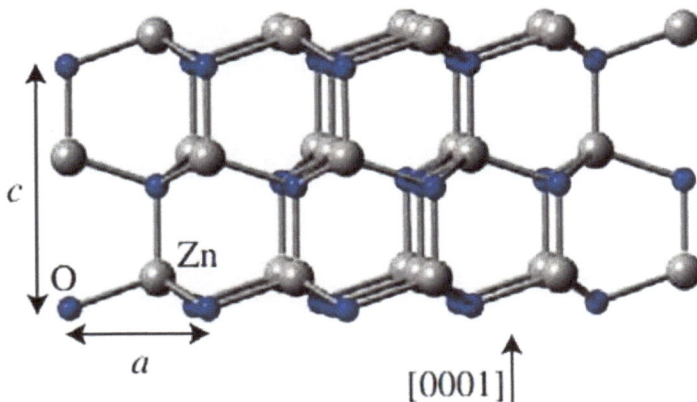

Fig. (12). Crystal structure of zinc oxide [51].

Physical Properties of Zinc Oxide

Zinc oxide (ZnO) exhibits several physical properties that make it highly desirable for use in semiconductor devices:

1. Wide Bandgap: ZnO possesses a wide bandgap energy of approximately 3.3 eV at room temperature. This wide bandgap allows ZnO to efficiently absorb and emit light in the Ultraviolet (UV) region, making it suitable for UV-based applications, such as UV photodetectors and UV Light-emitting Diodes (LEDs).

2. High Electron Mobility: ZnO exhibits high electron mobility, which is advantageous for semiconductor applications requiring fast electron transport. This property makes ZnO suitable for high-frequency electronic devices, such as FETs and high-speed integrated circuits.

3. Piezoelectricity: ZnO is piezoelectric, meaning it generates an electric charge in response to mechanical stress or pressure. This property is exploited in various sensor applications, including pressure sensors, acoustic wave sensors, and strain sensors.

4. Optical Transparency: ZnO is transparent in the visible spectrum, particularly in the range of 350-700 nm. This transparency makes ZnO suitable for Transparent Conductive Oxide (TCO) applications, such as transparent electrodes in displays, touchscreens, and solar cells.

5. Thermal Stability: ZnO exhibits good thermal stability, allowing it to maintain its structural and electrical properties at high temperatures. This property is important for semiconductor processing techniques, such as sputtering, CVD, and annealing.

6. Chemical Stability: ZnO is chemically stable under ambient conditions, with resistance to oxidation and corrosion. This stability contributes to the long-term reliability of semiconductor devices incorporating ZnO components.

7. Low Toxicity: Compared to other semiconductor materials, like cadmium telluride (CdTe) and lead sulfide (PbS), ZnO is relatively non-toxic, making it environmentally friendly and suitable for biomedical applications, such as biosensors and drug delivery systems.

Various Methods for Synthesizing Zinc Oxide Semiconductors

Zinc oxide (ZnO) can be synthesized using various methods, each offering different advantages in terms of scalability, purity, and control over properties. Here are some common methods for synthesizing ZnO semiconductors:

Deposition Methods

Physical Vapor Deposition (PVD): In this method, zinc oxide is deposited onto a substrate through the vapor phase. Techniques include thermal evaporation, sputtering, and Pulsed Laser Deposition (PLD).

Chemical Vapor Deposition (CVD): ZnO can be synthesized by reacting vapor-phase zinc precursors with oxygen or water vapor in the presence of a substrate at elevated temperatures. CVD methods offer precise control over film thickness and composition, making them suitable for thin film deposition.

Solution-based Methods

Hydrothermal Synthesis: ZnO nanoparticles or thin films are synthesized by reacting aqueous solutions of zinc salts (*e.g.*, zinc nitrate) with a hydroxide source (*e.g.*, sodium hydroxide) under high-temperature and high-pressure conditions in a closed vessel. This method allows for control over particle size and morphology.

Sol-gel Process: ZnO can be synthesized by hydrolyzing and condensing zinc alkoxide precursors in a solution to form a sol, followed by gelation and solidification. This method offers control over composition and morphology and is suitable for thin film deposition and coating applications.

Mechanical Methods

Ball Milling: ZnO nanoparticles can be synthesized by milling bulk ZnO powder in the presence of a grinding medium (*e.g.*, ceramic balls) and a surfactant. Ball milling allows for control over particle size and crystallinity and is a scalable method for large-scale production.

Ultrasonication: ZnO nanoparticles can be synthesized by dispersing bulk ZnO powder in a liquid medium and subjecting the dispersion to ultrasonic waves. Ultrasonication promotes the breakup of aggregates and the formation of nanoparticles.

Template-assisted Methods

Template-assisted Growth: ZnO nanostructures with controlled size, shape, and orientation can be synthesized using templates, such as porous membranes, nanowires, or nanopatterned substrates. The template directs the growth of ZnO structures through deposition or growth from solution.

Combustion Synthesis

Solution Combustion: ZnO nanoparticles can be synthesized by igniting a mixture of zinc-containing precursors and a fuel source (*e.g.*, urea or glycine). Combustion releases energy, driving the formation of ZnO nanoparticles.

Cadmium Arsenide

Cadmium arsenide (Cd_3As_2) is indeed a semiconductor compound. It belongs to the III-V semiconductor family, which comprises elements from groups III and V of the periodic table, specifically cadmium (Cd) and arsenic (As). Cadmium arsenide has a zinc-blende crystal structure.

Physical Properties of Cadmium Arsenide

Zinc oxide (ZnO) is a wide bandgap semiconductor with numerous physical properties that make it suitable for various semiconductor applications. Here are listed some key physical properties of zinc oxide relevant to its use in semiconductor devices:

1. Bandgap: Zinc oxide has a wide bandgap energy of around 3.37 eV at room temperature. This makes it suitable for applications requiring transparent conductive films, Ultraviolet (UV) optoelectronics, and high-frequency devices.

2. Crystal Structure: Zinc oxide typically crystallizes in the wurtzite structure, a hexagonal close-packed atom arrangement. This crystal structure contributes to its piezoelectric and pyroelectric properties, which are advantageous for sensor applications.

3. Thermal Stability: Zinc oxide exhibits good thermal stability at high temperatures, making it suitable for applications in harsh environments and high-temperature processes, such as sputtering or chemical vapor deposition.

4. Electrical Properties: ZnO is an n-type semiconductor under intrinsic conditions, meaning it has excess electrons. It can also be doped to modify its electrical properties. Various dopants, such as aluminum (Al), gallium (Ga), and indium (In), can be used to control the conductivity and other electronic properties of ZnO.

5. Transparent Conductivity: Zinc oxide thin films can exhibit excellent transparency in the visible spectrum, while maintaining good electrical conductivity. This property is highly desirable for applications, such as transparent electrodes in solar cells, touchscreens, and displays.

6. Piezoelectricity: ZnO is piezoelectric, which generates an electric charge in response to mechanical stress. This property is utilized in various sensor applications, including pressure sensors, accelerometers, and acoustic devices.

7. Optical Properties: Zinc oxide exhibits optical transparency over various wavelengths, from the ultraviolet to the visible spectrum. It is transparent to visible light while absorbing UV radiation, making it useful for UV-blocking coatings and UV photodetectors.

8. Chemical Stability: ZnO is chemically stable under normal operating conditions, which enhances its reliability in semiconductor devices. However, it can be susceptible to corrosion in certain aggressive environments.

Various Methods for Synthesizing Cadmium Arsenide Semiconductors

Synthesizing cadmium arsenide (Cd_3As_2) semiconductors typically involves numerous methods, each with its own advantages and disadvantages. Here are several common methods used for synthesizing cadmium arsenide:

Chemical Vapor Transport (CVT)

In this method, high-purity cadmium and arsenic precursors are heated in a closed ampoule under a controlled atmosphere. The reactants vaporize, and the cadmium arsenide product is deposited on a cooler substrate within the reaction vessel. CVT is a widely used method for synthesizing single crystals of cadmium arsenide with high purity and quality.

Molecular Beam Epitaxy (MBE)

MBE is a technique used to grow thin films of crystalline materials on a substrate. In a vacuum chamber, cadmium arsenide, high-purity cadmium, and arsenic beams are directed onto a heated substrate. The atoms deposit onto the substrate surface, forming a crystalline film of cadmium arsenide. MBE allows precise control over the growth parameters, resulting in high-quality thin films with controlled thickness and composition.

Metal-organic Chemical Vapor Deposition (MOCVD)

MOCVD is another technique for growing thin films of semiconductor materials. In this method, volatile metalorganic precursors containing cadmium and arsenic are introduced into a reaction chamber along with a carrier gas. These precursors decompose at elevated temperatures, depositing cadmium arsenide onto a heated substrate. MOCVD offers scalability and can be used to deposit cadmium arsenide films over large areas for industrial applications.

Hydrothermal Synthesis

Hydrothermal synthesis involves the growth of crystals from a solution at elevated temperatures and pressures. In the case of cadmium arsenide, a solution containing cadmium and arsenic precursors is heated under pressure in a sealed vessel. Cadmium arsenide crystals nucleate and grow from the solution over time. Hydrothermal synthesis can produce bulk cadmium arsenide crystals and is suitable for growing large single crystals.

Solid-state Reaction

Solid-state synthesis involves heating and reacting solid precursors to form the desired compound. In the case of cadmium arsenide, high-purity cadmium and arsenic powders are mixed and heated to high temperatures in a sealed ampoule. The precursors react to form cadmium arsenide, which can be collected and purified. Solid-state reaction methods are relatively simple and can synthesize cadmium arsenide powders or polycrystalline materials.

Zinc Phosphide

Zinc phosphide (Zn_3P_2) has a crystal structure that belongs to the zinc blende or sphalerite crystal structure type, a common structure for many binary compounds. In this structure, zinc (Zn) and phosphorus (P) atoms are arranged in a cubic close-packed arrangement, with zinc atoms occupying all the tetrahedral sites and phosphorus atoms occupying all the octahedral sites.

Physical Properties of Zinc Phosphide

Zinc phosphide (Zn_3P_2) has several physical properties that make it suitable for applications in semiconductor technology. Zinc phosphide is typically doped with impurities to modify its electrical properties when used as a semiconductor. Here are listed some of the key physical properties of zinc phosphide relevant to its use in semiconductors:

1. Bandgap: Zinc phosphide has a relatively wide bandgap, typically around 1.5 to 2.0 electron volts (eV), depending on the crystal structure and doping level. This band gap suits optoelectronic devices, such as LEDs and photovoltaic cells.

2. Crystal Structure: Zinc phosphide crystallizes in a cubic crystal lattice structure, commonly adopting the zinc blende or sphalerite crystal structure type. This crystal structure provides the foundation for its semiconductor behavior and allows the controlled doping of impurities to modify its conductivity.

3. Electrical Conductivity: In its pure form, zinc phosphide exhibits intrinsic semiconductor behavior, meaning its electrical conductivity can be modified by doping with impurities. It can be doped with elements, such as nitrogen, arsenic, or indium, to create n-type or p-type semiconductors, allowing for the fabrication of various semiconductor devices.

4. Thermal Stability: Zinc phosphide exhibits good thermal stability, making it suitable for high-temperature semiconductor applications. Its wide bandgap also contributes to its thermal stability by reducing the thermal generation of charge carriers.

5. Optical Properties: Zinc phosphide has optical properties relevant to its use in semiconductor devices. Its wide bandgap results in transparency to visible light, which is advantageous for optoelectronic applications, such as LEDs and photodetectors.

6. Mechanical Properties: While not as critical for semiconductor applications as other properties, the mechanical properties of zinc phosphide, such as its hardness and brittleness, may influence device fabrication processes and the reliability of semiconductor devices.

7. Compatibility: Zinc phosphide is compatible with standard semiconductor fabrication techniques, such as epitaxial growth, ion implantation, and photolithography, allowing for integrating zinc phosphide-based components into semiconductor devices and circuits.

Various Methods for Synthesizing Zinc Phosphide Semiconductors

Zinc phosphide (Zn_3P_2) can be synthesized using various methods tailored to produce semiconductor-grade material suitable for specific applications. Here are mentioned several common methods for synthesizing zinc phosphide semiconductors:

Solid-state Reaction

In this method, zinc and phosphorus precursors are mixed together and heated to high temperatures (typically above 800°C) in an inert atmosphere, such as argon or nitrogen. The reaction typically proceeds according to the following equation:

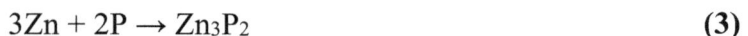

$$3Zn + 2P \rightarrow Zn_3P_2 \tag{3}$$

The resulting zinc phosphide powder can be further processed to produce semiconductor devices.

Chemical Vapor Transport (CVT)

In CVT synthesis, zinc and phosphorus reactants are heated in the presence of a transport agent, typically iodine (I_2), in a sealed ampoule under a vacuum. At elevated temperatures (typically around 700-900°C), the volatile zinc species react with phosphorus to produce zinc phosphide. The transport agent helps migrate the reaction products to form single crystals of zinc phosphide.

Hydride Vapor Phase Epitaxy (HVPE)

HVPE is a method commonly used to grow zinc phosphide thin films on suitable substrates. In this process, zinc and phosphorus precursors are reacted in a reactor chamber at high temperatures (typically 600-800°C) in the presence of hydrogen chloride (HCl) or hydrogen bromide (HBr) gas. The reaction produces volatile zinc and phosphorus species, which react on the substrate surface to deposit zinc phosphide thin films.

Solution-based Methods

Solution-based methods involve the synthesis of zinc phosphide using precursor solutions containing zinc and phosphorus sources. One example is solvothermal synthesis, where zinc and phosphorus precursors are dissolved in a suitable solvent, and the solution is heated under high pressure for a period of time. Another method is chemical bath deposition, where zinc phosphide thin films are deposited onto substrates by immersing them in a bath containing zinc and phosphorus precursors.

Hydrothermal Synthesis

Hydrothermal synthesis involves the reaction of zinc and phosphorus precursors in an aqueous solution at elevated temperatures and pressures. Under controlled conditions, zinc phosphide nanoparticles or thin films can be synthesized by this method.

Electrochemical Deposition

Zinc phosphide thin films can also be electrodeposited onto conductive substrates from electrolyte solutions containing zinc and phosphorus ions. By controlling the deposition parameters, such as voltage, current density, and electrolyte composition, the properties of the deposited zinc phosphide can be tailored for specific applications.

Zinc Antimonide

Zinc antimonide (ZnSb) is a semiconductor material with unique properties that make it suitable for various electronic and optoelectronic applications. Zinc antimonide crystallizes in the zinc blende crystal structure, which is a cubic crystal lattice similar to that of the diamond. In this structure, zinc and antimony atoms occupy alternating positions in the crystal lattice, as shown in Fig. (13) [52]. Zinc antimonide has a relatively narrow band gap, typically around 0.5 to 0.7 electron volts (eV) at room temperature. This makes it suitable for infrared optoelectronic devices and thermoelectric applications. Like other semiconductor materials, zinc antimonide can be doped with impurities to modify its electrical properties. Doping with elements, such as silicon (Si) or tin (Sn), can alter the carrier concentration and conductivity of the material.

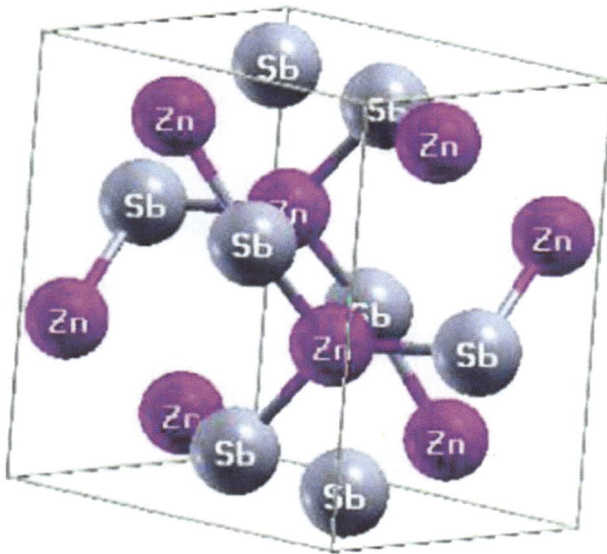

Fig. (13). Crystal structure of zinc antimonide [52].

Physical Properties of Zinc Antimonide

Zinc antimonide (ZnSb) possesses several physical properties that make it suitable for applications in semiconductor technology. Here are some of the key physical properties of zinc antimonide relevant to its use in semiconductors:

1. Crystal Structure: Zinc antimonide crystallizes in the zinc blende crystal structure, a cubic crystal lattice similar to diamond. In this structure, zinc and antimony atoms occupy alternating positions in the crystal lattice.

2. Bandgap: At room temperature, zinc antimonide has a relatively narrow bandgap, typically around 0.5 to 0.7 electron volts (eV). This band gap is smaller than other semiconductor materials, making zinc antimonide suitable for infrared optoelectronic devices.

3. Thermoelectric Properties: Zinc antimonide exhibits high thermoelectric efficiency, which can efficiently convert heat energy into electrical energy and *vice versa*. This property is attributed to its high Seebeck coefficient and relatively low thermal conductivity.

4. Electrical Conductivity: Zinc antimonide can be doped with impurities to modify its electrical properties. Doping with elements, such as silicon (Si) or tin (Sn), can alter the carrier concentration and conductivity of the material.

5. Carrier Mobility: Zinc antimonide has high electron mobility, which is desirable for electronic devices, such as transistors and high-speed circuits.

6. Optical Properties: Zinc antimonide is primarily used in infrared optoelectronic devices due to its narrow bandgap. It is transparent to visible light, but absorbs and emits light in the infrared spectrum.

7. Thermal Stability: Zinc antimonide exhibits good thermal stability, allowing it to withstand high temperatures without significantly degrading its properties. This makes it suitable for use in devices operating in harsh environments.

8. Thin Film Growth: Zinc antimonide thin films can be deposited using Molecular Beam Epitaxy (MBE), Metal-organic Chemical Vapor Deposition (MOCVD), and sputtering. These thin films find applications in infrared photodetectors and optoelectronic devices.

9. Mechanical Properties: While not as critical for semiconductor applications as other properties, the mechanical properties of zinc antimonide, such as its hardness and brittleness, may influence device fabrication processes and the reliability of semiconductor devices.

10. Chemical Stability: Zinc antimonide is generally stable under normal operating conditions and is not reactive with common semiconductor processing chemicals.

Various Methods for Synthesizing Zinc Antimonide Semiconductors

Synthesizing zinc antimonide (ZnSb) semiconductor material involves several methods tailored to produce high-quality material suitable for specific applications. Here are some common methods used for synthesizing zinc antimonide semiconductors:

Solid-state Reaction

In this method, zinc and antimony precursors are mixed together in stoichiometric proportions and heated to high temperatures (typically above 500°C) in an inert atmosphere, such as argon or nitrogen.

The reaction proceeds to form zinc antimonide according to the equation:

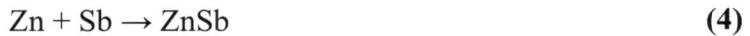

$$Zn + Sb \rightarrow ZnSb \tag{4}$$

The resulting zinc antimonide material can be further processed into bulk crystals or thin films for semiconductor device fabrication.

Chemical Vapor Transport (CVT)

CVT synthesis involves using a vapor transport agent to grow single crystals of zinc antimonide. Zinc and antimony precursors are placed in a sealed ampoule with a transport agent, such as iodine (I2), and heated to high temperatures (typically around 800-1000°C) in a gradient furnace. The volatile zinc and antimony species produced react at a cooler zone in the furnace to form single crystals of zinc antimonide.

Hydride Vapor Phase Epitaxy (HVPE)

HVPE is commonly used to grow zinc antimonide thin films on suitable substrates. In this process, zinc and antimony precursors are reacted in a reactor chamber at high temperatures (typically 500-700°C) in the presence of hydrogen chloride (HCl) or hydrogen bromide (HBr) gas. The reaction produces volatile zinc and antimony species, which react on the substrate surface to deposit zinc antimonide thin films.

Molecular Beam Epitaxy (MBE)

MBE is used to grow thin epitaxial films of zinc antimonide with precise control over film thickness and composition. In MBE, zinc and antimony atoms are evaporated from separate sources in ultra-high vacuum conditions and deposited onto a heated substrate. Controlling the deposition rate and substrate temperature allows epitaxial zinc antimonide layers with well-defined crystallographic orientations to be grown.

Chemical Solution Deposition

Solution-based methods involve the synthesis of zinc antimonide thin films using precursor solutions containing zinc and antimony sources. These solutions can be spin-coated, dip-coated, or sprayed onto substrates, followed by annealing at elevated temperatures to form zinc antimonide thin films.

Electrochemical Deposition

Zinc antimonide thin films can also be electrodeposited onto conductive substrates from electrolyte solutions containing zinc and antimony ions. By controlling the deposition parameters, such as voltage, current density, and electrolyte composition, the properties of the deposited zinc antimonide can be tailored for specific applications.

CONCLUSION

The semiconductor industry is growing daily due to its high demand in the electronic industry. So, the required characteristics of a semiconductor depend upon its material. Thus, to achieve better performance from the semiconductor devices, more research has to be done on the material part. Here, this chapter has delivered some fundamentals of semiconductor devices and basic materials used in semiconductor devices and described the classification of semiconductors. The final part of the chapter has focused on compound semiconductors, providing a clear understanding of various hybrid semiconductor materials used to date. Moreover, this chapter can enable the readers to identify the best semiconductor material for their application.

KEY POINTS

- This chapter delves into the necessity of semiconductor devices.
- A wide range of applications associated with semiconductor devices are highlighted.
- The chapter discloses various basic materials, such as Si, Ge, *etc.*, to manufacture semiconductor devices.
- Advancements in semiconductor devices to improve their performance and reliability with the help of compound materials are explored.
- Moreover, this chapter provides the merits and demerits of all the basic and compound materials used in the design of semiconductor devices.

ACKNOWLEDGEMENTS

The authors thank the O.P. Jindal University for providing the resources and environment necessary for their research.

REFERENCES

[1]   J. Millan, P. Godignon, X. Perpina, A. Perez-Tomas, and J. Rebollo, "A survey of wide bandgap power semiconductor devices", *IEEE Trans. Power Electron.,* vol. 29, no. 5, pp. 2155-2163, 2014.
[http://dx.doi.org/10.1109/TPEL.2013.2268900]

[2]   J. Rabkowski, D. Peftitsis, and H.P. Nee, "Recent advances in power semiconductor technology", *Power Electron. Renew. Energy Syst. Transp. Ind. Appl.,* vol. 9781118634035, pp. 69-106, 2014.

[3]   A. Narendra, "V. Naik N, and A. K. Panda, "Design of power electronic devices in the domain of energy storage,"", *Emerg. Trends Energy Storage Syst. Ind. Appl.,* no. Jan, pp. 27-65, 2023.

[4]   S. Nakamura, "Basics of Semiconductor", *Polym. Mater. Electron. Packag.,* no. Aug, pp. 1-7, 2023.

[5]   A. Narendra, N. Venkataramana Naik, A.K. Panda, and N. Tiwary, "A Comprehensive review of PV driven electrical motors", *Sol. Energy,* vol. 195, pp. 278-303, 2020.
[http://dx.doi.org/10.1016/j.solener.2019.09.078]

[6]   M. Gasik, V. Dashevskii, and A. Bizhanov, ""Metallurgy of silicon and silicon carbide," Top. Mining, Metall", *Mater. Des.,* pp. 35-55, 2020.

[7]   L.M. Soltys, I.F. Mironyuk, I.M. Mykytyn, I.D. Hnylytsia, and L.V. Turovska, "Synthesis and properties of silicon carbide (review)", *Physics and Chemistry of Solid State,* vol. 24, no. 1, pp. 5-16, 2023.
[http://dx.doi.org/10.15330/pcss.24.1.5-16]

[8]   H. Abderrazak, E. Selmane, and B. H. Hmida, "Silicon carbide: Synthesis and properties", *Prop. Appl. Silicon Carbide,* 2011.
[http://dx.doi.org/10.5772/15736]

[9]   Y.W. Kim, Y.H. Kim, and K.J. Kim, "Electrical properties of liquid-phase sintered silicon carbide ceramics: a review", *Crit. Rev. Solid State Mater. Sci.,* vol. 45, no. 1, pp. 66-84, 2020.
[http://dx.doi.org/10.1080/10408436.2018.1532394]

[10]   C. Zhou, C. Lai, C. Zhang, G. Zeng, D. Huang, M. Cheng, L. Hu, W. Xiong, M. Chen, J. Wang, Y. Yang, and L. Jiang, "Semiconductor/boron nitride composites: Synthesis, properties, and photocatalysis applications", *Appl. Catal. B,* vol. 238, pp. 6-18, 2018.
[http://dx.doi.org/10.1016/j.apcatb.2018.07.011]

[11]   N. Izyumskaya, D.O. Demchenko, S. Das, Ü. Özgür, V. Avrutin, and H. Morkoç, "Recent development of boron nitride towards electronic applications", *Adv. Electron. Mater.,* vol. 3, no. 5, p. 1600485, 2017.
[http://dx.doi.org/10.1002/aelm.201600485]

[12]   A. Nagakubo, H. Ogi, H. Sumiya, K. Kusakabe, and M. Hirao, "Elastic constants of cubic and wurtzite boron nitrides", *Appl. Phys. Lett.,* vol. 102, no. 24, p. 241909, 2013.
[http://dx.doi.org/10.1063/1.4811789]

[13]   A.V. Kurdyumov, V.L. Solozhenko, and W.B. Zelyavski, "Lattice parameters of boron nitride polymorphous modifications as a function of their crystal-structure perfection", *J. Appl. Cryst.,* vol. 28, no. 5, pp. 540-545, 1995.
[http://dx.doi.org/10.1107/S002188989500197X]

[14]   R.M. Chrenko, "Ultraviolet and infrared spectra of cubic boron nitride", *Solid State Commun.,* vol. 14, no. 6, pp. 511-515, 1974.
[http://dx.doi.org/10.1016/0038-1098(74)90978-8]

[15]   K. Watanabe, T. Taniguchi, and H. Kanda, "Direct-bandgap properties and evidence for ultraviolet lasing of hexagonal boron nitride single crystal", *Nat. Mater,* vol. 3, no. 6, pp. 404-409, 2004.
[http://dx.doi.org/10.1038/nmat1134]

[16]   G.A. Slack, "Nonmetallic crystals with high thermal conductivity", *J. Phys. Chem. Solids,* vol. 34, no. 2, pp. 321-335, 1973.

[http://dx.doi.org/10.1016/0022-3697(73)90092-9]

[17] G.A. Slack, and S.F. Bartram, "Thermal expansion of some diamondlike crystals", *J. Appl. Phys.*, vol. 46, no. 1, pp. 89-98, 1975.
[http://dx.doi.org/10.1063/1.321373]

[18] Z. Tong, D. Yang, Z. Li, Y. Nan, F. Ding, Y. Shen, and Z. Jiang, "Thylakoid-inspired multishell g-C 3 N 4 Nanocapsules with enhanced visible-light harvesting and electron transfer properties for high-efficiency photocatalysis", *ACS Nano*, vol. 11, no. 1, pp. 1103-1112, 2017.
[http://dx.doi.org/10.1021/acsnano.6b08251] [PMID: 28032986]

[19] D. Lee, B. Lee, K.H. Park, H.J. Ryu, S. Jeon, and S.H. Hong, "Scalable exfoliation process for highly soluble boron nitride nanoplatelets by hydroxide-assisted ball milling", *Nano Lett.*, vol. 15, no. 2, pp. 1238-1244, 2015.
[http://dx.doi.org/10.1021/nl504397h] [PMID: 25622114]

[20] Y. Xia, P. Yang, Y. Sun, Y. Wu, B. Mayers, B. Gates, Y. Yin, F. Kim, and H. Yan, "One-dimensional nanostructures: synthesis, characterization, and applications", *Adv. Mater.*, vol. 15, no. 5, pp. 353-389, 2003.
[http://dx.doi.org/10.1002/adma.200390087]

[21] B. Gates, Y. Yin, and Y. Xia, "A solution-phase approach to the synthesis of uniform nanowires of crystalline selenium with lateral dimensions in the range of 10-30 nm", *J. Am. Chem. Soc.*, vol. 122, no. 50, pp. 12582-12583, 2000.
[http://dx.doi.org/10.1021/ja002608d]

[22] W. Huang, M. Wang, L. Hu, C. Wang, Z. Xie, and H. Zhang, "Recent advances in semiconducting monoelemental selenium nanostructures for device applications", *Adv. Funct. Mater.*, vol. 30, no. 42, p. 2003301, 2020.
[http://dx.doi.org/10.1002/adfm.202003301]

[23] Xing. C, Xie. Z, Liang. Z, Liang. W, Fan. T, Ponraj. J. S & Zhang. H, "2D nonlayered selenium nanosheets: facile synthesis, photoluminescence, and ultrafast photonics", *Adv. Opt. Mater.*, vol. 5, no. 24, p. 1700884, 2017.

[24] Z.Y. Jiang, Z.X. Xie, S.Y. Xie, X.H. Zhang, R-B. Huang, and L.S. Zheng, "High purity trigonal selenium nanorods growth *via* laser ablation under controlled temperature"., *Chem. Phys. Lett.*, vol. 368, no. 3-4, pp. 425-429, 2003.
[http://dx.doi.org/10.1016/S0009-2614(02)01918-8]

[25] S. Chaudhary, and S.K. Mehta, "Selenium nanomaterials: applications in electronics, catalysis and sensors", *J. Nanosci. Nanotechnol.*, vol. 14, no. 2, pp. 1658-1674, 2014.
[http://dx.doi.org/10.1166/jnn.2014.9128] [PMID: 24749448]

[26] Y. Kumashiro, "Refractory semiconductor of boron phosphide", *J. Mater. Res.*, vol. 5, no. 12, pp. 2933-2947, 1990.
[http://dx.doi.org/10.1557/JMR.1990.2933]

[27] T.V. Vu, A.I. Kartamyshev, N.V. Hieu, T.D.H. Dang, S.N. Nguyen, N.A. Poklonski, C.V. Nguyen, H.V. Phuc, and N.N. Hieu, "Structural, elastic, and electronic properties of chemically functionalized boron phosphide monolayer", *RSC Advances*, vol. 11, no. 15, pp. 8552-8558, 2021.
[http://dx.doi.org/10.1039/D1RA00576F] [PMID: 35423400]

[28] J.S. Kang, M. Li, H. Wu, H. Nguyen, and Y. Hu, "Basic physical properties of cubic boron arsenide", *Appl. Phys. Lett.*, vol. 115, no. 12, p. 122103, 2019.
[http://dx.doi.org/10.1063/1.5116025]

[29] F. Tian, K. Luo, C. Xie, B. Liu, X. Liang, L. Wang, G.A. Gamage, H. Sun, H. Ziyaee, J. Sun, Z. Zhao, B. Xu, G. Gao, X-F. Zhou, and Z. Ren, "Mechanical properties of boron arsenide single crystal", *Appl. Phys. Lett.*, vol. 114, no. 13, p. 131903, 2019.
[http://dx.doi.org/10.1063/1.5093289]

[30]   S. Wang, S.F. Swingle, H. Ye, F.R.F. Fan, A.H. Cowley, and A.J. Bard, "Synthesis and characterization of a p-type boron arsenide photoelectrode", *J. Am. Chem. Soc.,* vol. 134, no. 27, pp. 11056-11059, 2012.
       [http://dx.doi.org/10.1021/ja301765v] [PMID: 22720867]

[31]   H.M. Manasevit, and W.I. Simpson, "The use of metal-organics in the preparation of semiconductor materials", *J. Electrochem. Soc.,* vol. 116, no. 12, p. 1725, 1969.
       [http://dx.doi.org/10.1149/1.2411685]

[32]   B. Schurink, W.T.E. van den Beld, R.M. Tiggelaar, R.W.E. van de Kruijs, and F. Bijkerk, "Synthesis and characterization of boron thin films using chemical and physical vapor depositions", *Coatings,* vol. 12, no. 5, p. 685, 2022.
       [http://dx.doi.org/10.3390/coatings12050685]

[33]   A. Iqbal, and F. Mohd-Yasin, "Reactive sputtering of aluminum nitride (002) thin films for piezoelectric applications: A review", *Sensors,* vol. 18, no. 6, p. 1797, 2018.
       [http://dx.doi.org/10.3390/s18061797]

[34]   N. Li, C.P. Ho, S. Zhu, Y.H. Fu, Y. Zhu, and L.Y.T. Lee, "Aluminium nitride integrated photonics: a review", *Nanophotonics,* vol. 10, no. 9, pp. 2347-2387, 2021.
       [http://dx.doi.org/10.1515/nanoph-2021-0130]

[35]   H. Yamashita, K. Fukui, S. Misawa, and S. Yoshida, "Optical properties of AlN epitaxial thin films in the vacuum ultraviolet region", *J. Appl. Phys.,* vol. 50, no. 2, pp. 896-898, 1979.
       [http://dx.doi.org/10.1063/1.326007]

[36]   C. Xiong, W.H.P. Pernice, X. Sun, C. Schuck, K.Y. Fong, and H.X. Tang, "Aluminum nitride as a new material for chip-scale optomechanics and nonlinear optics", *New J. Phys.,* vol. 14, no. 9, p. 095014, 2012.
       [http://dx.doi.org/10.1088/1367-2630/14/9/095014]

[37]   J.A. Carruthers, T.H. Geballe, H.M. Rosenberg, and J.M. Ziman, "The thermal conductivity of germanium and silicon between 2 an d 300° K", *Proc. R. Soc. London. Ser. A. Math. Phys. Sci.,* vol. 238, 1957no. 1215, pp. 502-514

[38]   J. Lähnemann, A. Ajay, M.I. Den Hertog, and E. Monroy, "Near-infrared intersubband photodetection in GaN/AlN nanowires", *Nano Lett.,* vol. 17, no. 11, pp. 6954-6960, 2017.
       [http://dx.doi.org/10.1021/acs.nanolett.7b03414] [PMID: 28961016]

[39]   N. Watanabe, T. Kimoto, and J. Suda, "The temperature dependence of the refractive indices of GaN and AlN from room temperature up to 515 °C", *J. Appl. Phys.,* vol. 104, no. 10, p. 106101, 2008.
       [http://dx.doi.org/10.1063/1.3021148]

[40]   R. Shinde, S.S.R.K.C. Yamijala, and B.M. Wong, "Improved band gaps and structural properties from Wannier–Fermi–Löwdin self-interaction corrections for periodic systems", *J. Phys. Condens. Matter,* vol. 33, no. 11, p. 115501, 2021.
       [http://dx.doi.org/10.1088/1361-648X/abc407] [PMID: 33091890]

[41]   J. Whitaker, "Electrical properties of n-type aluminium arsenide", *Solid-State Electron.,* vol. 8, no. 8, pp. 649-652, 1965.
       [http://dx.doi.org/10.1016/0038-1101(65)90032-8]

[42]   I.C. Kizilyalli, A.P. Edwards, O. Aktas, T. Prunty, and D. Bour, "Vertical power p-n diodes based on bulk GaN", *IEEE Trans. Electron Dev.,* vol. 62, no. 2, pp. 414-422, 2015.
       [http://dx.doi.org/10.1109/TED.2014.2360861]

[43]   G. Sabui, P.J. Parbrook, M. Arredondo-Arechavala, and Z.J. Shen, "Modeling and simulation of bulk gallium nitride power semiconductor devices", *AIP Adv.,* vol. 6, no. 5, p. 055006, 2016.
       [http://dx.doi.org/10.1063/1.4948794]

[44]   E. Gil, Y. André, R. Cadoret, and A. Trassoudaine, "Hydride vapor phase epitaxy for current III–V and nitride semiconductor compound issues", *Handb. Cryst. Growth Thin Film. Ep. Second Ed.,* vol. 3,

pp. 51-93, 2015.
[http://dx.doi.org/10.1016/B978-0-444-63304-0.00002-0]

[45]   C.T. Foxon, T.S. Cheng, D. Korakakis, S.V. Novikov, R.P. Campion, I. Grzegory, S. Porowski, M. Albrecht, and H.P. Strunk, "Homo- and hetero-epitaxial gallium nitride grown by molecular beam epitaxy", *MRS Internet J. Nitride Semicond. Res.,* vol. 4, no. 1, pp. 484-489, 1999.
[http://dx.doi.org/10.1557/S1092578300002933]

[46]   B. Wang, and M.J. Callahan, "Ammonothermal synthesis of III-nitride crystals", *Cryst. Growth Des.,* vol. 6, no. 6, pp. 1227-1246, 2006.
[http://dx.doi.org/10.1021/cg050271r]

[47]   N.N. Anua, R. Ahmed, M.A. Saeed, A. Shaari, and B.U. Haq, "DFT investigations of structural and electronic properties of gallium arsenide (GaAs)", *AIP Conf. Proc.,* vol. 1482, no. 1, pp. 64-68, 2012.
[http://dx.doi.org/10.1063/1.4757439]

[48]   D.C. Look, "Recent advances in ZnO materials and devices", *Mater. Sci. Eng. B,* vol. 80, no. 1-3, pp. 383-387, 2001.
[http://dx.doi.org/10.1016/S0921-5107(00)00604-8]

[49]   Ü. Özgür, Y.I. Alivov, C. Liu, A. Teke, M.A. Reshchikov, S. Doğan, V. Avrutin, S-J. Cho, and H. Morkoç, "A comprehensive review of ZnO materials and devices", *J. Appl. Phys.,* vol. 98, no. 4, p. 041301, 2005.
[http://dx.doi.org/10.1063/1.1992666]

[50]   V.A. Coleman, and C. Jagadish, "Basic properties and applications of ZnO, zinc oxide bulk", *Thin Film. Nanostructures Process. Prop. Appl.,* no. Jan, pp. 1-20, 2006.

[51]   A. Janotti, and C.G. Van de Walle, "Fundamentals of zinc oxide as a semiconductor", *Rep. Prog. Phys.,* vol. 72, no. 12, p. 126501, 2009.
[http://dx.doi.org/10.1088/0034-4885/72/12/126501]

[52]   S. Malki, and L. El Farh, "Structural and electronic properties of zinc antimonide ZnSb", *Mater. Today Proc.,* vol. 31, pp. S41-S44, 2020.
[http://dx.doi.org/10.1016/j.matpr.2020.05.598]

CHAPTER 3

A Comprehensive Overview of the Foundations of Semiconductor Materials

Agnibha Dasgupta[1,*], Soumya Sen[2], Prabhat Singh[3] and Ashish Raman[3]

[1] *GE Vernova T&D India Limited, GE VERNOVA; Services Specialist - DIG Grid Support, New Delhi, India*

[2] *University of Engineering and Management, Jaipur, Rajasthan, India*

[3] *Department of Electronics and Communication Engineering, Dr. B. R. Ambedkar National Institute of Technology, Jalandhar, Punjab, India*

Abstract: In the recent era, the semiconductor industry, which plays a pivotal role in powering today's cutting-edge technologies, relies heavily on a broad spectrum of materials, entailing of silicon and rare earth elements. These materials serve as the backbone for crucial components, such as solar cells, transistors, IoT sensors, and the intricate circuits found in self-driving cars. Consequently, there is a notable surge in demand for these devices, marking a paradigm shift in the technological landscape.

The first section of this comprehensive exploration delves deeply into semiconductor materials. Understanding their profound impact on electronic devices and the intricacies of the manufacturing process is fundamental for anyone seeking a comprehensive grasp of this dynamic industry. Moving forward, the second part focuses on the properties and physics governing semiconductor materials. The electronic conductivity of these materials is of paramount importance, and the chapter unravels the challenges involved in the efficient and cost-effective large-scale manufacturing of new materials with these crucial properties.

Segment three navigates through the vast realm of semiconductor applications, shedding light on their pivotal role in various electronic devices and cutting-edge technologies. It accentuates the unique electrical properties that make semiconductors indispensable in industrial settings.

In the fourth section, attention is paid to the present market scenario, where the semiconductor market stands out for its stability across diverse industrial sectors. The chapter meticulously examines the production expenses associated with different materials, ranging from the widely used silicon to the more exotic rare earth metals.

Essentially, this chapter guides readers through the complex trends in the semiconductor industry, offering a concise overview of material development and

---

* **Corresponding author Agnibha Dasgupta:** GE Vernova T&D India Limited, GE VERNOVA; Services Specialist - DIG Grid Support, New Delhi, India; E-mail: dasgupta.rony7@gmail.com

Ashish Raman, Prabhat Singh, Naveen Kumar & Ravi Ranjan (Eds.)

influential factors. It also encourages the exploration of innovative solutions to propel the Very Large Scale Integration (VLSI) industry toward unprecedented advancements.

Keywords: Electrical conductivity, IoT sensors, RREs, Semiconductor materials.

INTRODUCTION

Semiconductor devices stand as the cornerstone of modern electronic systems, propelling the technology that shapes our daily lives. The efficacy and performance of these devices are intricately tied to the design elements incorporated into their construction. The pursuit of smaller, faster, and more energy-efficient semiconductor devices has fuelled continuous advancements in the realm of science dedicated to these intricate structures. This introduction provides an overarching view of the key components instrumental in shaping semiconductor devices, shedding light on their roles, properties, and the pivotal influence they wield over device functionality.

The foundation of electronic components lies in semiconductor materials, facilitating the controlled flow of electric current. Silicon, renowned for its remarkable properties, has historically served as the primary component for semiconductor devices. Its crystalline structure, stability, and capacity to form a robust oxide layer make it an ideal choice for manufacturing integrated circuits. Nevertheless, as technology progresses, novel alternatives emerge to address the limitations inherent in conventional options [1 - 4].

Types of Semiconductor Materials

Beyond silicon, compound semiconductors, such as gallium arsenide (GaAs), gallium nitride (GaN), and indium phosphide (InP), have ascended to prominence. These compounds manifest distinctive electrical and thermal attributes that render them suitable for specialized deployments, including high-frequency devices, power amplifiers, and optoelectronics.

In addition to the aforementioned, the significance of dielectric materials, essential for insulating and isolating different components of a device, cannot be overstated. Insulating materials, like silicon dioxide (SiO_2) and various high-k dielectrics, play a critical role in ameliorating power consumption and enhancing the overall performance of semiconductor devices.

Furthermore, the interconnect elements utilized in the fabrication of integrated circuits play a pivotal role in determining the speed and efficiency of signal transmission. Copper and, more recently, low-k dielectric materials have become integral in surmounting the challenges associated with the increasing complexity and miniaturization of semiconductor devices.

Considering the demand involving quicker, petite, and economical electronic devices, researchers and engineers are exploring creative design elements and manufacturing methods. They are striving to discover fresh solutions, such as 2D materials, like graphene and transition metal dichalcogenides, in order to achieve the goal of advancing semiconductor technology. This endeavour aims to unlock fresh opportunities for the future evolution of electronic devices.

In the expansive field of semiconductor materials, III-V compounds occupy a distinctive and pivotal position, offering a unique array of properties that significantly contribute to the relentless advancement of electronic devices. The nomenclature "III-V" denotes elements derived from groups III and V of the periodic table, notably including gallium (Ga), indium (In), aluminum (Al) from group III, and nitrogen (N), phosphorus (P), arsenic (As), and antimony (Sb) from group V. This class of semiconductors has become a focal point of research and application due to its exceptional electrical, optical, and thermal characteristics.

Among the standout constituents of III-V semiconductors is gallium arsenide (GaAs), renowned for its direct bandgap feature, high electron mobility, and impeccable thermal stability. The direct bandgap represents a distinctive class of semiconductor materials that exhibit a specific electronic property crucial for various optoelectronic applications. This term refers to the alignment of the conduction band minimum and valence band maximum occurring at the unvarying momentum in the electronic band structure. Fig. (**1**) shows the energy *versus* momentum curve for the direct and indirect bandgap materials. This unique characteristic makes direct bandgap semiconductors highly efficient in radiative recombination processes, enabling the direct conversion of electrical energy into light or *vice versa*. GaAs is extensively employed in a myriad of electronic and optoelectronic applications, finding its niche in high-frequency devices, like microwave transistors and high-speed field-effect transistors, courtesy of its superior electron transport properties.

Indium phosphide (InP) is another luminary in the realm of III-V semiconductors, boasting a direct bandgap that renders it particularly apt for optoelectronic devices. Its applications span a wide spectrum, encompassing high-speed electronic and optoelectronic devices, such as high-frequency transistors, lasers, and photodetectors. The distinctive bandgap characteristics of InP make it especially valuable in the realms of telecommunications and high-speed data transmission [2 - 9, 15 - 18].

Gallium nitride (GaN) has emerged prominently in recent years, lauded for its wide gap, high breakdown voltage, and stellar thermal conductivity. This trifecta of properties positions it as a preferred choice for applications involving prominent power and frequencies, including power electronics, radar systems, and the efficient illumination provided by Light-emitting Diodes (LEDs). GaN's ability to function adeptly under elevated temperatures and voltages further propels its adoption in diverse electronic and power applications.

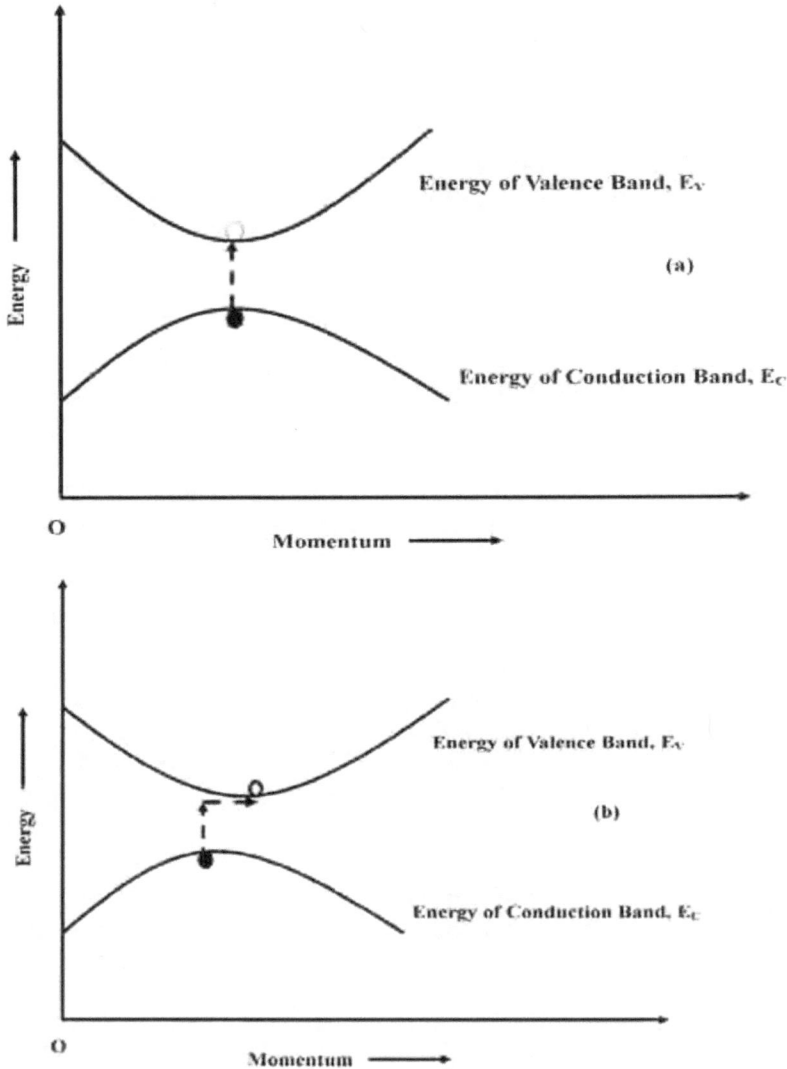

Fig. (1). Energy *vs.* momentum curve. **(a)** Direct bandgap semiconductors; **(b)** indirect bandgap semiconductors.

The ternary III-V compound aluminum gallium arsenide (AlGaAs) emerges as a decisive force in the intricate world of semiconductor technology, particularly in the construction of heterojunction devices, like bipolar transistors and photodiodes. This compound is distinguished by its inherent ability to present a tunable energy gap, a quality that grants engineers and researchers unparalleled control over the characteristics of electronic and optoelectronic devices. The versatility of the compound enables precise adjustments to the energy band structure, facilitating the optimization of device performance. This unique attribute positions AlGaAs as a linchpin in the development of high-performance electronic components, where the fine-tuning of energy gaps plays a pivotal role in achieving efficiency, speed, and reliability.

Adding to the symphony of innovative III-V semiconductor materials is indium gallium nitride (InGaN), a versatile alloy that finds widespread applications in the realm of LEDs, laser diodes, and photodetectors. The distinctive feature of InGaN lies in its malleability, allowing engineers to modulate the composition of indium and gallium within the alloy. This tunability facilitates precise adjustments to the energy bands, providing a powerful tool for tailoring the emission characteristics of light. The ability to control the energy bands in InGaN is particularly crucial in the context of LED technology, where a rich array of colours can be produced by harnessing the inherent flexibility of this semiconductor alloy. InGaN's versatility positions it as an indispensable material for the production of vibrant and efficient lighting solutions.

In all, the role of AlGaAs and InGaN in the semiconductor landscape is one of mastery over energy band structures, exemplifying the potential for tailored control in electronic and optoelectronic devices [10 - 22]. Table **1** summarizes the physics of silicon along with its counterpart III-V semiconductors.

Table 1. A comparison of the properties of silicon (Si) with III-V counterparts.

Properties	Silicon (Si)	GaAs	InP	GaN	AlGaAs	InGaN
Bandgap type	Indirect (1.12 eV)	Direct	Direct	Direct	Direct (AlGaAs portion can vary)	Direct
Bandgap energy (eV)	1.12	1.42	1.35	~3.4	Varies based on Al composition	Varies
Carrier mobility	Moderate (around 1500 $cm^2/V \cdot s$)	High (8500 $cm^2/V \cdot s$)	High (4000 $cm^2/V \cdot s$)	High (around 1500 $cm^2/V \cdot s$)	Moderate to high	Moderate to high

(Table 1) cont.....

Properties	Silicon (Si)	GaAs	InP	GaN	AlGaAs	InGaN
Thermal conductivity (W/m·K)	High (149)	~46	~80	~200	Varies based on Al composition	Varies
Breakdown voltage	Moderate	Varies	Varies	High	Varies based on Al composition	Varies
Crystal structure	Diamond cubic	Zinc blende	Zinc blende	Wurtzite	Zinc blende	Wurtzite
Applications	Integrated circuits, solar cells	Opto and high-frequency devices, solar cells	High-speed devices for telecommunication and photodetectors	Power electronics, LEDs, and Lasers	Heterojunction devices	LEDs and LASERs
Cost	Inexpensive	Typically more expensive than silicon, but costs decrease	Typically more expensive than silicon, but costs decrease	Varies	Varies based on Al composition	Varies
Processing technology	Mature	In progress, but not as mature as silicon	In progress, but not as mature as silicon	Advancing	Advancing	Advancing
Temperature stability	Stable	Moderate to high	Moderate to high	High	Varies based on Al composition	Varies

While AlGaAs serve as a linchpin in the development of high-performance heterojunction devices, InGaN's adaptability enables a spectrum of colours to be harnessed in LED technology. These materials, with their unique properties, continue to propel the evolution of semiconductor technology, offering unprecedented control and customization in the pursuit of cutting-edge electronic and optoelectronic advancements. In all, comprehending the intricacies of the design elements incorporated into semiconductor devices is fundamental to advancing electronic technology. The perpetual exploration of inventive components and cutting-edge fabrication processes is imperative for meeting the ever-evolving demands of the electronics industry, propelling progress toward more efficient, compact, and powerful semiconductor devices.

Each semiconductor material possesses a distinct set of characteristics tailored for specific applications. Silicon, characterized by its well-established processing technology and cost-effectiveness, remains the cornerstone of integrated circuits and solar cells. GaAs, featuring a direct bandgap, excels in optoelectronic and high-frequency devices. InP, another semiconductor with a direct bandgap, finds its specialization in telecommunications and photodetectors. GaN, renowned for its high breakdown voltage, takes the lead in power electronics, LEDs, and lasers. AlGaAs, a ternary compound, facilitates tunable bandgaps, playing a pivotal role in heterojunction devices. InGaN, boasting a versatile bandgap, proves essential for generating a diverse spectrum of colours in LEDs and also finds applications in lasers and photodetectors. The selection of a semiconductor material hinges on the precise demands of the application, considering elements, such as bandgap, carrier mobility, thermal conductivity, and cost. As technology progresses, the distinctive properties of III-V semiconductors continue to propel innovation across a wide range of electronic and optoelectronic applications [23 - 29].

PROPERTIES OF SEMICONDUCTOR MATERIALS

Semiconductor materials form the bedrock of modern electronics, enabling the creation of intricate devices that power our technological world. Among these materials, silicon and III-V semiconductors stand out for their unique properties and diverse applications. Silicon, a quintessential element in semiconductor technology, has long been the workhorse of the industry. On the other hand, III-V semiconductors, encompassing compounds, like gallium arsenide (GaAs), indium phosphide (InP), gallium nitride (GaN), aluminum gallium arsenide (AlGaAs), and indium gallium nitride (InGaN), introduce a new dimension with their distinctive characteristics. This exploration delves into the individual properties of silicon and III-V semiconductors, uncovering the underlying physics and culminating in a consideration of their potential synergy when blended [30 - 33].

SILICON

Silicon's crystalline structure, notably the diamond cubic crystal lattice, forms the foundation of its stability and uniformity. This arrangement ensures the dependability and repeatability of semiconductor devices. Silicon is characterized by a moderate bandgap of approximately 1.12 electron volts (eV), marking it as an indirect bandgap semiconductor. This property influences its electronic transitions, where energy is not efficiently radiated as light.

Its processing technology has reached maturity, allowing for the fabrication of intricate Integrated Circuits (ICs) and electronic components. The well-established infrastructure contributes to cost-effectiveness and widespread availability. Its stability and the emergence of a native oxide layer on its surface make silicon an excellent choice for applications requiring insulation or protection against environmental factors.

The band structure of silicon can be expressed mathematically through the energy dispersion relationship, commonly known as the E-k diagram. In an indirect band gap semiconductor, like silicon, the relationship between energy (E) and the electron momentum (k) differs for the Valence Band (VB) and the Conduction Band (CB). The energy dispersion relation for silicon near the band edges is given by:

$$E_{VB}(k) = E_{VB,0} - \frac{\hbar^2 k^2}{2m_h} \tag{1}$$

$$E_{CB}(k) = E_{CB,0} + \frac{\hbar^2 k^2}{2m_e} \tag{2}$$

Here, $E_{VB,0}$ and $E_{CB,0}$ represent the energy minima of the valence and conduction bands, respectively. m_h and m_e are the effective masses of holes and electrons, and h is the reduced Planck's constant. The indirect bandgap nature of silicon implies that the modest energy point in the conduction band does not align with the optimum energy point in the valence band. Consequently, an electron transitioning from the valence band to the conduction band requires an additional momentum change (Δk) to conserve both energy and momentum. Mathematically, this can be expressed as:

$$\Delta k = \frac{\sqrt{2m_e(E_{CB,0} - E_{VB,0})}}{\hbar} \tag{3}$$

This additional momentum requirement influences the efficiency of radiative recombination during electronic transitions, affecting the emission of light.

In terms of electronic performance, it exhibits a moderate carrier mobility of around 1500 cm²/V·s and a high thermal conductivity of 149 W/m·K, aiding in efficient heat dissipation. Silicon's versatility is evident in its widespread use in integrated circuits, microprocessors, memory devices, and solar cells, where it converts sunlight into electrical energy.

III-V Semiconductor (*e.g.*, GaAs) - Direct Band Gap Material

Contrastingly, III-V semiconductors, like GaAs, possess a direct bandgap, resulting in a more favorable scenario for radiative recombination. The energy dispersion relation near the band edges for GaAs is given by:

$$E_{VB}(k) = E_{VB,0} - \frac{\hbar^2 k^2}{2m_h} \tag{4}$$

$$E_{CB}(k) = E_{CB,0} - \alpha k^2 \tag{5}$$

In this case, the conduction band energy (E_{CB}) includes a term with a negative coefficient ($-\alpha$), leading to a direct transition of electrons between energy bands without the need for additional momentum. The coefficient α determines the curvature of the conduction band. The high electron mobility in GaAs is closely allied to the direct transition of electrons between energy bands. The electron mobility (μ) is mathematically expressed as:

$$\mu = \frac{e\tau}{m_e} \tag{6}$$

Here, e is the elementary charge, τ is the electron relaxation time, and m_e is the effective mass of electrons. The direct transition nature of GaAs allows for the swift movement of charge carriers, contributing to the high electron mobility observed in this material.

The mathematical expressions for the band structures of silicon and III-V semiconductors provide a quantitative foundation for understanding their electronic properties. The indirect bandgap nature of silicon introduces additional momentum requirements for electronic transitions, influencing its radiative efficiency. In contrast, the direct bandgap nature of III-V semiconductors, like GaAs, facilitates more efficient radiative recombination, influencing properties, such as electron mobility. These mathematical representations are integral to comprehending the underlying physics governing the behaviour of semiconductor materials [34 - 43].

Heterostructures and Quantum Wells

The synergistic approach of blending semiconductor materials, often realized through heterostructures, introduces a wealth of possibilities for enhancing the execution of electronic and optoelectronic devices. This blending is underpinned by mathematical principles, particularly exemplified by Vegard's Law, which governs the prediction of alloy properties based on the weighted average of constituent materials. Let's explore the mathematical foundations of blending semiconductor materials, emphasizing on the context of III-V semiconductors and the creation of alloys, like InGaN.

The electronic properties of hetero devices, formed by combining different semiconductor materials, can be mathematically described through the Schrödinger equation, which characterizes the quantum mechanics of particles. In the context of quantum wells within heterojunctions, the energy levels (E) for electrons can be modelled using the finite potential well equation:

$$E_n = \frac{\hbar^2 \pi^2 n^2}{2m^* L^2} \tag{7}$$

Here, E_n represents the energy level, h is the reduced Planck's constant, m^* is the effective mass of the electron, n is the quantum number, and L is the width of the quantum well. The quantization of energy levels in quantum wells is a key aspect impacting the electronic properties of unified semiconductor materials.

Vegard's Law for Blending Alloys

Vegard's law is a fundamental principle applied when blending different materials to create alloys. Mathematically, it can be expressed as:

$$Property_{alloy} = \Sigma_i x_i . Property_{material} \tag{8}$$

Here, Property$_{alloy}$ is the property of the alloy, x_i is the fraction by volume or weight of material i in the alloy, and Property$_{material}$ is the property of the individual material. Within the framework of III-V semiconductors, this law is crucial for predicting the bandgap of alloys, such as InGaN, where different proportions of indium (In), gallium (Ga), and nitrogen (N) are blended. The below example depicts the tailoring of bandgap in InGaN alloys. The bandgap energy E_g of an alloy, like InGaN, crucial for its applications in LEDs, can be expressed using the empirical Bowing parameter equation:

$$E_g(In_x Ga_{1-x} N) = x . E_g(InN) + (1-x) . E_g(GaN) - x(1-x) . B \tag{9}$$

Here, E_g*InN* and E_g*(GaN)* are the bandgaps of indium nitride and gallium nitride, respectively. The term B represents the bowing parameter, influencing the non-linearity in the relationship between alloy compositions (x). By varying x within the alloy, the bandgap of InGaN can be precisely tailored, enabling the emission of light across a spectrum in LEDs. Overall, this mathematical framework is instrumental in engineering alloys with desired electronic and optical characteristics, showcasing the potential of blended semiconductor materials in advancing device performance [44 - 62].

APPLICATION OF SEMICONDUCTOR MATERIALS

Contemporary electronic gadgets rely heavily on semiconductor materials, especially silicon and III-V semiconductors developing as essential components. Group IV member silicon has served as a mainstay of the semiconductor trade because of its plentiful supply and widely recognized production methods. On the other hand, III-V semiconductors, which are made up of elements from the periodic table's groups III and V, offer unique qualities that increase the potential applications of semiconductor technology. With a focus on silicon and III-V semiconductors, we have examined the uses of these semiconductor materials in this inquiry, backed by mathematical reasoning that explains the fundamentals of their operation [63 - 66].

Silicon Semiconductors

Integrated Circuits

Silicon-based Integrated Circuits (ICs) have transformed the landscape of electronics, playing a pivotal role in advancing modern technology. The prevalence of silicon in ICs, shown in Fig. (**2**), can be attributed to its outstanding semiconductor properties, making it an optimal material for constructing transistors, the fundamental components that underpin electronic devices.

The transistor, governed by Ohm's Law,

$$V = I.R \tag{10}$$

holds a central position in manufacturing. This equation delineates the entangled relationship among voltage (V), current (I), and resistance (R) within a transistor. Manipulating these variables allows engineers to finely regulate and modulate electronic signals, thereby facilitating the precise functionality of integrated circuits. Silicon's semiconductor prowess allows it to transition between conducting and insulating states, rendering it an ideal candidate for crafting electronic components. Transistors, serving as the foundational building blocks of ICs, assume various roles, such as amplifiers, switches, and signal modulators, thanks to silicon's capacity to form stable and dependable transistors that enable the creation of complex circuits with diverse functionalities.

Transistors, serving as the basic building blocks of ICs, assume various roles, such as amplifiers, switches, and signal modulators. These versatile components are crucial for the precise control and modulation of electronic signals within integrated circuits. By manipulating the variables outlined in Ohm's Law, engineers can finely regulate the behavior of transistors, thereby enabling the creation of complex circuits with diverse functionalities.

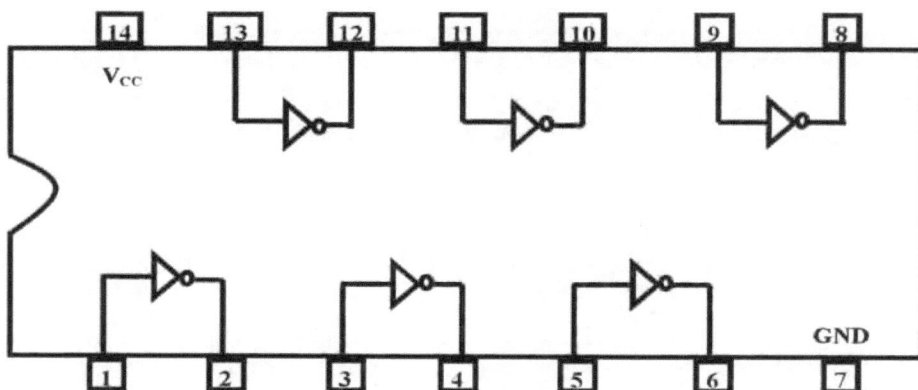

Fig. (2). Silicon-made IC 7404 for NOT gate.

The exceptional semiconductor properties of silicon render it an optimal material for crafting transistors that exhibit stability, reliability, and high performance. These properties enable precise control of electron flow within transistors, facilitating functions, such as amplification, switching, and modulation of electronic signals. The ability to form stable and dependable transistors is crucial for designing and manufacturing ICs that power a wide array of electronic devices, spanning from smartphones and computers to automotive systems and industrial equipment.

Additionally, its compatibility with established manufacturing processes and its abundance in the earth's crust has led to its widespread adoption in the semiconductor industry. Decades of research and development have refined fabrication techniques specific to silicon, allowing for the high-yield production of complex ICs at relatively low costs. Semiconductor foundries and fabrication facilities worldwide specialize *in silicon*-based processes, leveraging economies of scale to meet the growing demand for electronic devices.

Photovoltaic Solar Cell

Silicon-based Photovoltaic (PV) solar cells, depicted in Fig. (3), serve as pivotal components in the conversion of sunlight into electrical energy. Comprising monocrystalline or polycrystalline silicon, these semiconductor devices leverage the photovoltaic effect, where incident photons create electron-hole pairs within the material. The solar cell structure includes a p-n junction, creating an internal electric field that facilitates the disjunction of these charge carriers. As electrons move toward the n-type region and holes toward the p-type region, an electric current is generated. Metal contacts at the cell's surface collect this current, forming the basis for electricity production. The potency of silicon solar cells is quantified by the ratio of electrical power output to incident solar power, typically ranging from 15% to 25% in real-world applications. The mathematical equation representing efficiency (η) is expressed as:

$$\eta = \frac{V_{oc}I_{sc}}{P_{in}}$$

(11)

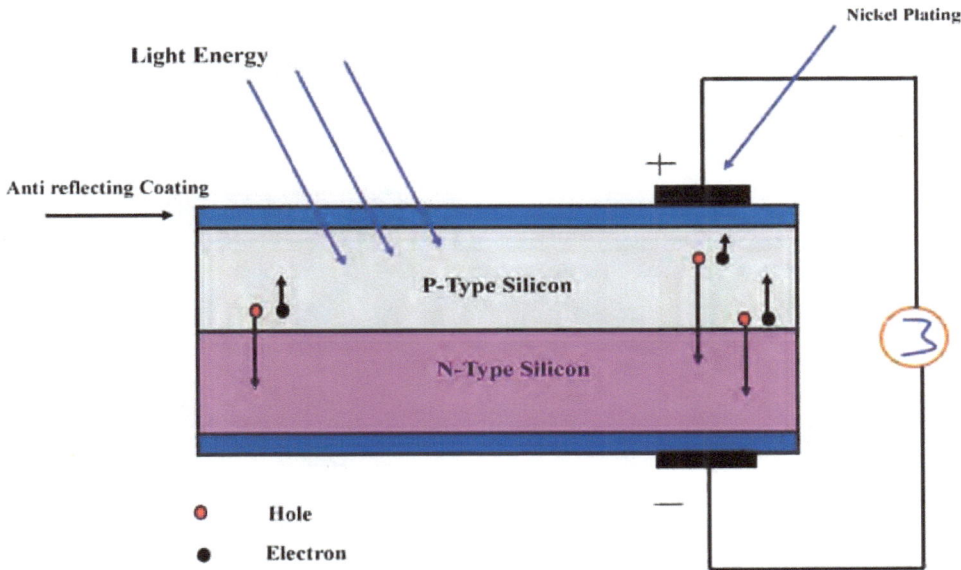

Fig. (3). Silicon-made photovoltaic/solar cell.

Where, V_{oc} is the open-circuit voltage, I_{sc} is the short-circuit current density, and P_{in} is the incident power density. These devices play a crucial role in advancing sustainable energy solutions, and ongoing research focuses on improving their efficiency and cost-effectiveness for broader integration into global energy systems.

They capitalize on the semiconductor properties of silicon to generate electricity. The process commences with incepting silicon wafers that act as the foundation for the solar cell. These wafers are typically produced using highly purified silicon ingots, derived from silica sand through processes, such as the Siemens process or the Czochralski method. The resulting silicon wafers are then doped with specific impurities to create regions of positive (p-type) and negative (n-type) conductivity, essential for the operation of the solar cell.

The photovoltaic effect occurs within the silicon structure of the solar cell when photons from sunlight strike the surface of the cell. The energy from these photons is conveyed to electrons in the atoms, causing them to disengage from their atomic bonds, which culminates in the establishment of electron-hole pairs. The electric field within the solar cell, generated by the p-n junction, then separates these electron-hole pairs, causing them to move in opposite directions. The separated electrons flow towards the n-type region, while the holes migrate towards the p-type region, creating a potential difference or voltage across the solar cell. This flow of electrons constitutes an electric current, which can be harnessed externally to power electrical devices or be stored in batteries for later use. By connecting multiple solar cells in series and parallel configurations, photovoltaic modules or solar panels can be created, capable of generating higher voltages and power outputs.

These devices offer several advantages that have contributed to their widespread adoption in the solar energy industry. Firstly, silicon is abundantly available in the earth's crust, making it a cost-effective and scalable material for large-scale solar cell production. Additionally, silicon is highly efficient at converting sunlight into electricity, with modern silicon solar cells achieving efficiency levels exceeding 20%.

Furthermore, silicon solar cells exhibit excellent long-term reliability and durability, with a typical lifespan ranging from 25 to 30 years or more. This longevity, coupled with minimal maintenance requirements, makes silicon-based photovoltaic systems a practical and reliable investment for residential, commercial, and utility-scale solar installations. Silicon solar cells also demonstrate versatility in terms of manufacturing techniques and form factors. They can be fabricated using various methods, including monocrystalline, polycrystalline, and thin-film deposition processes, allowing for flexibility in design and application. Monocrystalline silicon solar cells, produced from single-crystal silicon ingots, offer high efficiency and power output, making them ideal for rooftop installations and space-constrained environments. Polycrystalline silicon solar cells, manufactured from multi-crystalline silicon wafers, provide a cost-effective alternative with slightly lower efficiency but comparable performance. Thin-film silicon solar cells, deposited onto flexible substrates, such as glass or plastic, offer lightweight and flexible solutions suitable for Building-integrated Photovoltaics (BIPV) and portable applications [62 - 68].

Microelectromechanical Systems (MEMS)

Microelectromechanical Systems (MEMS), depicted in Fig. (4), represent a groundbreaking intersection of electronics and mechanical engineering, enabled by the remarkable compatibility of silicon with advanced microfabrication techniques. Silicon's unique properties and well-established processes have revolutionized the design and manufacturing of MEMS, paving the way for a myriad of innovative applications across various industries.

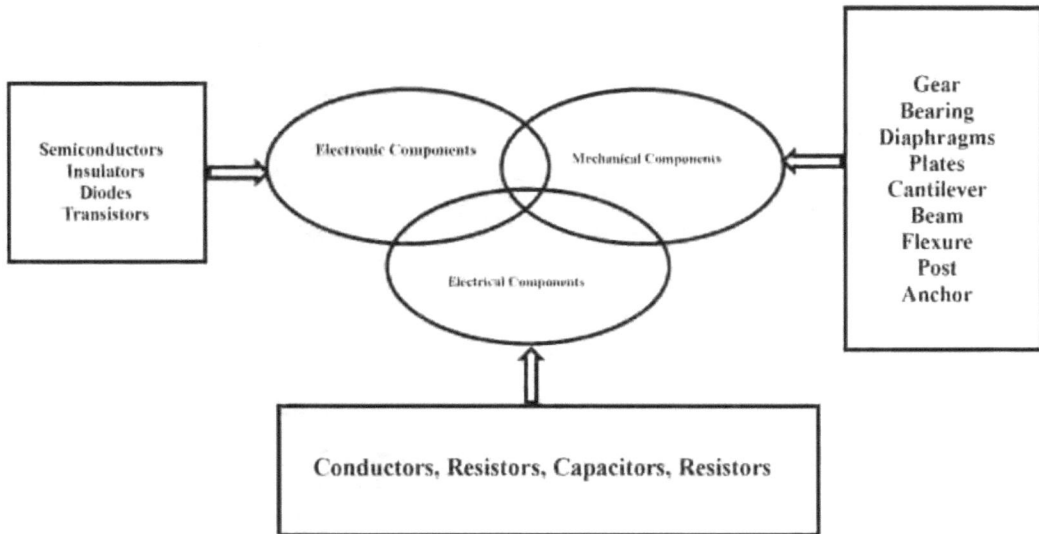

Fig. (4). MEMs technology.

It possesses characteristics that make it an ideal substrate for MEMS. Its mechanical stability, thermal conductivity, and electrical properties contribute to the reliability and performance of the miniature devices produced. Moreover, silicon's widespread use in the semiconductor industry has prompted the founding of sophisticated fabrication processes, such as photolithography and etching, which can be seamlessly applied to create intricate MEMS structures. These techniques allow for the precise and cost-effective production of devices on a miniature scale. They typically consist of tiny mechanical components, sensors, actuators, and electronic circuits blended into a single silicon chip. This integration is achieved through a series of intricate steps, including deposition, patterning, and etching, which enable the creation of intricate and precisely defined structures with dimensions on the micrometer or nanometre scale. The versatility of MEMS technology has given rise to a diverse array of applications, ranging from consumer electronics and healthcare to automotive and aerospace. In consumer electronics, their accelerometers and gyroscopes enable the motion sensing capabilities in smartphones and gaming controllers. In healthcare, these devices are utilized in miniaturized sensors for monitoring vital signs or drug delivery systems. Automotive applications include MEMS sensors for airbag deployment and tire pressure monitoring, while in aerospace, the devices find applications in navigation systems and aerospace instrumentation.

The ability to mass-produce miniature yet highly functional devices has opened doors to new possibilities, driving advancements in various fields and enhancing the capabilities of electronic systems [51 - 60].

III-V Semiconductors

Light Emitting Diodes (LEDs)

III-V semiconductors, like gallium nitride (GaN), have a critical impact on LED technology. The mathematical proof linking the emission wavelength (λ) to the bandgap energy E_g is given in the following equation:

$$\lambda = \frac{c}{f} = \frac{h}{E_g} \tag{12}$$

Here, c is the speed of light, f is the frequency, and h is the Planck's constant. This equation enables precise control over the emitted light's characteristics, influencing applications in displays, lighting, and telecommunications.

This equation, derived from fundamental principles in quantum mechanics and semiconductor physics, provides a mathematical proof of the inverse relationship between the emission wavelength and the bandgap energy of a semiconductor material. As the bandgap energy increases, corresponding to a wider energy gap between the valence and conduction bands of the material, the emitted photons have higher energy and shorter wavelengths. Conversely, a smaller bandgap results in lower energy photons with longer wavelengths.

Gallium nitride (GaN) and other III-V semiconductors possess unique properties that make them highly suitable for LED applications. One of the key advantages is its relatively wide bandgap energy, which allows for the emission of photons in the blue and ultraviolet regions of the spectrum. This property enables the production of high-efficiency blue and white LEDs, which are crucial components in a wide range of lighting and display technologies. Furthermore, III-V semiconductors exhibit superior electron mobility and thermal conductivity compared to traditional materials, like silicon, making them well-suited for high-power and high-frequency applications [62-70]. This makes GaN-based LEDs, as shown in Fig. (5), particularly attractive for applications requiring high brightness, such as automotive lighting, backlighting for displays, and solid-state lighting solutions. The mathematical relationship between emission wavelength and bandgap energy elucidated by the aforementioned equation serves as a guiding principle in the design and optimization of the devices [70-82]. By carefully selecting semiconductor materials with appropriate bandgap energies, engineers can tailor the emission characteristics of LEDs to meet specific application requirements, whether it is for general illumination, display technology, or specialized lighting applications [83 - 90].

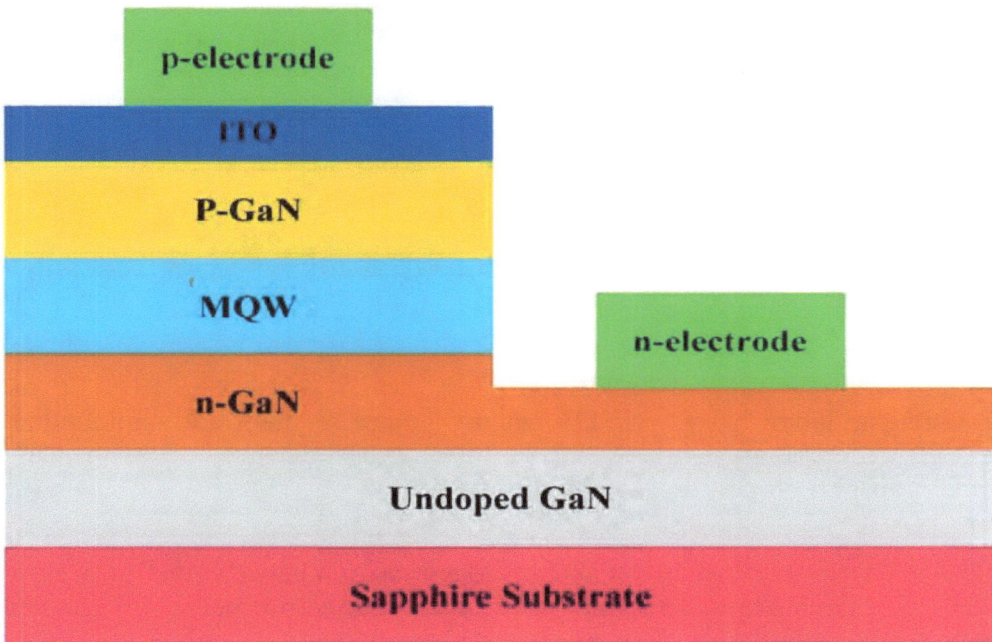

Fig. (5). GaN-oriented light emitting diodes.

High Electron Mobility Transistors (HEMTs)

High Electron Mobility Transistors (HEMTs), also known as HFETs (Heterostructure Field-effect Transistors), are semiconductor devices that offer high-speed, high-frequency operation, making them essential components in various electronic systems, including telecommunications, radar, and satellite communications. III-V semiconductor materials, such as gallium nitride (GaN) and indium phosphide (InP), are commonly used to fabricate HEMTs due to their unique physical properties and superior performance [42 - 54].

III-V compound semiconductors exhibit several advantageous properties that make them well-suited for HEMT applications, which are listed as follows:

Material Properties

The physics of the material characteristics has a significant contribution to the working principle of the compound semiconductor-based HEMTs.

    a. Wide Bandgap - GaN and InP have wide bandgaps, enabling operation at high temperatures and high-power densities. This property reduces self-heating effects and enhances device reliability, making III-V HEMTs suitable for exalted power and frequency applications.

    b. High Electron Mobility - III-V materials have high electron mobility compared to silicon, resulting in faster electron transport and higher electron velocities. This characteristic is crucial for achieving high-speed operation and low noise performance.

    c. High Breakdown Voltage – They exhibit high breakdown voltages, allowing HEMTs to operate at higher voltages without experiencing a catastrophic breakdown. This property is advantageous for power amplifier and RF switching applications where high voltage handling capability is required.

Device Structure

III-V semiconductor-based HEMTs usually comprise manifold semiconductor layers grown epitaxially on a substrate, forming a heterostructure with distinct electron transport properties.

    a. Buffer Layer - The buffer layer serves as a transition between the substrate and the active device layers, reducing defects and lattice mismatch. It is typically made of materials, like AlGaN or InAlAs, to provide a smooth growth surface for subsequent layers.

    b. Channel Layer - The channel layer is where electron conduction occurs in the HEMT. It is usually made of a narrow bandgap material, such as GaN or InGaAs, to confine electrons near the device interface and facilitate high electron mobility.

    c. Barrier Layer - The barrier layer is a wide-bandgap material, such as AlGaN or InAlN, which forms a potential barrier at the channel interface, confining electrons within the channel and preventing leakage current. It also enhances electron confinement and improves device performance.

    d. Gate Electrode - The gate electrode administers the flow of electrons in the channel by modulating the channel conductivity through the application of a voltage. It is typically made of metal or heavily doped semiconductor material and is separated from the channel by a thin insulating layer (gate dielectric).

Operating Principles

The operation of III-V semiconductor-based HEMTs relies on the modulation of electron density in the channel region through the application of a gate voltage.

a) ON-state Operation - When a positive voltage is applied to the gate electrode, it creates an electric field that attracts electrons toward the channel interface, forming an electron gas with high electron density. This results in a low-resistance conducting channel between the source and drain electrodes, allowing current flow.

b) OFF-state Operation - When the gate voltage is reduced or reversed, the electric field in the channel weakens, reducing the electron density and depleting the channel of charge carriers. This increases the channel resistance, effectively turning off the device and preventing current flow between the source and drain electrodes.

c) High-frequency Operation - III-V HEMTs have the knack to respond at elevated frequencies due to their short channel lengths, low parasitic capacitance, and high carrier velocities. This makes them suitable for RF and microwave applications, encompassing power amplifiers, low-noise amplifiers, and RF switches.

III-V semiconductors, especially indium gallium arsenide, are vital for High Electron Mobility Transistors (HEMTs), as shown in Fig. (**6**). The mathematical relation for the power gain (G) of a transistor is given in equation 13.

$$(G) = \frac{P_{out}}{P_{in}} \tag{13}$$

Here, P_{out} is the output power and P_{in} is the input power. III-V HEMTs, with their high electron mobility, find applications in high-frequency and high-power scenarios, impacting telecommunications and radar systems.

SEMICONDUCTOR MATERIALS MARKET

The semiconductor materials market stands as the backbone of modern technological advancement, driving innovation across industries ranging from consumer electronics to telecommunications and healthcare. This market's dynamism stems from its ability to evolve continuously, adapting to changing demands for performance, efficiency, and functionality in semiconductor devices. Central to this evolution are various semiconductor materials, with silicon traditionally dominating the market, but increasingly facing competition from III-

V semiconductors and other emerging materials [91]. This section provides a comprehensive overview of the semiconductor materials market, exploring the dominance of silicon, the growing importance of heterojunction semiconductors, and the ongoing exploration of other semiconductor materials [92 - 99].

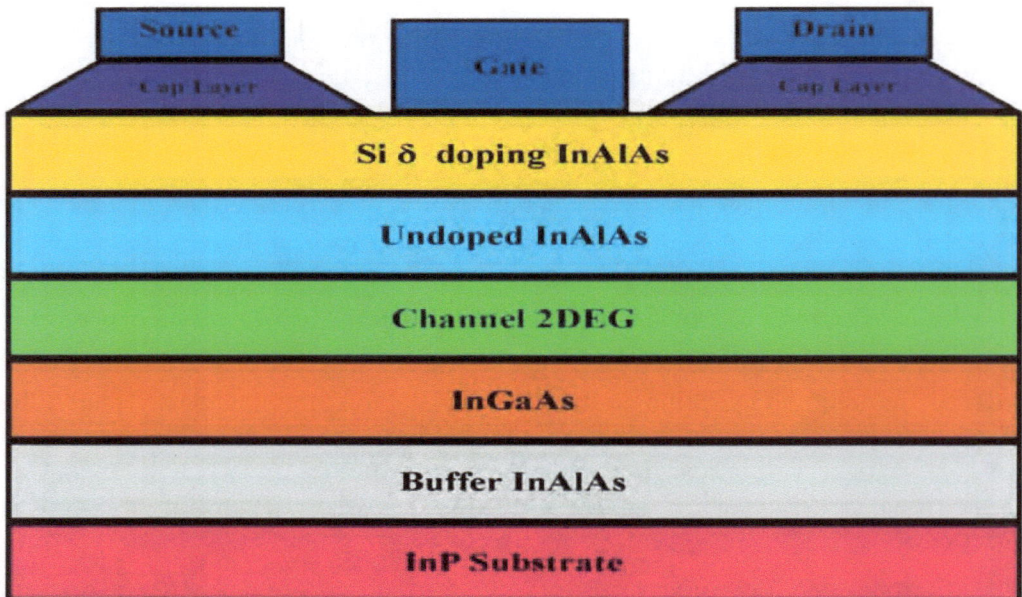

Fig. (6). III-V semiconductors-based High Electron Mobility Transistor (HEMT).

The semiconductor materials market is a pivotal sector within the broader electronics industry, deeply influenced by technological advancements and market demands. Historically, silicon has been the cornerstone of the semiconductor industry due to its abundance, cost-effectiveness, and well-established manufacturing processes. As a result, silicon has dominated the market, powering a wide array of electronic devices ranging from smart devices and computers to industrial automation systems [99 - 102].

However, the emergence of III-V semiconductors, such as gallium arsenide (GaAs) and gallium nitride (GaN), has introduced new dynamics and challenges to the market. They offer superior performance characteristics compared to silicon, including higher electron mobility and wider bandgaps, making them ideal for high-speed communication, power electronics, and optoelectronic applications. This has led to a surge in their demand, particularly in sectors, such as telecommunications, aerospace, and renewable energy [102].

The impact of silicon and III-V semiconductors on semiconductor device sales is significant. While silicon remains dominant in mainstream consumer electronics and industrial applications, the adoption of compound semiconductors in niche markets has reshaped the competitive landscape. This diversification has led to increased revenue streams in specialized segments, driving overall market growth. Silicon maintains a substantial market share due to its entrenched position and widespread adoption. However, the adoption of hetero devices is steadily increasing, challenging silicon's dominance and reshaping the market dynamics. This trend reflects the growing demand for high-performance semiconductor materials to meet evolving technological requirements [101].

Despite the advantages of III-V semiconductors, they also pose challenges to the semiconductor industry. One significant challenge is the complexity and cost associated with manufacturing compound semiconductor devices. Unlike silicon, which benefits from mature fabrication processes and infrastructure, III-V semiconductor manufacturing requires specialized techniques and facilities, driving up production costs. Additionally, integrating III-V semiconductors into existing silicon-based technologies presents compatibility and scalability challenges, further complicating their widespread adoption. The overall study is summarized in Table **2** .

Table 2. An overview of the silicon *vs.* compound semiconductors market.

Semiconductor Material	Market Share (%)	Main Applications	Growth Prospects
Silicon	70	Consumer electronics, industrial automation	Stable, incremental
III-V semiconductors	15	Telecommunications, aerospace, renewable energy	Rapid, niche-driven
Others	15	Emerging technologies, research and development	Variable, uncertain

This table provides a comprehensive overview of the market of semiconductor materials, highlighting the dominance of silicon, the growing importance of III-V semiconductors, and the ongoing exploration of other semiconductor materials. Overall, the semiconductor materials market is characterized by a delicate balance between tradition and innovation, as manufacturers strive to meet evolving demands for performance, efficiency, and functionality in semiconductor devices [70 - 92].

CONCLUSION

In conclusion, this chapter on materials used in the design of semiconductor devices has offered a comprehensive exploration of the fundamental properties, diverse applications, and market dynamics of semiconductor materials, with a focused examination of silicon and III-V semiconductors. Throughout this discussion, we have delved into the intricate details of these materials, ranging from their electrical conductivity and bandgap characteristics to their doping mechanisms, emphasizing the critical role of understanding these properties in optimizing semiconductor devices across a broad spectrum of industries, from microelectronics to optoelectronics.

At the heart of semiconductor technology lies the profound understanding of the materials that enable its functionality. Silicon, with its well-established infrastructure and favourable properties, stands as the cornerstone of the semiconductor industry. Its abundance, reliability, and scalability have made it the material of choice for integrated circuits powering the digital revolution. Through meticulous research and development, silicon-based technologies have achieved unprecedented levels of performance, driving innovation in fields ranging from computing and telecommunications to automotive and healthcare.

However, as technological demands evolve and the quest for enhanced performance continues, the semiconductor landscape is witnessing the emergence of alternative materials, chief among them being III-V semiconductors. These compounds, such as gallium arsenide (GaAs) and indium phosphide (InP), offer distinct advantages over silicon, including higher electron mobility, wider bandgaps, and superior performance at high frequencies. As a result, III-V semiconductors have found niche applications in specialized fields, such as high-frequency communication, advanced photovoltaics, and optoelectronics.

The discussion on the semiconductor materials market underscores the delicate balance between tradition and innovation. While silicon maintains its dominance in established markets, III-V semiconductors are gaining traction in niche applications where their superior properties confer a competitive edge. This dynamic interplay between traditional and emerging materials reflects the evolutionary nature of the semiconductor industry, where continuous innovation drives progress and propels technological advancements.

Moreover, the chapter highlights the wide-ranging applications of semiconductor materials, showcasing their pivotal role in revolutionizing modern technology. From microelectronics to optoelectronics, semiconductor devices have permeated every aspect of our lives, powering the digital infrastructure that underpins our interconnected world. Whether the microprocessors in our smartphones, the photovoltaic cells in solar panels, or the laser diodes in optical communication networks, semiconductor materials play a central role in shaping the technologies that define the 21st century.

In essence, this chapter has provided a holistic overview of the materials landscape shaping semiconductor device design and development. By elucidating the fundamental properties, diverse applications, and market dynamics of semiconductor materials, it has offered valuable insights into the evolving semiconductor ecosystem and its profound impact on technological advancement and innovation. As we navigate the complexities of the semiconductor materials market, it is essential to recognize the symbiotic relationship between tradition and innovation, where the fusion of established practices with emerging technologies can drive progress and shape the future of electronics.

KEY POINTS

- *Silicon as dominant semiconductor material:* Silicon is the most widely used semiconductor material in electronic devices due to its abundance, reliability, and well-understood properties.
- *Doping for control of electrical properties:* Doping with specific impurities, such as phosphorus or boron, is crucial for controlling the conductivity of silicon and creating P-type and N-type regions necessary for device functionality.
- *Integration of III-V compound semiconductors:* III-V compound semiconductors, like gallium arsenide (GaAs), are often integrated with silicon for specialized applications, taking advantage of their superior electronic properties for high-speed and high-frequency devices.
- *High-k dielectrics for advanced transistors:* In modern semiconductor device design, high-k dielectric materials are employed as gate insulators in transistors to reduce leakage current and enable further miniaturization.
- *Advanced packaging materials:* Packaging materials play a crucial role in protecting semiconductor devices from environmental factors and facilitating heat dissipation. Materials, like ceramics and organic substrates, are commonly used for this purpose.
- *Emerging materials for future devices:* Researchers are exploring various emerging materials, such as carbon nanotubes, graphene, and 2D transition metal dichalcogenides, for their potential in next-generation semiconductor devices with enhanced performance and novel functionalities.

ACKNOWLEDGMENT

The authors would like to express their profound gratitude to the Central Laboratory, Computer Science and Engineering Department, University of Engineering and Management (UEM), Jaipur, Rajasthan, India, for their invaluable support throughout the research process for this book chapter. Their state-of-the-art facilities and dedicated team have provided the perfect environment for conducting the research and analysis. Additionally, the authors extend their heartfelt thanks to the VLSI Design Laboratory, Electronics and Communication Engineering Department, Dr. B.R. Ambedkar National Institute of Technology (NIT), Jalandhar, Punjab, India. Their expertise and resources have been instrumental in enabling authors to delve deep into the complexities of VLSI design, significantly enriching the quality and depth of their work. Without the unwavering support and contributions from these esteemed institutions, the completion of this research could not have been possible. The authors are deeply grateful for their assistance and collaboration.

REFERENCES

[1]   S.M. Sze, Y. Li, and K.K. Ng, *Physics of Semiconductor Devices*. John Wiley & Sons, 2021.

[2]   H. Yoo, K. Heo, M.H.R. Ansari, and S. Cho, "Recent advances in electrical doping of 2D semiconductor materials: Methods, analyses, and applications", *Nanomaterials (Basel),* vol. 11, no. 4, p. 832, 2021.
        [http://dx.doi.org/10.3390/nano11040832] [PMID: 33805062]

[3]   G. Johannesson, and N. Salem, "Design structure of compound semiconductor devices and its applications", *Acta Energetica,* vol. 2, pp. 28-35, 2022.

[4]   Wenqi Shi, Canwen Zou, Yulian Cao, and Jianguo Liu, "The progress and trend of heterogeneous integration silicon/III-V semiconductor optical amplifiers", *Photonics,* vol. 10, no. 2, p. 161, 2023.
        [http://dx.doi.org/10.3390/photonics10020161]

[5]   Y. Taur, and T.H. Ning, *Fundamentals of Modern VLSI Devices*. Cambridge University Press, 2009.
        [http://dx.doi.org/10.1017/CBO9781139195065]

[6]   Y.B. Kim, "Challenges for nanoscale MOSFETs and emerging nanoelectronics", *Transactions on Electrical and Electronic Materials,* vol. 11, no. 3, pp. 93-105, 2010.
        [http://dx.doi.org/10.4313/TEEM.2010.11.3.093]

[7]   R. Parekh, and R. Dhavse, *Nanoelectronics: Physics*. Technology, and Applications, 2023.
        [http://dx.doi.org/10.1088/978-0-7503-4811-9]

[8]   S. Shekhar, W. Bogaerts, L. Chrostowski, J.E. Bowers, M. Hochberg, R. Soref, and B.J. Shastri, "Roadmapping the next generation of silicon photonics", *Nat. Commun.,* vol. 15, no. 1, p. 751, 2024.
        [http://dx.doi.org/10.1038/s41467-024-44750-0] [PMID: 38272873]

[9]   D.J. Lockwood, and L. Pavesi, *Silicon Photonics IV*. vol. Vol. 139. Springer International Publishing, 2021.
        [http://dx.doi.org/10.1007/978-3-030-68222-4]

[10]   A.C. Diebold, Ed., *Handbook of Silicon Semiconductor Metrology*. 1st ed. CRC Press, 2001. [Online]
        [http://dx.doi.org/10.1201/9780203904541]

[11]   P. Gargini, F. Balestra, and Y. Hayashi, "Roadmapping of nanoelectronics for the new electronics industry", *Appl. Sci. (Basel),* vol. 12, no. 1, p. 308, 2021.
        [http://dx.doi.org/10.3390/app12010308]

[12]   O.A. Oyubu, and O.U. Kazeem, "An overview of nanoelectronics and nanodevices", *Journal of Engineering Studies and Research,* vol. 26, no. 3, pp. 165-172, 2020.
        [http://dx.doi.org/10.29081/jesr.v26i3.220]

[13]   "Liang, Di, and John E. Bowers. "Recent progress in heterogeneous III-V-on-silicon photonic

integration." Light", *Adv. Manuf.,* vol. 2, no. 1, pp. 59-83, 2021.

[14] Davenport, M. L., Skendžić, S., Volet, N., Hulme, J. C., Heck, M. J., & Bowers, J. E. "Heterogeneous silicon/III–V semiconductor optical amplifiers", *IEEE Journal of Selected Topics in Quantum Electronics,* vol. 22, no. 6, no. 78, p. 88, 2016.

[15] C.P. Auth, and J.D. Plummer, "Scaling theory for cylindrical, fully-depleted, surrounding-gate MOSFET's", *IEEE Electron Device Lett.,* vol. 18, no. 2, pp. 74-76, 1997.
[http://dx.doi.org/10.1109/55.553049]

[16] I. Ferain, C.A. Colinge, and J.P. Colinge, "Multigate transistors as the future of classical metal–oxide–semiconductor field-effect transistors", *Nature,* vol. 479, no. 7373, pp. 310-316, 2011.
[http://dx.doi.org/10.1038/nature10676] [PMID: 22094690]

[17] M. H. Nayfeh, and C. R. Abernathy, *III-V Compound Semiconductors: Integration with Silicon-Based Microelectronics.,* 2010.

[18] A. Malinowski, and S.K. Mishra, "Challenges in performance improvement of silicon systems on Chip in advanced Nanoelectronics technology nodes",
[http://dx.doi.org/10.23919/MIXDES49814.2020.9155752]

[19] J.C. Bean, and K.I. Fujino, *Silicon Heterostructure Handbook: Materials.* Fabrication, Devices, Circuits, and Applications of SiGe and Si Strained-Layer Epitaxy, 2005.

[20] John Wiley & Sons, *III-V Compound Semiconductors: Integration with Silicon-Based Microelectronics.,* 2010.

[21] S. Hirano, and S. Somiya, *Silicon-Based Structural Ceramics for the New Millennium.,* 2000.

[22] D.A. Neamen, *Semiconductor Physics and Devices.,* 1992.

[23] M. Jamal Deen and Jaspreet Singh, *Silicon Photonics: Fundamentals and Devices.,* John Wiley & Sons, 2012.

[24] Stanley Wolf and Richard N. Tauber. "Silicon Processing for the VLSI Era, Vol. 1: Process Technology". *Lattice Press,* 1986.

[25] M. Willander, *III–V Semiconductor Materials and Devices.* Springer, 2013.

[26] T.C. Damen, *Epitaxial Growth of III-V Semiconductors: An Introduction to the Fundamental Principles.* North-Holland, 1992.

[27] Lorenzo Pavesi and David J. Lockwood, *Silicon Nanophotonics: Basic Principles, Present Status, and Perspectives.* CRC Press, 2009.

[28] M.A. Green, Y. Hishikawa, E.D. Dunlop, D.H. Levi, J. Hohl-Ebinger, and A.W.Y. Ho-Baillie, "Silicon materials for photovoltaic applications – A review." Progress in Materials Science", *Online (Bergh.),* 2018.
[http://dx.doi.org/10.1016/j.pmatsci.2018.02.001]

[29] J.E. Bowers, and M.J.W. Rodwell, "Recent progress in III-V semiconductors and devices on silicon", *Proceedings of the IEEE,* 2010.

[30] N. Grandjean, and J. Massies, "III-V semiconductors: Growth, properties and applications", *Journal of Crystal Growth,* 1998.
[http://dx.doi.org/10.1016/S0022-0248(97)00736-0]

[31] R. Saive, "Light trapping in thin silicon solar cells: A review on fundamentals and technologies", *Prog. Photovolt. Res. Appl.,* vol. 29, no. 10, pp. 1125-1137, 2021.
[http://dx.doi.org/10.1002/pip.3440]

[32] A.S. Algamili, M.H.M. Khir, J.O. Dennis, A.Y. Ahmed, S.S. Alabsi, S.S. Ba Hashwan, and M.M. Junaid, "A review of actuation and sensing mechanisms in MEMS-based sensor devices", *Nanoscale Res. Lett.,* vol. 16, no. 1, p. 16, 2021.
[http://dx.doi.org/10.1186/s11671-021-03481-7] [PMID: 33496852]

[33]   French, Paddy J., Gijs J. M. Krijnen, Sten Vollebregt, and Massimo Mastrangeli. Technology development for MEMS: A tutorial., *IEEE Sens. J.,* vol. 22, no. 11, pp. 10106-10125, 2021.

[34]   R. Hajare, V. Reddy, and R. Srikanth, "MEMS based sensors – A comprehensive review of commonly used fabrication techniques", *Mater. Today Proc.,* vol. 49, pp. 720-730, 2022.
[http://dx.doi.org/10.1016/j.matpr.2021.05.223]

[35]   M. Li, and T. Hu, "Research status and development trend of MEMS S&A devices: A review", *Defence Technology,* vol. 17, no. 2, pp. 450-456, 2021.
[http://dx.doi.org/10.1016/j.dt.2020.02.014]

[36]   W.A. Gill, I. Howard, I. Mazhar, and K. McKee, "A review of MEMS vibrating gyroscopes and their reliability issues in harsh environments", *Sensors (Basel),* vol. 22, no. 19, p. 7405, 2022.
[http://dx.doi.org/10.3390/s22197405] [PMID: 36236508]

[37]   C. Chircov, and A.M. Grumezescu, "Microelectromechanical systems (MEMS) for biomedical applications", *Micromachines (Basel),* vol. 13, no. 2, p. 164, 2022.
[http://dx.doi.org/10.3390/mi13020164] [PMID: 35208289]

[38]   M. Karimzadehkhouei, B. Ali, M. Jedari Ghourichaei, and B.E. Alaca, "Silicon nanowires driving miniaturization of microelectromechanical systems physical sensors: A review", *Adv. Eng. Mater.,* vol. 25, no. 12, p. 2300007, 2023.
[http://dx.doi.org/10.1002/adem.202300007]

[39]   A.K. Basu, A. Basu, S. Ghosh, S. Bhattacharya, Ed., *MEMS Applications in Biology and Healthcare.* AIP Publishing: Melville, 2021.
[http://dx.doi.org/10.1063/9780735423954]

[40]   T.B. Taha, A.A. Barzinjy, F.H.S. Hussain, and T. Nurtayeva, "Nanotechnology and computer science: Trends and advances", *Memories - Materials, Devices, Circuits and Systems,* vol. 2, p. 100011, 2022.
[http://dx.doi.org/10.1016/j.memori.2022.100011]

[41]   S. Raman, R.S. A, and S. M, "Advances in silicon nanowire applications in energy generation, storage, sensing, and electronics: a review", *Nanotechnology,* vol. 34, no. 18, p. 182001, 2023.
[http://dx.doi.org/10.1088/1361-6528/acb320] [PMID: 36640446]

[42]   Singh, P., Sharma, K., Puchades, I., & Agarwal, P. B. "A comprehensive review on MEMS-based viscometers", *Sensors and Actuators A: Physical,* vol. 338, p. 113456, 2002.

[43]   H. Helmers, "Advancing solar energy conversion efficiency to 47.6% and exploring the spectral versatility of III-V photonic power converters", *in Physics, Simulation, and Photonic Engineering of Photovoltaic Devices XIII,* vol. 12881, pp. 6-15, 2024.
[http://dx.doi.org/10.1117/12.3000352]

[44]   Spinelli, P., Ferry, V. E., Van de Groep, J., Van Lare, M., Verschuuren, M. A., Schropp, R. E. I., & Polman, A. Plasmonic light trapping in thin-film Si solar cells, *Journal of Optics,* vol. 14, no. 2, p. 024002, 2012.

[45]   A.S. Al-Ezzi, and M.N.M. Ansari, "Photovoltaic solar cells: a review", *Appl. Syst. Innov.,* vol. 5, no. 4, p. 67, 2022.
[http://dx.doi.org/10.3390/asi5040067]

[46]   E.T. Efaz, M.M. Rhaman, S.A. Imam, K.L. Bashar, F. Kabir, M.D.E. Mourtaza, S.N. Sakib, and F.A. Mozahid, "A review of primary technologies of thin-film solar cells", *Eng. Res. Express,* vol. 3, no. 3, p. 032001, 2021.
[http://dx.doi.org/10.1088/2631-8695/ac2353]

[47]   A. Udabe, I. Baraia-Etxaburu, and D.G. Diez, "Gallium nitride power devices: A state of the art review", *IEEE Access,* vol. 11, pp. 48628-48650, 2023.
[http://dx.doi.org/10.1109/ACCESS.2023.3277200]

[48]   S.N. Shuji Nakamura, "GaN growth using GaN buffer layer", *Jpn. J. Appl. Phys.,* vol. 30, no. 10A, pp.

L1705-L1707, 1991.
[http://dx.doi.org/10.1143/JJAP.30.L1705]

[49]   U.K. Mishra, Shen Likun, T.E. Kazior, and Yi-Feng Wu, "GaN-based RF power devices and amplifiers", *Proc. IEEE,* vol. 96, no. 2, pp. 287-305, 2008.
[http://dx.doi.org/10.1109/JPROC.2007.911060]

[50]   A. Jarndal, A.Z. Markos, and G. Kompa, "Improved modeling of GaN HEMTs on Si substrate for design of RF power amplifiers", *IEEE Trans. Microw. Theory Tech.,* vol. 59, no. 3, pp. 644-651, 2011.
[http://dx.doi.org/10.1109/TMTT.2010.2095034]

[51]   M. Wang, H. Zhu, G. Yang, J. Li, and L. Kong, "A generally reliable model for composition-dependent lattice constants of substitutional solid solutions", *Acta Mater.,* vol. 211, p. 116865, 2021.
[http://dx.doi.org/10.1016/j.actamat.2021.116865]

[52]   S.K. Behura, A. Miranda, S. Nayak, K. Johnson, P. Das, and N.R. Pradhan, "Moiré physics in twisted van der Waals heterostructures of 2D materials", *Emergent Mater.,* vol. 4, no. 4, pp. 813-826, 2021.
[http://dx.doi.org/10.1007/s42247-021-00270-x]

[53]   A. K. Pimachev, and S. Neogi, "First-principles prediction of electronic transport in fabricated semiconductor heterostructures *via* physics-aware machine learning", *npj Computational Materials,* vol. 7, p. 93, 2021.
[http://dx.doi.org/10.1038/s41524-021-00562-0]

[54]   J. Guo, R. Xiang, T. Cheng, S. Maruyama, and Y. Li, "One-dimensional van der Waals heterostructures: A perspective", *ACS Nanosci. Au,* vol. 2, no. 1, pp. 3-11, 2022.
[http://dx.doi.org/10.1021/acsnanoscienceau.1c00023] [PMID: 37101518]

[55]   C. Ballif, F-J. Haug, M. Boccard, P.J. Verlinden, and G. Hahn, "Status and perspectives of crystalline silicon photovoltaics in research and industry", *Nat. Rev. Mater.,* vol. 7, no. 8, pp. 597-616, 2022.
[http://dx.doi.org/10.1038/s41578-022-00423-2]

[56]   H.M. Tun, and M.S. Nwe, "Fabrication of high performance high electron mobility transistor design based on III-V compound semiconductor", *Proceedings of the 11th International Conference on Robotics, Vision, Signal Processing and Power Applications: Enhancing Research and Innovation through the Fourth Industrial Revolution,* pp. 376-381, 2022.
[http://dx.doi.org/10.1007/978-981-16-8129-5_59]

[57]   J. A. del Alamo, "The high electron mobility transistor: 40 years of excitement and surprises", *in 75th Anniversary of the Transistor,* pp. 253-262, 2023.
[http://dx.doi.org/10.1002/9781394202478.ch20]

[58]   E. Raghuveera, G.P. Rao, and T.R. Lenka, "Prospects of III–V semiconductor-based high electron mobility transistors (HEMTs) towards emerging applications", *International Conference on Micro/Nanoelectronics Devices, Circuits and Systems,* pp. 123-137, 2023.

[59]   V. Hemaja, and D.K. Panda, "A comprehensive review on high electron mobility transistor (HEMT) Based biosensors: recent advances and future prospects and its comparison with Si-based biosensor", *Silicon,* vol. 14, no. 5, pp. 1873-1886, 2022.
[http://dx.doi.org/10.1007/s12633-020-00937-w]

[60]   I. Berdalovic, M. Poljak, and T. Suligoj, "A comprehensive model and numerical analysis of electron mobility in GaN-based high electron mobility transistors", *J. Appl. Phys.,* vol. 129, no. 6, p. 064303, 2021.
[http://dx.doi.org/10.1063/5.0037228]

[61]   M. Haziq, S. Falina, A.A. Manaf, H. Kawarada, and M. Syamsul, "Challenges and opportunities for high-power and high-frequency AlGaN/GaN high-electron-mobility transistor (HEMT) applications: A review", *Micromachines (Basel),* vol. 13, no. 12, p. 2133, 2022.
[http://dx.doi.org/10.3390/mi13122133] [PMID: 36557432]

[62]   A. B. Khan, "Challenges and opportunities for high-power and high-frequency AlGaN/GaN high-

electron-mobility transistor (HEMT) applications: A review", *Micromachines,* vol. 13, no. 12, p. 2133, 2021.
[http://dx.doi.org/10.1002/9781119755104.ch5]

[63]   K. Biswas, R. Ghoshhajra, and A. Sarkar, ""High electron mobility transistor: Physics-based TCAD simulation and performance analysis," in HEMT Technology and Applications, Singapore, Springer", *Nat. Singap.,* pp. 155-179, 2022.

[64]   O. Odabaşı, D. Yilmaz, E. Aras, K.E. Asan, S. Zafar, B.C. Akoglu, B. Butun, and E. Ozbay, "AlGaN/GaN-based laterally gated high-electron-mobility transistors with optimized linearity", *IEEE Trans. Electron Dev.,* vol. 68, no. 3, pp. 1016-1023, 2021.
[http://dx.doi.org/10.1109/TED.2021.3053221]

[65]   E. Raghuveera, "Prospects of III-V Semiconductor-Based", *Micro and Nanoelectronics Devices, Circuits and Systems: Select Proceedings of MNDCS,* p. 1067, 2023.

[66]   S. Marcinkevičius, R. Yapparov, Y.C. Chow, C. Lynsky, S. Nakamura, S.P. DenBaars, and J.S. Speck, "High internal quantum efficiency of long wavelength InGaN quantum wells", *Appl. Phys. Lett.,* vol. 119, no. 7, p. 071102, 2021.
[http://dx.doi.org/10.1063/5.0063237]

[67]   G. Verma, K. Mondal, and A. Gupta, "Si-based MEMS resonant sensor: A review from microfabrication perspective", *Microelectronics J.,* vol. 118, p. 105210, 2021.
[http://dx.doi.org/10.1016/j.mejo.2021.105210]

[68]   G. Langfelder, M. Bestetti, and M. Gadola, "Silicon MEMS inertial sensors evolution over a quarter century", *J. Micromech. Microeng.,* vol. 31, no. 8, p. 084002, 2021.
[http://dx.doi.org/10.1088/1361-6439/ac0fbf]

[69]   O. Burkacky, M. de Jong, and J. Dragon, *Strategies to lead in the semiconductor world,* MGI, 2022.

[70]   Utmel,   "Analysis   of   Semiconductor   Wafers,"   Available   from: https://www.utmel.com/blog/categories/semiconductor/analysis-of-semiconductor-wafers Apr. 16, 2020.

[71]   O. Burkacky, J. Dragon, and N. Lehmann, *The semiconductor decade: A trillion-dollar industry.* vol. Vol. 1. McKinsey & Company, 2022.

[72]   O. Burkacky, J. Deichmann, P. Pfingstag, and J. Werra, *Semiconductor shortage: How the automotive industry can succeed.* McKinsey & Company, 2022.

[73]   C.S. Kuo, J.J. Yu, and F.C. Chang, "Revenue recognition and channel stuffing in the Taiwanese semiconductor industry", *Int. J. Discl. Gov.,* vol. 19, no. 3, pp. 352-361, 2022.
[http://dx.doi.org/10.1057/s41310-022-00144-6]

[74]   M. Boquet, *Key success factors in the semiconductor industry: Comparative analysis of the US, European and Chinese semiconductor industries,* 2021.

[75]   C. Richard, *"The Semiconductor Industry–Past, Present, and Future," in Understanding Semiconductors: A Technical Guide for Non-Technical People.* Apress: Berkeley, CA, 2022, pp. 175-210.

[76]   X. Chen, Research on marketing strategy of the semiconductor company C, 2023.

[77]   S.M. Khan, D. Peterson, and A. Mann, "The semiconductor supply chain", *CSET Issue Brief. Center for Security and Emerging Technology,* 2021.https://cset.georgetown.edu/publication/the-semiconductor-supply-chain

[78]   C. Mouré, "Technological change and strategic sabotage: A capital as power analysis of the US semiconductor business", *Real-World Economics Review,* vol. 103, pp. 26-55, 2023.

[79]   M. Mann, and V. Putsche, *Semiconductor: Supply Chain Deep Dive Assessment* USDOE Office of Policy (OP), Washington DC (United States),, 2022.
[http://dx.doi.org/10.2172/1871585]

[80]   A. Chikhalkar, "Interaction of charge carriers with defects at interfaces and grain boundaries in compound semiconductors". 2021.

[81]   A. K. Ahirwar, and V. Jangde, *A comprehensive study on indian semiconductor industry*, 2024.

[82]   J. Weinstein, *Semiconductors and the Calculation of the Balance of Power*, 2023.

[83]   Y. Liu, Y. Fang, D. Yang, X. Pi, and P. Wang, "Recent progress of heterostructures based on two dimensional materials and wide bandgap semiconductors", *J. Phys. Condens. Matter,* vol. 34, no. 18, p. 183001, 2022.
[http://dx.doi.org/10.1088/1361-648X/ac5310] [PMID: 35134786]

[84]   P.V. Pham, S.C. Bodepudi, K. Shehzad, Y. Liu, Y. Xu, B. Yu, and X. Duan, "2D heterostructures for ubiquitous electronics and optoelectronics: principles, opportunities, and challenges", *Chem. Rev.,* vol. 122, no. 6, pp. 6514-6613, 2022.
[http://dx.doi.org/10.1021/acs.chemrev.1c00735] [PMID: 35133801]

[85]   Y. Li, J. Zhang, Q. Chen, X. Xia, and M. Chen, "Emerging of heterostructure materials in energy storage: a review", *Adv. Mater.,* vol. 33, no. 27, p. 2100855, 2021.
[http://dx.doi.org/10.1002/adma.202100855] [PMID: 34033149]

[86]   P. Singh, and D.S. Yadav, "Assessment of temperature and ITCs on single gate L-shaped tunnel FET for low power high frequency application", *Eng. Res. Express,* vol. 6, no. 1, p. 015319, 2024.
[http://dx.doi.org/10.1088/2631-8695/ad32b0]

[87]   V. Zatko, S.M.M. Dubois, F. Godel, C. Carrétéro, A. Sander, S. Collin, M. Galbiati, J. Peiro, F. Panciera, G. Patriarche, P. Brus, B. Servet, J.C. Charlier, M.B. Martin, B. Dlubak, and P. Seneor, "Band-gap landscape engineering in large-scale 2D semiconductor van der Waals heterostructures", *ACS Nano,* vol. 15, no. 4, pp. 7279-7289, 2021.
[http://dx.doi.org/10.1021/acsnano.1c00544] [PMID: 33755422]

[88]   H. Xu, M.K. Akbari, and S. Zhuiykov, "2D semiconductor nanomaterials and heterostructures: Controlled synthesis and functional applications", *Nanoscale Res. Lett.,* vol. 16, no. 1, p. 94, 2021.
[http://dx.doi.org/10.1186/s11671-021-03551-w] [PMID: 34032946]

[89]   P. Singh, and D.S. Yadav, "Impact of work function variation for enhanced electrostatic control with suppressed ambipolar behavior for dual gate L-TFET", *Curr. Appl. Phys.,* vol. 44, pp. 90-101, 2022.
[http://dx.doi.org/10.1016/j.cap.2022.09.014]

[90]   P.J. Parbrook, B. Corbett, J. Han, T-Y. Seong, and H. Amano, "Micro-light emitting diode: from chips to applications", *Laser Photonics Rev.,* vol. 15, no. 5, p. 2000133, 2021.
[http://dx.doi.org/10.1002/lpor.202000133]

[91]   D. Liang, and J.E. Bowers, ""Recent progress in heterogeneous III-V-on-silicon photonic integration," Light", *Adv. Manuf.,* vol. 2, no. 1, pp. 59-83, 2021.

[92]   S. Sen, M. Khosla, and A. Raman, *"Design and Challenges in TFET,"* in *Advanced Field-Effect Transistors.* CRC Press, 2023, pp. 23-45.
[http://dx.doi.org/10.1201/9781003393542-2]

[93]   P. Singh, and D.S. Yadav, *Design and investigation of f-shaped tunnel fet with enhanced analog/rf parameters.* Silicon, 2021, pp. 1-16.

[94]   S. Sen, A. Raman, and M. Khosla, "A literature survey on tunnel field effect transistors", *AIJR Proceedings,* pp. 506-512, 2021.
[http://dx.doi.org/10.21467/proceedings.114.65]

[95]   D.S. Yadav, S.B. Rahi, S. Tirkey, Ed., *Advanced Field-Effect Transistors: Theory and Applications.* CRC Press, 2023.
[http://dx.doi.org/10.1201/9781003393542]

[96]   W. Cao, H. Bu, M. Vinet, M. Cao, S. Takagi, S. Hwang, T. Ghani, and K. Banerjee, "The future transistors", *Nature,* vol. 620, no. 7974, pp. 501-515, 2023.

[http://dx.doi.org/10.1038/s41586-023-06145-x] [PMID: 37587295]

[97]   P.A. Gargini, "Overcoming semiconductor and electronics crises with IRDS: Planning for the future", *IEEE Electron Devices Magazine,* vol. 1, no. 3, pp. 32-47, 2023.
[http://dx.doi.org/10.1109/MED.2023.3340123]

[98]   A. Chen, "Beyond-CMOS roadmap—from Boolean logic to neuro-inspired computing", *Japanese Journal of Applied Physics,* vol. 61, p. SM1003, 2022.
[http://dx.doi.org/10.35848/1347-4065/ac5d86]

[99]   N. Kumar, R. Dhar, C. Pascual Garcia, and V.P. Georgiev, "A novel computational framework for simulations of bio-field effect transistors", *ECS Trans.,* vol. 111, no. 1, pp. 249-260, 2023.
[http://dx.doi.org/10.1149/11101.0249ecst]

[100]   P. Singh, A. Raman, D.S. Yadav, N. Kumar, A. Dixit, and M.D.H.R. Ansari, "Ultra thin finger-like source region-based TFET: Temperature sensor", *IEEE Sens. Lett.,* vol. 8, no. 5, pp. 1-4, 2024.
[http://dx.doi.org/10.1109/LSENS.2024.3390689]

[101]   N. Kumar, C.P. García, A. Dixit, A. Rezaei, and V. Georgiev, "Charge dynamics of amino acids fingerprints and the effect of density on FinFET-based Electrolyte-gated sensor", *Solid-State Electron.,* vol. 210, p. 108789, 2023.
[http://dx.doi.org/10.1016/j.sse.2023.108789]

[102]   R. Ranjan, P. Kumar, and N. Kumar, *Demonstration and Performance Assessment of Dopant Free TFET Including Lattice Heating and Temperature Effects.* Silicon, 2024, pp. 1-8.

Innovative Materials Shaping the Future: A Deep Dive into the Design of Semiconductor Devices

Peeyush Phogat[1,*], **Shreya**[1], **Ranjana Jha**[1] and **Sukhvir Singh**[1]

[1] *Research Lab for Energy Systems, Department of Physics, Netaji Subhas University of Technology, New Delhi, India*

Abstract: The pursuit of advanced semiconductor materials drives innovations across various technological domains. This chapter explores cutting-edge materials essential for semiconductor device development. Key applications include solar cells, capacitors, supercapacitors, thermoelectric devices, sensors, and reactions, such as the Hydrogen Evolution Reaction (HER) and Oxygen Evolution Reaction (OER), also known as water splitting. For solar cells, the chapter highlights materials engineered to boost efficiency and durability, reflecting the evolving landscape of photovoltaic technologies. Capacitors and supercapacitors are analyzed for their energy storage capabilities, with a focus on novel materials promising improved performance and longevity. Thermoelectric materials are examined for their ability to convert waste heat into electrical energy. Sensor technologies are explored, emphasizing materials designed to enhance sensitivity, selectivity, and response times. The chapter also delves into electrocatalysis, specifically addressing semiconductor materials used in water splitting. As the demand for sustainable energy grows, understanding the role of semiconductor materials in these catalytic reactions becomes crucial. This comprehensive exploration provides researchers, engineers, and scientists with a deep understanding of the diverse semiconductor materials shaping the future of electronic and energy applications. Through a multidimensional perspective, it underscores the pivotal role of innovative materials in advancing semiconductor nanoscale devices toward new levels of performance and functionality.

Keywords: Electrocatalysis, Nanomaterials, Solar cells, Supercapacitors, Thermoelectric materials.

INTRODUCTION

The field of semiconductor devices has been a cornerstone of technological progress, driving innovation across a myriad of applications, ranging from electronics to renewable energy. As the demand for smaller, faster, and more

---

* **Corresponding author Peeyush Phogat:** Research Lab for Energy Systems, Department of Physics, Netaji Subhas University of Technology, New Delhi, India; E-mail: peeyush.phogat@gmail.com

Ashish Raman, Prabhat Singh, Naveen Kumar & Ravi Ranjan (Eds.)

energy-efficient devices continues to surge, researchers and engineers are increasingly turning their attention to the critical role that materials play in shaping the future of semiconductor devices. "Innovative Materials Shaping the Future: A Deep Dive into the Design of Semiconductor Devices" represents a significant contribution to this evolving landscape. The chapter emerges against the backdrop of the ever-expanding exploration of advanced semiconductor materials, which has become a focal point of contemporary research. The relentless pursuit of materials with enhanced properties has the potential to revolutionize the design and performance of semiconductor devices, paving the way for unprecedented technological advancements. The chapter aims to provide a comprehensive overview of the cutting-edge materials that are poised to redefine the landscape of semiconductor technology. By focusing on pivotal applications, such as solar cells, capacitors, supercapacitors, thermoelectric devices, sensors, and water-splitting materials, the chapter addresses the diverse facets of semiconductor materials' impact on various technological domains [1-6].

The intricate design challenges and opportunities associated with each application are explored, offering a deep dive into the complexities that researchers and engineers face in harnessing the full potential of innovative materials. From enhancing the efficiency and durability of solar cells to unlocking energy storage capabilities in capacitors and supercapacitors, the chapter navigates through the multifaceted world of semiconductor materials [7-9]. Moreover, the exploration extends to thermoelectric materials, shedding light on the unique properties that make certain semiconductors ideal for converting waste heat into valuable electrical energy [10, 11]. The examination of sensors emphasizes the role of semiconductor materials in amplifying sensitivity, selectivity, and response times, essential for a myriad of applications.

The chapter delves into the fascinating realm of electrocatalysis, addressing semiconductor materials employed in water splitting [12-14]. This becomes particularly pertinent as the global demand for sustainable energy sources intensifies, highlighting the importance of understanding semiconductor materials in catalysing these reactions. Lastly, the exploration extends to the role of semiconductor materials in electrochemical sensing, showcasing their significance in developing sensitive and reliable devices for real-time detection and monitoring.

The significance of advanced semiconductor materials in contemporary research and technology cannot be overstated. As the backbone of electronic devices and an essential component in various technological applications, semiconductors play a pivotal role in shaping the landscape of modern innovation. The relentless pursuit of advanced semiconductor materials stems from the ever-growing demand for more efficient, powerful, and compact electronic devices [15-18]. These materials are crucial in enhancing the performance of devices, such as transistors, integrated circuits, and sensors. The advent of nanotechnology has further underscored the importance of innovative semiconductor materials, as researchers explore nanoscale structures to unlock new functionalities and properties [19-21]. Additionally, in the context of renewable energy, semiconductor materials are key players in the development of solar cells, photodetectors, and energy storage devices. Their unique electronic properties make them indispensable in converting and manipulating electrical signals. The ongoing quest for materials with superior conductivity, stability, and other tailored properties is driving research forward, offering the promise of more sophisticated and sustainable technologies that will shape the future of electronics and energy systems [22 - 24]. Ultimately, the significance of advanced semiconductor materials lies in their transformative potential to revolutionize the design and performance of a wide array of devices, contributing to advancements that permeate every aspect of our interconnected, technology-driven world.

SOLAR CELLS: ENHANCING EFFICIENCY AND DURABILITY OVERVIEW OF SOLAR CELL TECHNOLOGIES

Solar cell technologies represent a critical frontier in the pursuit of sustainable energy sources. As the global demand for clean and renewable energy intensifies, solar cells play a pivotal role in harnessing the power of the sun and converting it into electricity. At present, there are six types of Solar Cells (SC) that use different semiconductor materials for enhancing the efficiency of solar cell applications, including crystalline silicon SC, thin-film SC, organic SC, perovskite SC, multi-junction SC, and tandem SC.

Among the most established solar cell technologies, crystalline silicon solar cells dominate the market. These cells are characterized by their high efficiency and reliability. Monocrystalline and polycrystalline silicon variants are the primary categories, with monocrystalline offering higher efficiency due to its uniform crystal structure. These cells are widely deployed in both residential and commercial solar installations [25]. While generally less efficient than crystalline silicon cells, thin-film technologies excel in specific applications where flexibility and lower production costs are paramount [26]. Organic solar cells, also known as Organic Photovoltaics (OPVs), leverage organic compounds as the active material. These cells offer the advantage of being lightweight, flexible, and potentially cost-effective. However, their efficiency is currently lower than traditional solar cell technologies, and research efforts focus on improving their performance to make them more competitive in the market [27]. Perovskite solar cells have garnered significant attention for their rapid advancements in efficiency. These cells utilize perovskite-structured materials, often made of lead halide compounds, as the light-absorbing layer. Although promising, challenges, such as the stability and toxicity of materials, need to be addressed. Ongoing research aims to overcome these hurdles and establish perovskite solar cells as a mainstream technology [28]. Multi-junction solar cells, commonly used in Concentrated Photovoltaics (CPV), consist of multiple semiconductor materials stacked on top of each other. This design allows the cell to capture a broader spectrum of sunlight, increasing overall efficiency. Multi-junction cells find applications in specialized scenarios, such as space-based solar arrays and solar concentrators [29 - 31]. Tandem solar cells combine multiple layers of different materials with complementary absorption spectra. This approach aims to enhance the overall efficiency by capturing a broader range of sunlight. Tandem structures often involve stacking traditional solar cell materials with emerging technologies, like perovskites, to achieve higher performance [32].

The field of solar cell technologies is dynamic, with ongoing research exploring novel materials and design concepts. Emerging trends include the integration of quantum dots, advanced nanomaterials, and the exploration of tandem structures with perovskite-silicon combinations. These endeavours seek to push the efficiency boundaries and drive the continuous evolution of solar cell technologies.

LATEST SEMICONDUCTOR MATERIALS IN SOLAR CELL DESIGN

Silicon emerged as a cornerstone material in the initial stages of solar energy exploration, maintaining its pivotal role in the sector today [33]. The prevalence of silicon in the Earth's crust, coupled with its exceptional semiconducting properties and thermal stability, makes it a prime candidate for solar cell applications. The research and commercial focus have predominantly centered on crystalline silicon solar cells, categorized into monocrystalline and polycrystalline variants [34]. Mono-Si cells, distinguished by a single crystal with good efficiency, incur higher costs compared to the more affordable polycrystalline silicon cells, characterized by multiple smaller crystals.

The journey of silicon-based solar cells began in the 1950s, with the discovery of silicon's photovoltaic effect. However, significant advancements materialized in the 1970s and 1980s, marked by intensified efforts to enhance performance, reliability, and cost-effectiveness. The introduction of the "p-n junction" structure, where silicon is doped to create distinct electron-rich and electron-deficient regions, proved transformative [35]. This structure, facilitating charge separation upon exposure to sunlight, laid the foundation for contemporary silicon solar cells. The evolution of silicon solar cell technology has predominantly focused on improving light absorption efficiency and charge carrier collection. Various strategies, including surface passivation, texturizing, and anti-reflective coatings, have been employed to minimize losses and enhance overall performance [36]. The popularity of silicon solar cells can be attributed to the extensive research history, allowing continuous refinement of design and manufacturing processes. The established network of suppliers and economies of scale, fueled by the semiconductor sector, have led to cost reductions and ensured abundant access to premium silicon resources.

Despite the impressive track record of silicon solar cells, challenges persist, such as relatively high production costs associated with the energy-intensive creation of silicon wafers [37]. The rigidity of silicon cells limits their integration into unconventional surfaces, favoring flexible and lightweight alternatives for applications, like building facades and wearable devices. However, ongoing research aims to address these challenges through innovative designs, novel production methods, and the incorporation of silicon with other materials to enhance performance. While silicon solar cell technology faces obstacles, its continued dominance in the market is anticipated due to its reliability, affordability, and consistent efficiency improvements. Silicon remains a foundational material in the pursuit of a sustainable and clean energy future, with ongoing research endeavors seeking to mitigate its drawbacks and further elevate its contribution to solar energy deployment. Although silicon has conventionally served as the predominant material for solar cells, ongoing research and scientific endeavors are actively exploring alternative materials to overcome certain constraints and further enhance the performance of solar cells [38]. This exploration is poised to expand the potential applications of solar photovoltaic technology, simultaneously reducing production costs and improving energy conversion efficiency.

One notable category of alternative materials is organic solar cells, utilizing organic materials, such as carbon materials, as their active layer components [39]. Organic materials offer benefits, including the possibility of economical production methods, lightweight and flexible designs, and suitability for diverse applications.

These materials are typically divided into small molecules and polymers, each possessing distinct advantages and drawbacks [40].

Perovskite solar cells constitute another promising alternative material, characterized by a specific crystal structure that excels in light absorption and charge transportation. Despite their remarkable efficiency, challenges related to long-term stability, water detection, and use of potentially harmful or erratic materials still need resolution [41]. QDs, nano-size semiconductors with unique optical and electrical properties, offer another avenue for exploration. Quantum dot solar cells, leveraging the size-dependent absorption properties of quantum dots, demonstrate benefits, such as the quantity of absorbing a broad range of sunlight and the potential for low-cost production techniques [33].

Tandem Solar Cells (TSCs), combining one or more materials with complementary absorption properties, represent a new way to enhance solar cell effectiveness. By optimizing bandgaps and layer arrangements, TSCs can outclass one-junction solar cells [42]. Further alternatives under investigation include nanomaterials, like CNT, graphene, and 2D materials, such as CdTe and CuInSe. Each material presents a unique combination of benefits and challenges, necessitating ongoing efforts to improve performance, stability, and scalability [41]. Overcoming obstacles related to long-term stability, reliability, and achieving high energy alteration effectiveness on a commercial scale remains a significant task. The achievement of substitute materials will hinge on achieving financial prudence of scale, optimizing engineering processes, and addressing cost considerations. Despite challenges, the ongoing research and development activities in the field are driving the advancement and potential commercialization of alternative materials. The exploration of organic materials, QDs, TSCs, and other emerging materials holds the promise of increasing energy conversion efficiency, lowering costs, and expanding the range of solar cell applications. As these materials continue to develop, they may play a crucial role in shaping the future of solar energy production, contributing to a cleaner and more sustainable energy future [43].

Organic Photovoltaic (OPV) cells have gained attention as potential substitutes for traditional Si solar cells. These cells utilize a carbon-based active layer, often referred to as organic semiconductors, for light absorption and charge generation. The appeal of organic materials lies in their potential for cost-effectiveness, flexibility, and compatibility with various substrates [44]. Organic semiconductors, characterized by polymer chains or molecules with π-e$^-$ delocalization, effectively absorb photons across a broad spectrum of solar wavelengths, generating excitons or electron-hole pairs [44]. To extract electrical energy, efficient charge separation at the p-type and n-type junction within the organic layer is crucial [45]. The architecture of organic solar cells often incorporates Bulk Heterojunction (BHJ) structures, combining a donor material with an acceptor material. This combination facilitates charge separation at the donor-acceptor interface, resulting in a charge-separated state that contributes to electrical energy extraction [46]. The manufacturing of organic solar cells commonly employs solution-based technologies, such as spin coating, inkjet printing, or roll-to-roll coating, enabling deposition on flexible or textured surfaces, and promising low-cost, large-scale production. An inherent advantage of organic solar cells lies in their tunability. This tunability allows for the optimization of stability and efficiency [46]. Duan *et al.* showed an effective method for enhancing device stability in Organic Solar Cells (OSCs) [47]. Various examples, such as the addition of an ultrathin Al layer and the use of Atomic Layer Deposition (ALD) with hafnium oxide (HfO2), illustrate improved performance and shelf-life stability. However, concerns are raised about the complexity introduced in the fabrication process, particularly in large-scale production. Despite its effectiveness, the bilayer carrier transport layer strategy may face challenges in compatibility with roll-to-roll fabrication, suggesting a cautious approach in its application for future research, as shown in Fig. (1).

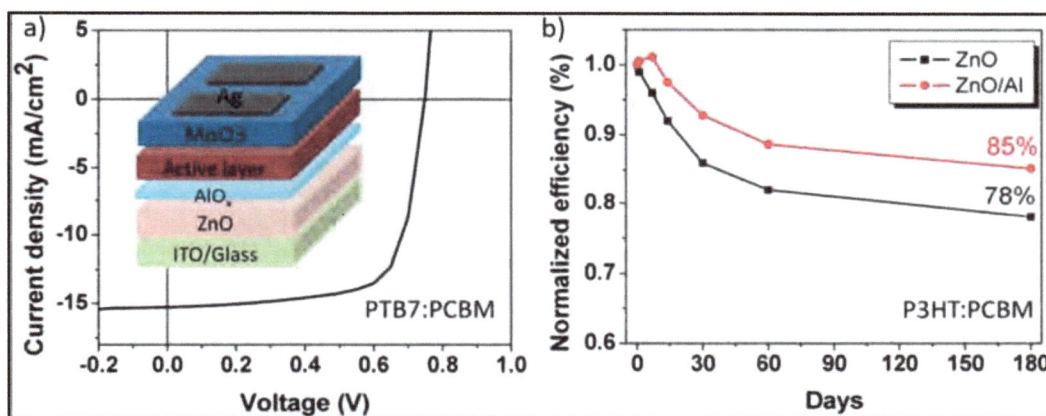

Fig. (1). (a) Device structure with J-V curve for OSC solar cell, **(b)** stability comparison between ZnO and ZnO/Al layer; reproduced from reference [48]. Copyright © 2017, American Chemical Society.

The review by Park *et al.* offered a comprehensive exploration of perovskite solar cells, an emerging photovoltaic technology rooted in the evolution of dye-sensitized solar cells. Initially, the liquid-based perovskite solar cell demonstrated a modest Power Conversion Efficiency (PCE) of 3–4%, but significant advancements were made by doubling the PCE through perovskite coating optimization within two years. However, stability concerns hindered the liquid-based approach, prompting the development of a more stable, long-term perovskite solar cell with a remarkable efficiency of ~10% in 2012. Subsequent years witnessed a rapid increase in efficiencies, reaching 18% within two years. With expectations of achieving PCE values surpassing 20% using cost-effective organometal halide perovskite materials, the review underscored the promising trajectory of perovskite solar cells as a notable photovoltaic technology. The comprehensive discussion encompassed the opto-electronic properties of perovskite materials, recent progress in perovskite solar cells, and insightful commentary on current challenges and future prospects [49]. Perovskite solar cells utilize organometal halide light absorbers, positioning them as a highly promising photovoltaic technology. Highlighting their remarkable PCE and cost-effectiveness, the text traced the evolution from a durable solid-state cell with a 9.7% PCE in 2012 to a groundbreaking 19.3% in 2014. Despite these achievements, the complex working principles of perovskite solar cells remain elusive, necessitating a deeper understanding. Fig. (2) shows the increasing efficiency over the years.

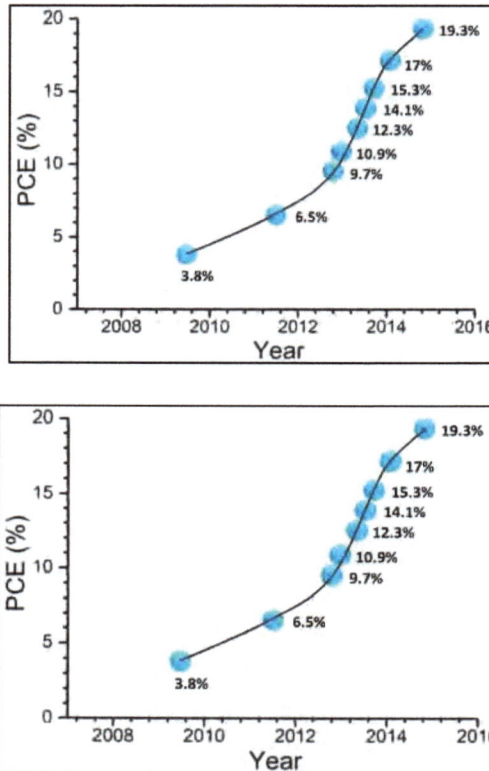

Fig. (2). Increase in the conversion efficiency of perovskite solar cells from 2009 to 2016; data taken from reference [49].

In the domain of photovoltaics, tandem solar cells have emerged as an advanced technology with the potential to significantly enhance energy conversion efficiency compared to conventional single-junction solar cells. This innovative approach involves combining multiple materials with complementary absorption characteristics to effectively capture a broader spectrum of sunlight. This section discusses the conceptual framework, operational principles, and transformative potential of tandem solar cells, which aim to overcome the limitations imposed by the Shockley-Queisser limit on single-junction solar cells, particularly those based on silicon [50]. The fundamental concept behind tandem solar cells involves stacking two or more subcells, each designed to absorb specific wavelengths of sunlight. This arrangement enables the overall voltage and power output to increase through the series connection of subcells, with the bottom subcell absorbing lower-energy photons and the top subcell targeting high-energy photons. However, achieving effective current matching between subcells poses a primary challenge, necessitating careful optimization of materials, bandgaps, and thicknesses for balanced absorption and efficient charge extraction [50]. The discussion highlights various techniques employed in constructing tandem solar cells, such as monolithic integration and mechanically stacked or hybrid approaches. Monolithic integration involves stacking subcells on top of each other using transparent interlayers, demanding meticulous control over material characteristics, interfaces, and optical coupling to minimize losses and enhance overall efficiency. Alternatively, mechanically stacked or hybrid tandem solar cells allow independent manufacturing of subcells, providing flexibility in material selection and bandgap combinations [51].

Tandem solar cells demonstrate superiority over single-junction counterparts by capturing a wider sunlight spectrum, making them highly effective in low-light conditions. The section emphasizes their ability to utilize each semiconductor material efficiently, assigning specific solar spectrum wavelengths to different subcells and avoiding losses due to the thermalization of high-energy photons in a single material [52]. Ongoing research has propelled tandem solar cells beyond the 30% power conversion efficiency mark in lab-scale developments, outperforming some commercial single-junction solar cells. The discussion also delves into the challenges hindering the commercialization of tandem solar cells, emphasizing the need for precise control over subcell thicknesses and compositions. Scalable fabrication processes, such as solution processing or roll-to-roll manufacturing, are deemed essential for cost-effective production. The stability, durability, and compatibility of subcells in tandem structures are addressed, highlighting the ongoing investigations into encapsulation techniques, passivation layers, and interface engineering strategies to ensure long-term performance and commercial viability. Fig. (3) illustrates that there is a continuous improvement in the PCE of single-junctions, particularly evident in the case of perovskites, and practical limits of around ≈27% on the cell level are expected to persist in the coming years [53]. This limitation has driven the emergence of TSC technology, demonstrating its ability to surpass the practical PCE limit associated with single-junctions by mitigating losses from thermalization and non-absorbed photons [54 - 57].

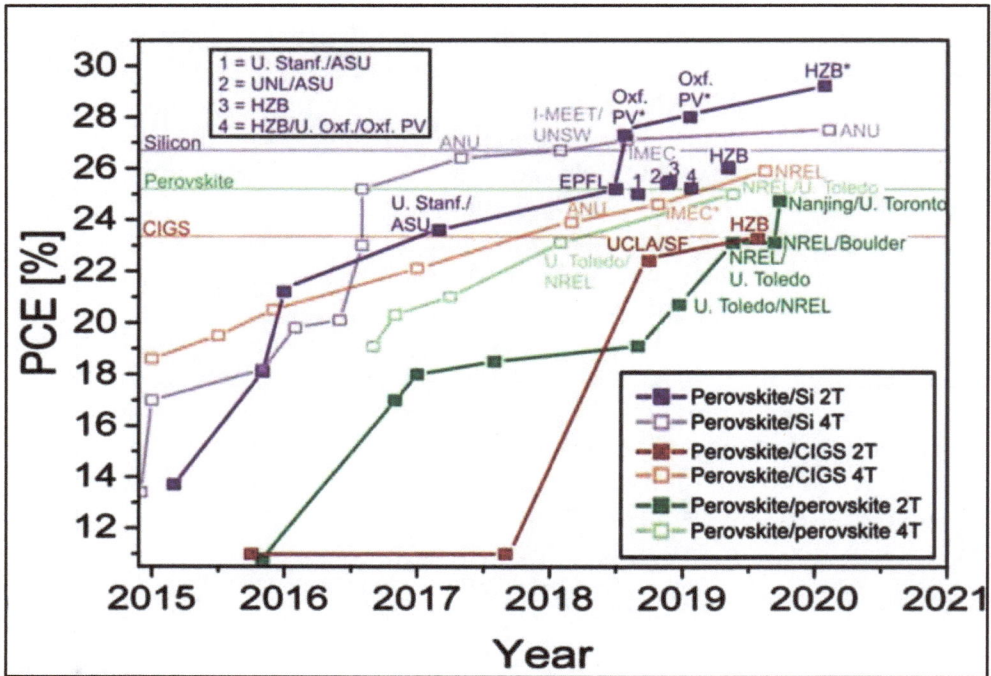

Fig. (3). The progress in efficiency for tandem solar cells based on perovskite, either through press releases (indicated with *) or published papers, has been examined. Reproduced from reference [58]. Copyright © 1999-2024 John Wiley & Sons, Inc.

INSIGHTS INTO THE EVOLVING LANDSCAPE OF PHOTOVOLTAIC TECHNOLOGIES

Nolden *et al.* critically examined the ongoing transformation of the energy system, emphasizing the pivotal role of citizens and communities in shaping the future of energy systems [59]. The traditional focus on citizen participation, particularly in the establishment of SC assemblages, is undergoing significant shifts as supportive national policies for small-scale renewable projects are phased out across European countries. In response, the review proposed social innovation and the adaptation of business models to support and enhance citizen engagement in this evolving policy environment. Focusing on public solar photovoltaic projects as a fundamental technology, the analysis observed the historical growth of public energy business models in England. Significantly, the research highlighted the emergence of new intermediary entities, which play a vital role in facilitating intricate and commercial community energy business models. The authors contended that this shift marks a notable departure from the past, potentially providing increased opportunities for community involvement in the transformation of the energy system, while also enabling some communities to maintain their grassroots character.

Fontaine *et al.* critically examined the nexus between renewable energy development and sustainability. While the rhetoric around sustainable transitions often places renewable energies at the forefront of low-carbon societies, the authors emphasized that the mere categorization as renewable does not guarantee sustainability [60]. The literature review has been found to challenge prevailing notions of sustainability and introduce a relational focus on resource construction processes, offering fresh analytical insights. Applying the concept of different modes of existence, the authors have demonstrated how certain energy resource configurations align with sustainability, while others fall short. In the concluding section, the paper has advocated for a relational framework as a catalyst for redefining and advancing more sustainable renewable energy systems, signaling a paradigm shift in the discourse on the intersection of renewable energy and sustainability [61]. Table **1** provides an insight into the different types of PV technologies and their output efficiency.

CAPACITORS AND SUPERCAPACITORS: UNLEASHING ENERGY STORAGE CAPABILITIES

Role of Semiconductor Materials in Energy Storage

The role of semiconductor materials in energy storage is pivotal to the development of advanced and efficient energy storage devices. Semiconductors play a crucial role in various energy storage technologies, influencing the performance, efficiency, and overall functionality of these devices [62]. In the realm of batteries, semiconductor materials are employed as electrodes, separators, and in some cases, as solid electrolytes. The unique electronic properties of semiconductors facilitate charge storage and transport within the energy storage system, contributing to the overall efficiency and responsiveness of the device. In lithium-ion batteries, for instance, semiconductor materials are commonly utilized in the electrodes, enabling the reversible intercalation and deintercalation of lithium ions during charging and discharging cycles [63]. Semiconductor nanoparticles are also being explored for their potential application in enhancing the conductivity of electrolytes, leading to improved charge/discharge rates and overall performance. Semiconductor materials are also integral to the development of supercapacitors, which offer rapid energy storage and release capabilities. The high surface area and excellent conductivity of certain semiconductors contribute to the enhanced electrochemical performance of supercapacitors. Moreover, semiconductor materials are crucial in the field of

Photoelectrochemical (PEC) energy storage, where they are employed in devices, like photoelectrochemical cells and solar fuel cells [64].

The exploration of semiconductor materials for energy storage extends to emerging technologies, such as lithium-sulfur batteries, where semiconductors are being investigated to mitigate issues related to the polysulfide shuttle effect and improve the overall stability and cycling performance of the battery. Semiconductor materials are indispensable in the landscape of energy storage, contributing to advancements in batteries, supercapacitors, and photoelectrochemical devices. Their unique properties are harnessed to optimize energy storage systems, paving the way for more sustainable and efficient solutions in the rapidly evolving field of energy storage technology [85].

Table 1. Types of PV systems along with their efficiency.

Type of PV Cell	Refs.	Efficiency %	Type of PV Cell	Refs.	Efficiency %
Monocrystalline	[65]	>25%	Quantum dot	[66]	12.1%
CdTe thin-film	[67]	22.1%	Perovskite–silicon tandem	[68]	25.2%
Copper indium Gallium selenide	[69]	23.35%	Ternary organic	[70]	17.5%
Perovskite	[71]	25.2%	Dye-sensitized	[72]	6.3%
Tandem	[51]	29.15%	Perovskite	[73]	20.3%
Silicon	[74]	25.7%	Silicon heterojunction	[75]	21.38%
Multi-junction	[76]	47.1%	III–V multi-junction	[77]	41.6%
Perovskite	[78]	25.2%	Graphene-based	[79]	17.1%
Gallium arsenide	[80]	28.3%	Silicon heterojunction	[81]	22.1%
Dye-sensitized	[82]	11.3%	Silicon heterojunction	[81]	17.4%
Organic–inorganic hybrid	[83]	11%	Perovskite–silicon tandem	[84]	30.7%

SCRUTINY OF CAPACITORS AND SUPERCAPACITORS

The scrutiny of capacitors and supercapacitors involves a comprehensive examination of these energy storage devices, delving into their design, performance, and applications. Capacitors, including supercapacitors or ultracapacitors, are essential components in electronic systems, offering rapid energy storage and release capabilities. Capacitors store electrical energy in an electric field between two conductive plates, separated by an insulating material. Supercapacitors, specifically, exhibit higher capacitance than traditional capacitors due to their unique electrode materials and electrochemical design [86]. In the scrutiny of capacitors, researchers and engineers focus on several key aspects. The capacitance, voltage rating, and leakage current are critical parameters that influence the overall performance of capacitors. Understanding the dielectric material used in capacitors is also crucial, as it significantly impacts the energy storage capacity and efficiency [33]. Additionally, advancements in materials science, including the exploration of novel dielectric materials, contribute to improving the energy density and overall performance of capacitors.

Supercapacitors, often known for their high power density and long cycle life, undergo detailed scrutiny to enhance their capabilities. Electrode materials, such as activated carbon or conductive polymers, play a pivotal role in determining the energy storage capacity and charge/discharge rates of supercapacitors [87, 88]. Researchers are continually investigating advanced materials and innovative designs to push the boundaries of supercapacitor performance. The scrutiny extends to the application-specific requirements of capacitors and supercapacitors. For instance, in electric vehicles and renewable energy systems, supercapacitors are scrutinized for their ability to provide rapid bursts of power during acceleration or absorb excess energy during braking. In electronic devices, capacitors are scrutinized for their stability, reliability, and ability to handle varying voltage levels [89, 90]. Ongoing research and advancements in materials science and design contribute to refining these energy storage devices, making them integral components in various technological domains. Fig. (**4**) shows the Ragone plot for charge storage devices.

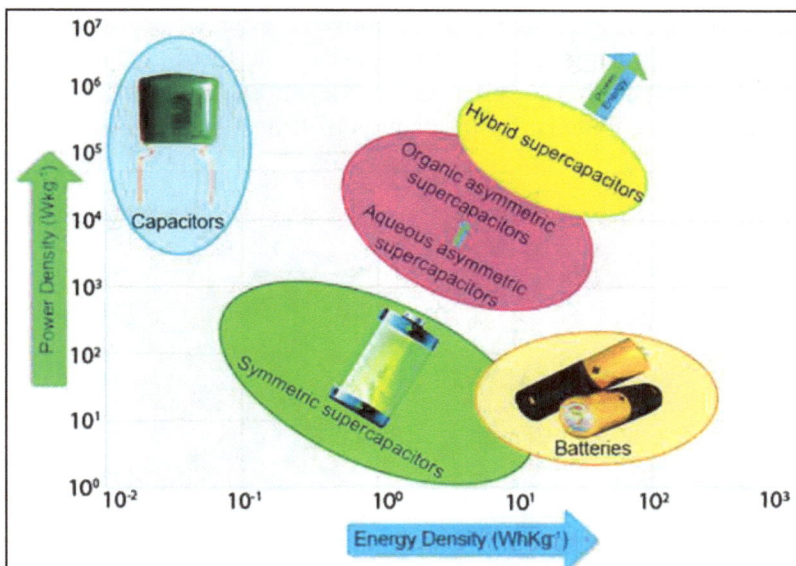

Fig. (4). Ragone plot for capacitors, supercapacitors, and batteries; reproduced from reference [91]. Copyright © 1996-2024 MDPI (Basel, Switzerland).

NOVEL SEMICONDUCTOR MATERIALS FOR HEIGHTENED PERFORMANCE AND LONGEVITY

There are several reasons for the advancement in the capacitor and supercapacitor applications rather than batteries. Various factors play distinct roles in the operation of batteries, capacitors, and supercapacitors. To begin, batteries exhibit considerably more weight compared to capacitors and supercapacitors, posing potential challenges in practical applications. The second factor involves the charge mechanism, where the charging and discharging process occurs upon applying a voltage to the battery terminals. Supercapacitors (SC) demonstrate significantly higher power density than both batteries and traditional capacitors, discharging rapidly, as indicated in Table **2**. While batteries boast greater energy density than capacitors, their lifespan is comparatively shorter. Supercapacitors offer the advantage of environmental friendliness, eliminating the need for chemical combustion seen in batteries [92]. Types of SC, such as Electrochemical Double-layer Capacitors (EDLCs), hybrid capacitors, and pseudocapacitors, display varied attributes. Capacitors can be classified into three categories. The initial form is the electrostatic capacitor utilizing a dry separator [93]. This type of capacitor demonstrates low capacitance and is utilized in applications, like frequency tuning and filtering. Electrostatic capacitors come in various sizes, ranging from picoF to microF. The second type, called the electrolytic capacitor, boasts higher capacitance when compared to its electrostatic counterpart [89]. These capacitors employ a wet separator between electrodes and are commonly employed in signal filtering and buffering tasks. The third and last category is the supercapacitor, known for its capacitance measured in farads, which surpasses that of an electrolytic capacitor by thousands of times (Fig. **5**).

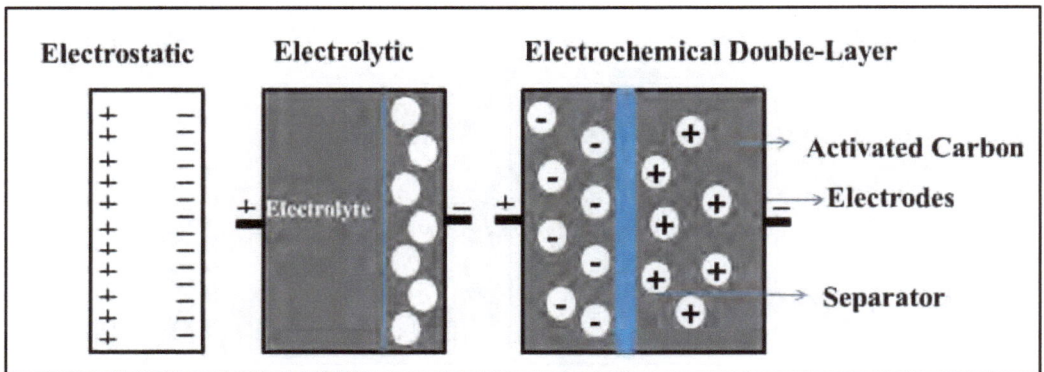

Fig. (5). Schematic diagram of a **(a)** dry capacitor, **(b)** electrolytic capacitor, and **(c)** supercapacitor.

Table 2. Differences in capacitors, batteries, and supercapacitors.

Parameters	Capacitors	Batteries	Supercapacitors
Charge method	Voltage across terminals	Current and voltage	Voltage across terminals
Charge/discharge time	Less	Large	Very less
Power delivered	Rapid discharge	Constant voltage for a larger time	Rapid discharge, linear or exponential voltage decay
Temperature sensitivity	Excellent temperature performance	Brilliant temperature performance	Outstanding temperature performance
Weight	1-100 gm	10 gm to 10 Kg	1-2 gm
Lifetime	> 100k cycles	1500 to 150 cycles	> cycles of 100 k

The utilization of various carbon-based materials in supercapacitor electrodes has been comprehensively studied due to their distinctive characteristics. Activated carbon emerges as a suitable choice owing to its large S/V ratio, excellent conductivity, cost-effectiveness, and abundant availability, making it a preferred material for supercapacitor fabrication [93]. However, challenges related to specific surface area in terms of capacitance have been acknowledged, particularly with graphene, where the rGO may result in a lower surface area and suboptimal pore size distribution, impacting capacitance negatively [93].

The performance of Electric Double-layer Capacitors (EDLC) is influenced not only by surface area, but also by factors, such as pore size distribution, structure, shape, surface functionality, and conductivity [94]. To optimize performance, enhanced porosity and active surface functionality become crucial [95]. Carbon nanotubes, classified as single-walled and multi-walled, exhibit remarkable electrochemical features, including high specific capacitance, low internal resistance, and stability under high current loads, making them excellent candidates for polarizable electrodes in supercapacitors [96]. The performance of carbon nanotubes is intricately linked to micropores and internal resistance, factors that can affect specific capacitance. Researchers are actively exploring flexible carbon fiber hybrid electrodes to leverage synergistic effects [97]. Graphene, with its remarkable properties, such as high surface area, electrical and thermal conductivity, light weight, and chemical stability, has been a focal point in electrode material research for energy storage, especially in supercapacitors [98]. Graphene oxide serves as a precursor for graphene-based materials, with the functionality of the graphene surface significantly influencing specific capacitance [99]. The difficulty lies in attaining complete electrochemical double layers on both sides of graphene because of van der Waals forces, but graphene's extensive surface area and robustness, along with its optical characteristics and excellent conductivity, provide the opportunity for enhanced electrodes

[99]. 3DGFs represent a promising class of carbon materials with unique properties, including an interconnected macroporous structure and excellent electrical conductivity [100]. These materials find applications in various forms, such as foams, aerogels, and sponges.

The utilization of metal oxides in supercapacitor electrodes is significant due to their high specific capacitance, making them favorable for efficient charge storage [101]. Examples of such metal oxides include RuO_2, IrO_2, and MnO_2, with a substantial focus on ruthenium oxide in much of the research due to its remarkable capacitance achieved through the insertion and removal of protons within its amorphous structure [102]. Despite the efficacy, the high cost of ruthenium oxide necessitates the exploration of alternatives, like manganese and ferrous oxides, which share similar properties. However, challenges arise as these metal oxides exhibit poor conductivity, leading to suboptimal supercapacitor performance. To address this limitation, research delves into composite materials, with cobalt oxide, nickel oxide, and various composites emerging as potential electrode materials [103].

The utilization of polymers in the fabrication of supercapacitors is due to their inherent insulating nature and the consequent adoption of doped polymers to enhance conductivity and improve overall properties [104]. The electrochemical formation mechanism has been highlighted, emphasizing the growth of conducting polymers on the current collector's surface to enable p-doping or n-doping, crucial for the charge/discharge process. Polyaniline (PANI) and polypyrrole emerge as exemplary conducting polymers for energy storage devices due to their unique characteristics, such as high conductivity, thermal stability, and facile electrochemical processes [105]. The study underlines the significance of a tailored fabrication process, citing the example of PANI nanostructures, which require a specific separation process to preserve their integrity [106].

THERMOELECTRIC DEVICES: CONVERTING WASTE HEAT TO ELECTRICAL ENERGY

Thermoelectricity and its Applications

Thermoelectricity is a phenomenon that involves the direct conversion of heat energy into electrical energy through the Seebeck effect. The Seebeck effect is based on the principle that a temperature difference across a material can generate an electric voltage. This unique property has led to the development and exploration of thermoelectric materials and devices with diverse applications

[107]. One prominent application of thermoelectricity is in power generation. Thermoelectric Generators (TEGs) (Fig. **6**) are devices that convert waste heat from industrial processes or other sources into electricity [108]. This is particularly valuable for enhancing energy efficiency and sustainability, as it allows the utilization of otherwise wasted thermal energy. Another significant application is in thermoelectric cooling, where the Peltier effect is employed [109]. Thermoelectric coolers use an electric current to transfer heat from one side of the device to the other, resulting in cooling on one side and heating on the opposite side. This technology finds applications in various cooling systems, such as in portable refrigeration units, electronic devices, and even in the aerospace industry for temperature control in spacecraft. In the automotive sector, thermoelectric materials are explored for waste heat recovery from exhaust systems. By converting the high-temperature exhaust heat into electricity, vehicles can improve fuel efficiency and reduce emissions. Moreover, thermoelectric devices are gaining traction in wearable technology, where they can be integrated into clothing to power small electronic devices using the body heat of the wearer. This has the potential to enhance the autonomy of wearable gadgets by providing a continuous and renewable power source.

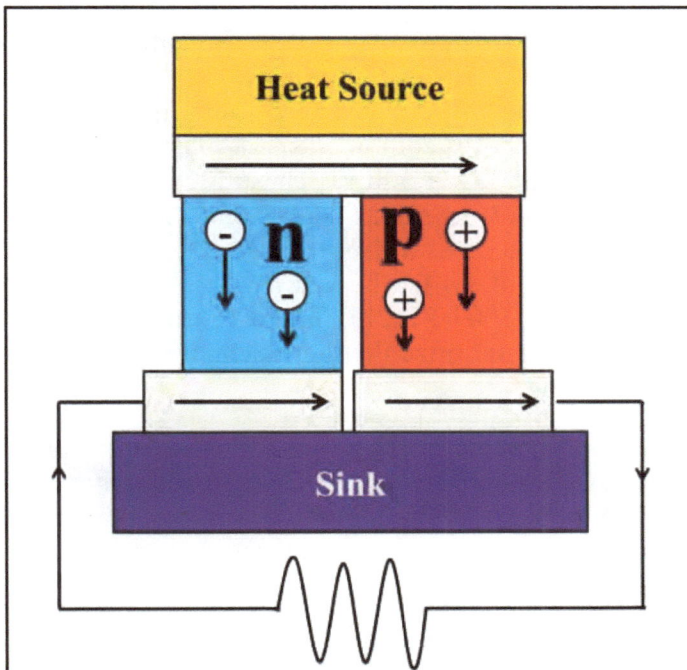

Fig. (6). Thermoelectric circuit for TEGs.

UNIQUE PROPERTIES OF SEMICONDUCTOR MATERIALS IN THERMOELECTRIC DESIGN

P-type Thermoelectric Materials

In recent years, noteworthy work has been dedicated to enhancing the thermoelectric performance of materials, particularly alloys of $(Bi, Sb)_2$ and $(Se, Te)_3$ are known for being thermoelectric semiconductors at 25 °C [110]. Various strategies have been employed, including adjusting the concentration of the carrier, band structure manipulation, defect manipulation, and nanostructure engineering. Duan *et al.* [111] explored the incorporation of $Ge_{0.5}Mn_{0.5}Te$ and $BiCl_3$ into $Bi_{0.38}Sb_{1.62}Te_3$ alloys, achieving an optimal ZT^- value of 1.22. High-pressure techniques, like hot pressing, have been applied, as seen in the work of Wang *et al.* [112], where $CuInTe_2$ inclusion in $Bi_{0.4}Sb_{1.6}Te_3$ led to a ZT^- of 1.15 at 300 K. Ge doping in $Bi_{0.4}Sb_{1.6}Te_3$, investigated by Wang *et al.* [113] demonstrated a ZT^- peak of 1.36 at 300 K. Spark plasma sintering was utilized by Chiu *et al.* [114] on $Sb^{2-}xInxTe_3$, resulting in an optimal ZT^- of 0.27 with $Sb_{1.85}In_{0.15}Te_3$ at 300 K. Cu doping in $Bi_{0.5}Sb_{1.5}Te_3$ *via* spark plasma sintering, as conducted by Hao *et al.* [115], achieved a ZT^- of 0.97. Guo and team [116] used extreme temperature and pressure for $Bi_{0.5}Sb_{1.5}Te_{3x}Se_x$ alloys, obtaining a ZT^- of 0.48 with $Bi_{0.5}Sb_{1.5}Te_{2.7}Se_{0.3}$. Alternative methods, such as ball milling and melting, were explored by Serrano-Sánchez *et al.* [117], yielding ZT^- of 0.8 with $Sb_{1.85}In_{0.15}Te_3$. Gao and team [118] used liquid-state manipulation on $Bi_{0.4}Sb_{1.6}Te_3$, achieving a ZT^- of 0.48. Kim and team [119] combined spark plasma sintering and excess Te addition, reaching a ZT^- of 1.7 at 300 K. These methods, while effective, have been associated with complex processes. Fan *et al.* [120] introduced high entropy alloys, achieving a ZT^- of 0.25 at 300 K. Liu *et al.* [41] explored Yb doping in MgAgSb, achieving a ZT^- of 1.14, while Zheng *et al.* [121] utilized Zn doping, reaching a ZT^- of 0.28 at 300 K [122]. These diverse strategies showcase the multifaceted approaches to optimize the thermoelectric performance of materials for various applications.

Achieving optimal Thermoelectric (TE) performance in conducting polymers requires a careful equilibrium between electrical conductivity and the Seebeck coefficient, both of which are interconnected. Numerous approaches have been investigated to boost charge mobility and regulate carrier concentration in conducting polymers to enhance their TE capabilities.

For instance, Fan *et al.* [123] employed sequential post-treatments with acids and bases on PEDOT:PSS films, resulting in significantly enhanced electrical conductivity and a slight increase in the Seebeck coefficient. Similarly, Jung *et al.* [124] focused on high carrier mobility matrices and minimal dopant concentration, achieving superior TE performance in PDPP3T compared to P3HT. Flexible and mechanically robust conducting polymers have been prepared through electrochemical polarization. Zhang *et al.* [125] synthesized PEDOT:S-PHE films, adjusting polymerization conditions to simultaneously increase electrical conductivity and the Seebeck coefficient. Chemically or electrochemically doping semiconducting polymers introduce charge carriers, impacting electrical conductivity and structural changes. Patel *et al.* [126] compared vapor deposition and immersion doping methods, demonstrating the superior performance of vapor-doped PBTTT over solution-doped counterparts. Lim *et al.* [127] utilized vapor doping to minimize morphological alterations in P3HT films, resulting in higher electrical conductivity and overall power factor compared to solution doping. Li *et al.* [128] emphasized the role of dopant type and polymer structure in power factor, revealing the influence of these factors on the Seebeck coefficient and electrical conductivity. Polymer side chains were found to significantly impact TE performance, with polar side chains enhancing power factor effectively [129].

N-type Thermoelectric Materials

The quest for efficient thermoelectric materials continues to drive research, especially in improving the performance of n-type bismuth telluride (Bi_2Te_3) alloys, which exhibit high performance around room temperature. Strategies, such as doping with excessive tellurium and iodine (I), have been explored to mitigate these effects. Another approach involves high-pressure synthesis of Bi_2Se_3 alloys, where varying sintering pressures influence the thermoelectric properties, with the highest ZT value achieved at 1 GPa [130]. Efforts to improve the TE characteristics of $Bi_2Te_{2.7}Se_{0.3}$ semiconductors for wearable TEGs have been explored using multiple techniques. Among these, the combination of microwave processing and re-spark plasma sintering has proven most effective, yielding a maximum ZT value of 0.76 at 300 K [131]. However, the technique's drawbacks, including extended processing time, has led to the suggestion of incorporating a liquid state manipulation method with directionally solidified texturing [132]. This innovative approach has demonstrated a significant improvement, yielding an optimal ZT value of 0.81 at 300 K [132]. The resulting hierarchical microstructures and improved carrier concentration have led to a maximum ZT value of 0.33 at 300 K [133]. Comparing fabrication methods, solvothermal synthesis offers better control over nanograin characteristics. However, challenges arise in the removal of surfactants, impacting TE performance. Hong *et al.* synthesized $Bi_2Te_{3-x}Se_x$ within a shorter timeframe. The experiments revealed that limiting Se concentration to 0.4 wt% yielded the highest ZT value of 0.68 at 300 K [134]. Despite these advancements, inorganic TE materials still face challenges, such as high manufacturing costs, toxicity, and processing difficulties.

Furthermore, they are inherently rigid, hindering their application in flexible and stretchable wearable TEGs [135]. As research progresses, addressing these challenges and exploring alternative materials remain crucial for advancing thermoelectric technology.

Significant progress has been observed in solution-processable and air-stable Thermoelectric (TE) polymers, leading to notable improvements in p-type polymers. However, the development of n-type TE polymers has been relatively sluggish, encountering obstacles, such as reduced air stability, limited solubility, and comparatively lower charge carrier mobility. To address these issues, various strategies involving n-doping have been explored. Madan and team [136] employed Na-SG as a reducing factor to improve the TE properties of polymers. The study demonstrated that the incorporation of Na-SG salts significantly improved the Power Factor (PF) of the polymers, attributed to three orders of magnitude increase in electrical conductivity compared to pristine polymers. In a similar vein, Un *et al.* [136] investigated the impact of different dopants on the TE properties of FBDPPV, highlighting a converse relationship between dopant mol% and the Seebeck coefficient. Molecular dopants, particularly $(N-DMBI)_2$, exhibited the best electrical conductivity and highest PF, emphasizing the importance of dopant characteristics, including planar shape and efficient pathways, for electron transfer. Addressing the need for electron doping in soluble n-type conducting polymers, electron-withdrawing modification of the donor moiety led to changes in polymer packing orientation and an increased PF. These studies have collectively showcased the diverse strategies and considerations in enhancing the TE properties of n-type polymers, ranging from the selection of suitable dopants to the careful engineering of polymer structures for improved electron affinity and mobility. These advancements have contributed significantly to the broader goal of optimizing polymer materials for efficient thermoelectric applications.

In the quest to enhance the performance of n-type organic Thermoelectric (TE) films, stable doping emerges as a primary challenge. The intrinsic lower power factor of n-type films, exacerbated by exposure to air, necessitates innovative approaches. Montgomery *et al.* [137] adopted a solution doping strategy to create n-type films in CNT/polymer composites. Addressing the influence of oxygen adsorption during the fabrication of n-type films under air exposure, Zhou *et al.* [138] used a transformed SWNT film from p-type to n-type by applying a solution of PEI through drop-casting. The application of PEI molecules onto the CNT bundles enhanced the electron concentration and prevented oxygen doping, preserving the n-type properties for up to 3 months. This strategy resulted in n-type TE films achieving a remarkable power factor of 1500 μW/mK2, underscoring the influence of PEI concentration. Horike *et al.* [139] demonstrated the transformation of a p-type SWCNT into an n-type through polymer doping, using various polymer solutions. The choice of polymer, including PVC, PVP, PVA, and PVAc, significantly influenced the negative Seebeck coefficients of the SWCNTs.

CONVERTING WASTE HEAT INTO VALUABLE ELECTRICAL ENERGY

Converting waste heat into valuable electrical energy through thermoelectric materials represents a promising avenue for enhancing energy efficiency and sustainability. The key to the success of thermoelectric materials lies in their ability to efficiently convert heat differentials into electricity. Researchers focus on materials with high thermoelectric efficiency, characterized by a high Seebeck coefficient, electrical conductivity, and low thermal conductivity. This combination allows for the effective generation of electrical power from modest temperature differences. In industrial applications, such as power plants or manufacturing facilities, where significant amounts of waste heat are produced, TEGs can be integrated to capture and convert this heat into electrical energy. This process not only reduces the overall energy consumption of the facility, but also contributes to a more sustainable energy landscape by utilizing what would otherwise be wasted. The automotive industry is another sector exploring thermoelectric materials for waste heat recovery. Incorporating thermoelectric modules into the exhaust systems of vehicles enables the conversion of heat from the engine into electricity, improving fuel efficiency and reducing emissions. Moreover, thermoelectric materials find applications in various electronic devices and wearable technologies, where they can efficiently convert the body heat of users into electrical power, enhancing the autonomy and sustainability of these devices. While there are challenges, such as the need for materials with improved efficiency and cost-effectiveness, the ongoing research and development in the field of thermoelectric materials hold great promise for transforming waste heat into a valuable and sustainable source of electrical energy. As advancements continue, the widespread implementation of thermoelectric technologies has the potential to significantly contribute to the global pursuit of cleaner and more efficient energy systems.

SENSORS: AMPLIFYING SENSITIVITY, SELECTIVITY, AND RESPONSE TIMES

Importance of Semiconductor Materials in Sensor Technologies

Converting semiconductor materials play a pivotal role in sensor technologies, contributing to the development of highly sensitive and efficient devices that are integral to various industries. The importance of semiconductor materials in sensor technologies can be outlined in several key aspects:

Sensitivity and Precision: Semiconductor materials, particularly those with unique electronic properties, enable the creation of sensors with high sensitivity and precision. The ability of semiconductors to undergo changes in conductivity or voltage in response to external stimuli makes them ideal for detecting subtle variations in the environment, such as temperature, pressure, or chemical composition.

Diversity of Sensor Types: Semiconductors are versatile and can be tailored for different sensor applications. They are employed in a wide range of sensor types, including temperature sensors, humidity sensors, and many more (gas sensors, as shown in Fig. (7). This adaptability allows for the development of sensors suitable for specific requirements in industries, such as healthcare, environmental monitoring, and manufacturing.

Miniaturization and Integration: Semiconductor technologies enable the miniaturization of sensors, making them compact and suitable for integration into various devices and systems. This feature is crucial for modern electronics, wearable devices, and Internet of Things (IoT) applications where space constraints are a consideration.

Response Time: Semiconductors often exhibit rapid response times, allowing sensors to provide real-time data. This is particularly important in applications, such as automotive sensors, medical devices, and industrial monitoring systems where timely and accurate information is critical.

Selectivity and Specificity: Semiconductor materials can be engineered to exhibit selectivity and specificity in detecting particular substances. This is essential for sensors used in applications, such as chemical analysis, environmental monitoring, and biological sensing.

Energy Efficiency: Semiconductor-based sensors can be designed to operate with low power depletion, making them power-affective and appropriate for battery devices. This is advantageous in applications where continuous monitoring is required over extended periods.

The importance of semiconductor materials in sensor technologies lies in their ability to provide sensitivity, precision, versatility, miniaturization, rapid response times, selectivity, energy efficiency, compatibility with electronics, and cost-effectiveness. These characteristics make semiconductor-based sensors indispensable for a wide range of applications, driving advancements in technology and contributing to the development of smarter and more efficient systems.

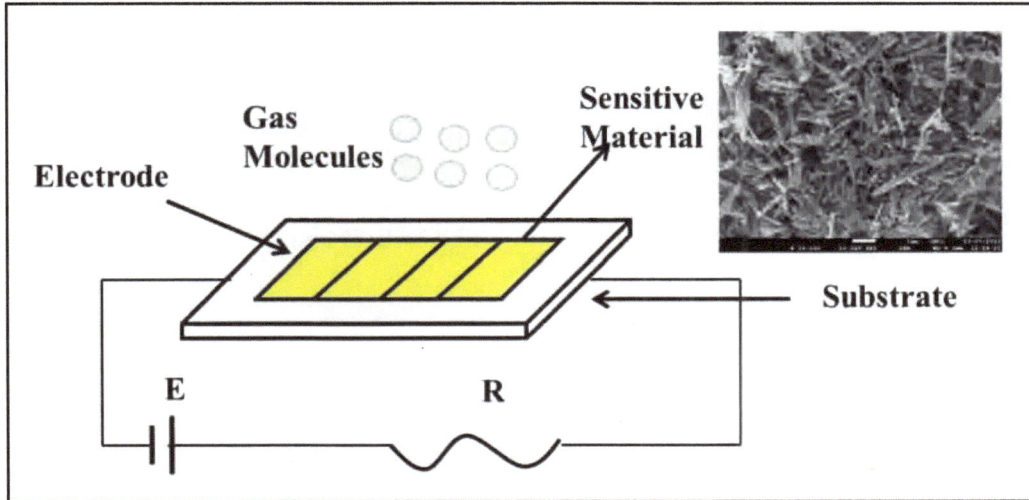

Fig. (7). Gas detection sensor using MoO_3 semiconductor nanorods.

EXAMINATION OF SENSORS FOR VARIOUS APPLICATIONS

Sensors, devices that detect and convert stimuli into measurable electronic signals, find diverse applications in various fields. In the realm of smart textiles, wearable electronic systems, or electronic textiles (e-textiles), integrate sensors into textiles to meet advanced applications in military, safety, healthcare, and sports [140]. These textiles incorporate capacitive, resistive, and optical sensors, allowing them to sense stimuli, such as strain, pressure, temperature, and humidity [141]. Wearable sensors play a crucial role in monitoring physiological parameters in humans and animals. For instance, a wearable system has been developed for monitoring the physiological characteristics of sheep, including heart rate and skin temperature, based on an IoT multi-sensor platform [142]. Biosensors, capable of measuring various biochemical parameters, have applications in healthcare for monitoring calcium, lithium, lactate, cholesterol, *etc* [142]. The application of wearable sensors extends to diverse fields, such as sports performance improvement, where they monitor heartbeat rate during activities, like resting, walking, and running. Wearable textile sensors, equipped with optical sensors, accelerometers, and pressure sensors, detect vital signs and record their time-based history. These sensors are available for various body parts, including the finger, ear, chest, and wrist (Fig. **8**), contributing to sports performance enhancement [143]. Furthermore, wearable textile sensors play a crucial role in healthcare, monitoring vital signs, like blood oxygen saturation levels and body temperature. These sensors are designed to detect disturbances in the immunological system, providing valuable information for patient care. The sensors' high sensitivity, passive wireless operation, multitasking capability, low cost, and easy installation have garnered attention for their utility in measurement applications across industries [144]. Wearable sensors have become essential for monitoring complex human movements and have been integrated into electronic textiles for applications, such as energy storage, monitoring body movements, and strain detection of angular displacements [145]. The wearable textile sensors have become indispensable for monitoring and enhancing various aspects of human and animal health, sports performance, and general well-being. Their versatility, ease of use, and integration into textiles make them valuable tools for real-time data acquisition and analysis in diverse applications.

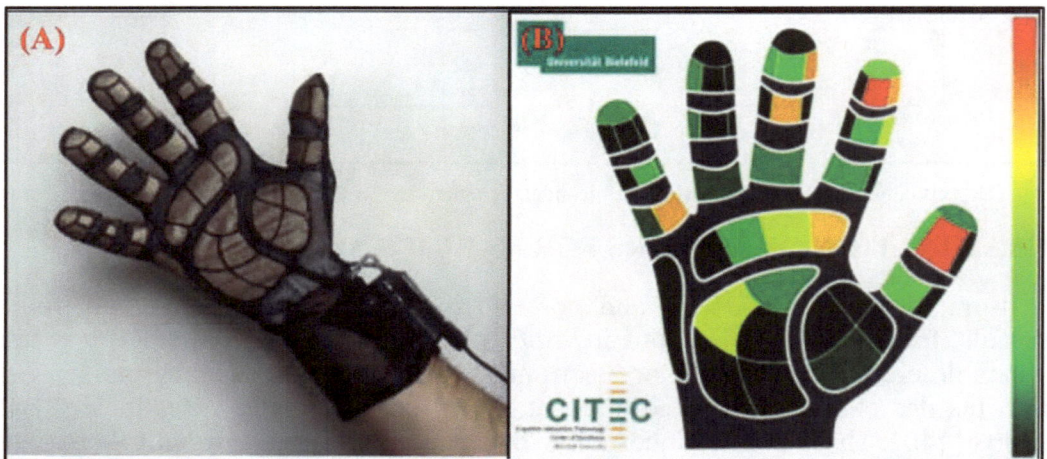

Fig. (8). (a) Tecticle sensor glove as wearable electronics, **(b)** sensor output; reproduced from reference [146]. Copyright © 2014 The Authors. Published by Elsevier B.V.

SEMICONDUCTOR MATERIALS DESIGNED FOR ENHANCED SENSING CAPABILITIES

Various semiconductor materials have been extensively studied and utilized for gas-sensing applications. Among the investigated metal oxide semiconductors, n-type oxides, such as SnO_2, TiO_2, ZnO, $\alpha\text{-}Fe_2O_3$, and WO_3, have received considerable attention [16, 23]. Additionally, p-type semiconducting oxides, including CuO, NiO, Cr_2O_3, and Co_3O_4, have been explored to a lesser extent [147]. Complex metal oxides, like perovskites, have also been investigated. Doping has been a common approach to improve the sensing properties of metal oxides, with examples including Sr-doped Fe_2O_3, Sm-doped $CoFe_2O_4$, and Nb-doped TiO_2 [148 - 161]. Metal Oxide Semiconductors (MOS) provide benefits, like affordability, simple manufacturing processes, and the capability to detect a wide range of gases, including hazardous and combustible ones. Nonetheless, they come with drawbacks, such as limited specificity, reduced sensitivity to lower gas levels, increased energy consumption, baseline resistance fluctuations, and elevated operational temperatures. Semiconductor materials play a crucial role in the development of sensors for various applications, offering unique properties that enable sensitive and selective detection of target substances. Several semiconductor materials have been extensively studied and employed in sensor technologies, each with distinct characteristics and advantages.

Tin Oxide (SnO_2): Tin oxide is a widely studied MOS known for its gas-sensing capabilities. It exhibits a wide bandgap and stable chemical properties. SnO_2-based sensors operate based on changes in conductivity upon interaction with reducing or oxidizing gases. Doping SnO_2 with noble metals or metal ions, such as Pt, Pd, Au, Ni, Fe, and Cu, has been shown to enhance sensitivity and selectivity to various gases, including CO, CH_4, NO_2, and ethanol [162 - 170].

Zinc Oxide (ZnO): Zinc oxide has gained attention in sensor applications due to its direct bandgap and high exciton binding energy. ZnO nanostructures, such as nanorods, nanowires, and nanobelts, offer large surface areas for gas interaction, leading to improved sensitivity. Surface modification, doping, and light activation techniques further enhance ZnO-based sensor performance, particularly in detecting flammable and explosive gases [171 - 173].

Copper Oxide (CuO): CuO is a p-type MOS, which has been examined for gas sensing applications, especially for catalyzing the oxidation of volatile organic compounds. CuO sensors exhibit high selectivity to specific gases and tolerate humidity well. Various CuO morphologies, including urchin-like, fiber-like, and nanorod structures, have been synthesized to tailor sensor properties [164 - 186].

Nickel Oxide (NiO): Nickel oxide, another p-type MOS with a wide bandgap, shows promise in detecting hazardous and toxic gases. NiO thin films and nanowires exhibit high sensitivity and fast response times, particularly towards gases like NH3, toluene, ethanol, and acetone. Doping and surface decoration with noble metal catalysts further enhance NiO sensor performance [167 - 191].

Chromium Oxide (Cr_2O_3): Chromium oxide, also a p-type MOS, has been explored for gas sensing, particularly towards neurotoxic volatile organic compounds, like toluene. Surface decoration with noble metal nanoparticles improves the sensitivity and selectivity of Cr_2O_3 sensors, making them effective for detecting specific gases in complex environments [192 - 203].

ELECTROCATALYSIS: SEMICONDUCTOR MATERIALS IN HER AND OER

Overview of Electrocatalysis

Electrocatalysis is a pivotal process in electrochemical reactions, essential for accelerating chemical transformations at electrode surfaces. It involves the use of catalysts to facilitate the conversion of chemical energy into electrical energy, or *vice versa*, in electrochemical cells. The primary objective of electrocatalysis is to enhance reaction kinetics, lower overpotential, and improve the efficiency of energy conversion processes, such as fuel cells, electrolyzers, and batteries. Catalysts employed in electrocatalysis can modify reaction pathways, adsorb reaction intermediates, and provide active sites for electron transfer, thus enabling faster and more selective reactions. Common electrocatalysts include metals, metal oxides, metal alloys, and carbon-based materials, like graphene and carbon nanotubes. The design and optimization of electrocatalysts involve considerations, such as catalytic activity, stability, selectivity, and compatibility with specific reaction environments. Recent advances in electrocatalysis have led to the development of efficient and sustainable catalysts for renewable energy technologies, such as water splitting for hydrogen production and carbon dioxide reduction for fuel synthesis. Understanding the fundamental principles of electrocatalysis and advancing catalyst design are crucial for realizing clean energy solutions and mitigating environmental challenges associated with traditional energy sources.

SEMICONDUCTOR MATERIALS IN WATER SPLITTING

Semiconductor materials play a crucial role in advancing sensor technologies, offering diverse properties and functionalities essential for detecting and analyzing various environmental parameters. These materials are extensively utilized in the fabrication of sensors due to their unique electronic characteristics and responsiveness to external stimuli [176, 177, 200]. The landscape of semiconductor materials for sensors encompasses a wide range of elements and compounds, each tailored to specific sensing applications [178 - 201]. Noble metals, such as platinum, gold, and silver, along with transition metals, like titanium, iron, and nickel, are commonly employed in sensor fabrication, offering excellent conductivity and stability [179]. Nonmetals and metalloids, such as silicon, carbon, and sulfur, also play significant roles, providing distinct electronic properties suitable for sensor functionalities [180]. In the realm of sensor technologies, the choice of semiconductor material depends on the desired sensing parameters and environmental conditions. For instance, titanium dioxide (TiO_2) serves as a benchmark material in photocatalysis due to its stability and versatility, although its limited absorption of solar radiation poses challenges in utilizing sunlight efficiently [181]. Similarly, iron oxide (Fe_2O_3) exhibits favorable bandgap characteristics for photoelectrochemical water splitting, but its slow kinetics hinders overall efficiency [182 - 202]. Other semiconductor materials, like cadmium sulfide (CdS), gallium nitride (GaN), and silicon carbide (SiC), offer promising capabilities for water-splitting applications, each addressing specific challenges related to light absorption, charge transfer kinetics, and stability. Despite significant progress, ongoing research aims to address the limitations of semiconductor materials in sensor technologies. Efforts are focused on enhancing light harvesting efficiency, improving charge transfer kinetics, and extending the operational lifetime of semiconductor-based sensors [183, 184, 202]. By exploring a diverse array of semiconductor materials (Fig. **9**) and optimizing their properties, researchers strive to develop sensors with enhanced sensitivity, selectivity, and reliability, paving the way for innovative sensing solutions across various industries.

Fig. (9). Popular highlighted elements known for water splitting application.

CATALYZING REACTIONS FOR SUSTAINABLE ENERGY SOURCES

Semiconductor materials play a crucial role in various sensor applications, leveraging their unique electronic properties to detect and respond to external stimuli. Fundamentally, semiconductors possess an energy band structure consisting of a valence band and a conduction band, separated by an energy gap. When exposed to photons, as in photoelectrolysis processes, semiconductor materials generate electron-hole pairs, initiating electrochemical reactions [185]. For instance, in the Photoelectrochemical Hydrogen Evolution Reaction (PEC-HER), photons induce electron-hole pairs, which are then separated and participate in the catalysis of water to produce hydrogen and oxygen gases (Fig. **10**) [186].

The reaction for water splitting is given as follows [187, 188]:

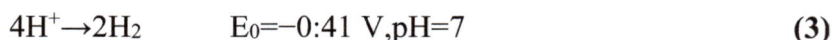

$$2H_2O \rightarrow 2H_2 + O_2 \quad E_0 = 1{:}23 \text{ V} \tag{1}$$

$$2H_2O \rightarrow 4H^+ + O_2 \quad E_0 = +0{:}82 \text{ V,pH}=7 \tag{2}$$

$$4H^+ \rightarrow 2H_2 \quad E_0 = -0{:}41 \text{ V,pH}=7 \tag{3}$$

By connecting semiconductor electrodes, the efficiency of this process is enhanced as electron-hole recombination is minimized, and external potentials can be applied to manipulate electron or hole flow, facilitating selective reduction or oxidation reactions [189]. For instance, holes in the valence band oxidize water to form oxygen, while electrons in the conduction band reduce water to produce hydrogen [190]. To optimize the performance of such processes, semiconductor materials with bandgap energies close to the redox potential of water are preferred, with techniques, like doping and quantum size control, used to adjust bandgap energies [191]. Transition Metal Dichalcogenides (TMDs) have emerged as promising candidates due to their bandgap alignment with water redox potentials, making them suitable for water-splitting applications. Additionally, co-catalyst combinations further enhance their performance [191]. This underscores the importance of semiconductor materials in sensor technologies, particularly in facilitating efficient electrochemical reactions for various applications.

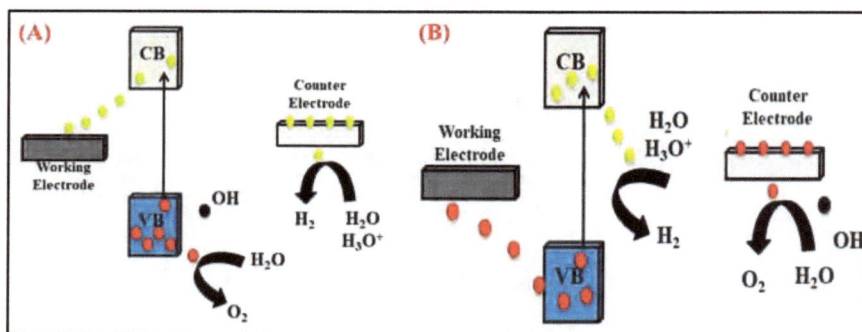

Fig. (10). Redox mechanism for water splitting application of **(a)** n-type semiconductor, **(b)** p-type semiconductor.

CONCLUSION

In conclusion, semiconductor materials serve as fundamental building blocks across a vast spectrum of technological domains, encompassing energy conversion, storage, sensing, and beyond. Throughout our exploration, we have delved into the myriad ways in which semiconductor materials catalyze advancements in renewable energy, sensor technologies, and beyond. Within the realm of renewable energy, semiconductor devices, notably solar cells and thermoelectric generators, leverage the distinctive characteristics of semiconductor materials to effectively convert sunlight and waste heat into invaluable electrical energy. These innovations hold tremendous potential in mitigating global energy challenges and facilitating the transition toward more sustainable energy ecosystems. Furthermore, semiconductor materials form the bedrock of sensor technologies, empowering the creation of highly sensitive and efficient devices tailored for applications ranging from environmental monitoring to healthcare diagnostics. Their adaptability, sensitivity, and seamless integration with electronic systems render semiconductor-based sensors indispensable in modern society's fabric. As we continue to push the frontiers of semiconductor science and engineering, we are met with both challenges and opportunities. Innovations in material synthesis, device fabrication, and system integration stand poised to propel further advancements, ushering in an era of more efficient and sustainable technologies. Undoubtedly, the significance of semiconductor materials in shaping the trajectory of technology cannot be overstated. With sustained research and development endeavors, semiconductor-based innovations will persistently drive progress toward a more sustainable, efficient, and technologically advanced future.

KEY POINTS

1. Exploration of Innovative Materials: The chapter delves into the cutting-edge materials crucial for semiconductor device design, offering insights into their applications in diverse fields.
2. Comprehensive Coverage: The chapter provides a comprehensive exploration of semiconductor materials, covering areas, such as solar cells, capacitors, thermoelectric devices, sensors, and electrocatalysis.
3. Design Challenges and Opportunities: The chapter addresses the intricate design challenges and opportunities associated with semiconductor materials, offering a multidimensional perspective on their utilization in various electronic and energy applications.
4. Solar Cell Technologies: Detailed discussions on solar cell technologies, including tandem solar cells, highlight advancements in efficiency and performance, along with strategies to overcome design limitations.

5. Energy Storage Solutions: Capacitors and supercapacitors are scrutinized for their energy storage capabilities, with an emphasis on novel semiconductor materials and fabrication techniques for enhancing performance and durability.
6. Thermoelectric Materials: Insights into thermoelectric materials elucidate their role in converting waste heat into electrical energy, addressing the growing demand for sustainable energy solutions.
7. Sensor Technologies: Semiconductor materials play a vital role in sensor technologies, enabling the development of highly sensitive and reliable devices for applications, such as environmental monitoring and healthcare diagnostics.
8. Electrocatalysis and Energy Conversion: Discussions on electrocatalysis explore the utilization of semiconductor materials in hydrogen and oxygen evolution reactions, paving the way for advancements in clean energy production.
9. Future Directions: The chapter highlights the ongoing research efforts and potential future directions in semiconductor material design, fabrication, and integration, aiming to propel semiconductor nanoscale devices to unprecedented levels of performance and functionality.

ACKNOWLEDGEMENTS

The authors extend their deepest gratitude to Prof. Anand Srivastava, VC, NSUT, for his unwavering support and encouragement throughout this research endeavor.

REFERENCES

[1]   V.D. Silva, R.A. Raimundo, T.A. Simões, F.J.A. Loureiro, D.P. Fagg, M.A. Morales, D.A. Macedo, and E.S. Medeiros, "Nonwoven Ni–NiO/carbon fibers for electrochemical water oxidation", *Int. J. Hydrogen Energy,* vol. 46, no. 5, pp. 3798-3810, 2021.
[http://dx.doi.org/10.1016/j.ijhydene.2020.10.156]

[2]   Y. Cao, T. Lai, F. Teng, C. Liu, and A. Li, "Highly stretchable and sensitive strain sensor based on silver nanowires/carbon nanotubes on hair band for human motion detection", *Prog. Nat. Sci.,* vol. 31, no. 3, pp. 379-386, 2021.
[http://dx.doi.org/10.1016/j.pnsc.2021.05.005]

[3]   T. Hao, S. Wang, H. Xu, X. Zhang, J. Xue, S. Liu, Y. Song, Y. Li, and J. Zhao, "Stretchable electrochromic devices based on embedded WO3@AgNW Core-Shell nanowire elastic conductors", *Chem. Eng. J.,* vol. 426, p. 130840, 2021.
[http://dx.doi.org/10.1016/j.cej.2021.130840]

[4]   C. A. Foss, A. Z. M. S. Rahman, K. A. Eldressi, and S. A. Elsheikhi, "Optical Properties of Nanoparticle Pair Structures", *in Reference Module in Materials Science and Materials Engineering,* Elsevier, 2019.

[5]   J. Meng, T. Wang, H. Zhu, L. Ji, W. Bao, P. Zhou, L. Chen, Q.Q. Sun, and D.W. Zhang, "Integrated in-sensor computing optoelectronic device for environment-adaptable artificial retina perception application", *Nano Lett.,* vol. 22, no. 1, pp. 81-89, 2022.
[http://dx.doi.org/10.1021/acs.nanolett.1c03240] [PMID: 34962129]

[6]   S. Lee, J. Jang, T. Park, Y.M. Park, J.S. Park, Y.K. Kim, H.K. Lee, E.C. Jeon, D.K. Lee, B. Ahn, and C.H. Chung, "Electrodeposited silver nanowire transparent conducting electrodes for thin-film solar

cells", *ACS Appl. Mater. Interfaces,* vol. 12, no. 5, pp. 6169-6175, 2020.
[http://dx.doi.org/10.1021/acsami.9b17168] [PMID: 31933356]

[7]   P.P. Dipti, P. Phogat, Shreya, D. Kumari, and S. Singh, "Fabrication of tunable band gap carbon based zinc nanocomposites for enhanced capacitive behaviour", *Phys. Scr.,* vol. 98, no. 9, p. 095030, 2023.
[http://dx.doi.org/10.1088/1402-4896/acf07b]

[8]   P. Shreya, Phogat, R. Jha, and S. Singh, "Elevated refractive index of mos2 amorphous nanoparticles with a reduced band gap applicable for optoelectronics bt - recent advances in mechanical engineering.*Recent Advances in Mechanical Engineering.,* B. Sethuraman, P. Jain, M. Gupta, Eds., Springer Nature Singapore: Singapore, 2023, pp. 431-439.
[http://dx.doi.org/10.1007/978-981-99-2349-6_39]

[9]   P. Phogat, R. Shreya, R. Jha, and S. Singh, "Electrochemical analysis of thermally treated two dimensional zinc sulphide hexagonal nano-sheets with reduced band gap", *Phys. Scr.,* vol. 98, no. 12, p. 125962, 2023.
[http://dx.doi.org/10.1088/1402-4896/ad0d93]

[10]   A. Yadav, "Preliminary Observations of synthesized ws2 and various synthesis techniques for preparation of nanomaterials", In: *in Advances in Manufacturing Technology and Management* Springer Science and Business Media Deutschland GmbH, 2023, pp. 546-556.

[11]   T. Kumar, P. Shreya, P. Phogat, V. Sahgal, and R. Jha, "Surfactant-mediated modulation of morphology and charge transfer dynamics in tungsten oxide nanoparticles", *Phys. Scr.,* vol. 98, no. 8, p. 085936, 2023.
[http://dx.doi.org/10.1088/1402-4896/ace566]

[12]   A. Wolcott, W.A. Smith, T.R. Kuykendall, Y. Zhao, and J.Z. Zhang, "Photoelectrochemical study of nanostructured Zno thin films for hydrogen generation from water splitting", *Adv. Funct. Mater.,* vol. 19, no. 12, pp. 1849-1856, 2009.
[http://dx.doi.org/10.1002/adfm.200801363]

[13]   H.X. Sang, X.T. Wang, C.C. Fan, and F. Wang, "Enhanced photocatalytic H 2 production from glycerol solution over ZnO/ZnS core/shell nanorods prepared by a low temperature route", *Int. J. Hydrogen Energy,* vol. 37, no. 2, pp. 1348-1355, 2012.
[http://dx.doi.org/10.1016/j.ijhydene.2011.09.129]

[14]   S.H. Hong, Y.K. Kim, S.H. Hwang, H.J. Seo, and S.K. Lim, "Effect of morphology of ZnO on colorimetric hydrogen sensitivity of PdO@ZnO hybrids", *Int. J. Hydrogen Energy,* vol. 57, pp. 717-726, 2024.
[http://dx.doi.org/10.1016/j.ijhydene.2024.01.087]

[15]   S. Sharma, P. Phogat, R. Jha, and S. Singh, "Electrochemical and optical properties of microwave assisted mos2 nanospheres for solar cell application", *International Journal of Smart Grid and Clean Energy,* vol. 12, no. 3, pp. 66-72, 2023.
[http://dx.doi.org/10.12720/sgce.12.3.66-72]

[16]   P. Shreya, P. Phogat, R. Jha, and S. Singh, "Microwave-synthesized γ-WO3 nanorods exhibiting high current density and diffusion characteristics", *Trans. Met. Chem. (Weinh.),* vol. 48, no. 3, pp. 167-183, 2023.
[http://dx.doi.org/10.1007/s11243-023-00533-y]

[17]   P. Phogat, and R. Shreya, Jha, and S. Singh, "Phase transition of thermally treated polyhedral nano nickel oxide with reduced band gap", *MATEC Web Conf.,* vol. 393, 2024.
[http://dx.doi.org/10.1051/matecconf/202439301001]

[18]   S. Rai, and P. Shreya, Phogat, R. Jha, and S. Singh, "Hydrothermal synthesis and characterization of selenium-doped MoS2 for enhanced optoelectronic properties", *MATEC Web Conf.,* vol. 393, 2024.
[http://dx.doi.org/10.1051/matecconf/202439301008]

[19]   D. Kumari, P. Shreya, P. Phogat, Dipti, S. Singh, and R. Jha, "Enhanced electrochemical behavior of C@CdS Core-Shell heterostructures", *Mater. Sci. Eng. B,* vol. 301, p. 117212, 2024.

[http://dx.doi.org/10.1016/j.mseb.2024.117212]

[20]   P. Phogat, Shreya, R. Jha, and S. Singh, "Diffusion controlled features of microwave assisted zns/zno nanocomposite with reduced band gap", *ECS J. Solid State Sci. Technol.,* vol. 12, no. 3, p. 034004, 2023.
[http://dx.doi.org/10.1149/2162-8777/acc426]

[21]   P. Phogat, S. Shreya, R. Jha, and S. Singh, "Impedance study of zinc sulphide quantum dots *via* one step green synthesis", *Mater. Sci. Forum,* vol. 1099, pp. 119-125, 2023.
[http://dx.doi.org/10.4028/p-G1CCxq]

[22]   A. Shreya, Yadav, R. Khatri, N. Jain, A. Bhandari, and N. K. Puri, "Double zone thermal CVD and plasma enhanced cvd systems for deposition of films/coatings with eminent conformal coverage bt - advances in manufacturing technology and management.*Advances in Manufacturing Technology and Management.,* R.M. Singari, P.K. Jain, H. Kumar, Eds., Springer Nature Singapore: Singapore, 2023, pp. 273-283.
[http://dx.doi.org/10.1007/978-981-16-9523-0_31]

[23]   P. Phogat, R. Shreya, and S. Singh, "Optical and microstructural study of wide band gap zno@zns core--shell nanorods to be used as solar cell applications", In: *in Recent Advances in Mechanical Engineering,* 2023, pp. 419-429.

[24]   L. Baia, R. Stefan, J. Popp, S. Simon, and W. Kiefer, "Vibrational spectroscopy of highly iron doped B2O3–Bi2O3 glass systems", *J. Non-Cryst. Solids,* vol. 324, no. 1-2, pp. 109-117, 2003.
[http://dx.doi.org/10.1016/S0022-3093(03)00227-8]

[25]   Y. Zhou, L. Zhang, W. Liu, H. Zhang, S. Huang, S. Lan, H. Zhao, H. Fu, A. Han, Z. Li, K. Jiang, X. Yu, D. Zhao, R. Li, F. Meng, and Z. Liu, "Improved interface microstructure between crystalline silicon and nanocrystalline silicon oxide window layer of silicon heterojunction solar cells", *Sol. Energy Mater. Sol. Cells,* vol. 265, p. 112652, 2024.
[http://dx.doi.org/10.1016/j.solmat.2023.112652]

[26]   A. Hosen, S. Yeasmin, K.M.S. Bin Rahmotullah, M.F. Rahman, and S.R.A. Ahmed, "Design and simulation of a highly efficient CuBi2O4 thin-film solar cell with hole transport layer", *Opt. Laser Technol.,* vol. 169, p. 110073, 2024.
[http://dx.doi.org/10.1016/j.optlastec.2023.110073]

[27]   L. Wen, H. Mao, L. Zhang, J. Zhang, Z. Qin, L. Tan, and Y. Chen, "Achieving desired pseudo-planar heterojunction organic solar cells *via* binary-dilution strategy", *Adv. Mater.,* vol. 36, no. 3, p. 2308159, 2024.
[http://dx.doi.org/10.1002/adma.202308159] [PMID: 37831921]

[28]   R. Tian, S. Zhou, Y. Meng, C. Liu, and Z. Ge, "Material and device design of flexible perovskite solar cells for next-generation power supplies", *Adv. Mater.,* 2024.
[http://dx.doi.org/10.1002/adma.202311473]

[29]   L.M. Nieto Nieto, J.P. Ferrer Rodríguez, R.M. Campos, and P.J. Pérez Higueras, "Multi-junction solar cell measurements at ultra-high irradiances for different temperatures and spectra", *Sol. Energy Mater. Sol. Cells,* vol. 266, p. 112651, 2024.
[http://dx.doi.org/10.1016/j.solmat.2023.112651]

[30]   P. Shreya, P. Phogat, R. Jha, and S. Singh, "Enhanced electrochemical performance and charge-transfer dynamics of 2D MoS 2 /WO 3 nanocomposites for futuristic energy applications", *ACS Appl. Nano Mater.,* vol. 7, no. 8, pp. 8593-8611, 2024.
[http://dx.doi.org/10.1021/acsanm.3c06017]

[31]   A. Sharma, and P. Shreya, Phogat, R. Jha, and S. Singh, "Hydrothermally Synthesized NiS2 and NiSO4(H2O)6 Nanocomposites and its Characterizations", *MATEC Web Conf.,* vol. 393, 2024.
[http://dx.doi.org/10.1051/matecconf/202439301016]

[32]   E. Aydin, "Pathways toward commercial perovskite/silicon tandem photovoltaics", *Science,* vol. 383, no. 6679, p. eadh3849, 2024.

[http://dx.doi.org/10.1126/science.adh3849]

[33] B. Hoex, J.J.H. Gielis, M.C.M. van de Sanden, and W.M.M. Kessels, "On the c-Si surface passivation mechanism by the negative-charge-dielectric Al2O3", *J. Appl. Phys.,* vol. 104, no. 11, p. 113703, 2008.
[http://dx.doi.org/10.1063/1.3021091]

[34] J. Benick, B. Hoex, M.C.M. van de Sanden, W.M.M. Kessels, O. Schultz, and S.W. Glunz, "High efficiency n-type Si solar cells on Al2O3-passivated boron emitters", *Appl. Phys. Lett.,* vol. 92, no. 25, p. 253504, 2008.
[http://dx.doi.org/10.1063/1.2945287]

[35] T.G. Allen, J. Bullock, X. Yang, A. Javey, and S. De Wolf, "Passivating contacts for crystalline silicon solar cells", *Nat. Energy,* vol. 4, no. 11, pp. 914-928, 2019.
[http://dx.doi.org/10.1038/s41560-019-0463-6]

[36] M. Hermle, F. Feldmann, M. Bivour, J.C. Goldschmidt, and S.W. Glunz, "Passivating contacts and tandem concepts: Approaches for the highest silicon-based solar cell efficiencies", *Appl. Phys. Rev.,* vol. 7, no. 2, p. 021305, 2020.
[http://dx.doi.org/10.1063/1.5139202]

[37] J. Nelson, *The Physics of Solar Cells.* Published by imperial college press and distributed by world scientific publishing co, 2003.
[http://dx.doi.org/10.1142/p276]

[38] G. Coletti, P.C.P. Bronsveld, G. Hahn, W. Warta, D. Macdonald, B. Ceccaroli, K. Wambach, N. Le Quang, and J.M. Fernandez, "Impact of metal contamination in silicon solar cells", *Adv. Funct. Mater.,* vol. 21, no. 5, pp. 879-890, 2011.
[http://dx.doi.org/10.1002/adfm.201000849]

[39] N.E. Grant, F.E. Rougieux, D. Macdonald, J. Bullock, and Y. Wan, "Grown-in defects limiting the bulk lifetime of p -type float-zone silicon wafers", *J. Appl. Phys.,* vol. 117, no. 5, p. 055711, 2015.
[http://dx.doi.org/10.1063/1.4907804]

[40] A.Y. Liu, and D. Macdonald, "Impurity gettering effect of atomic layer deposited aluminium oxide films on silicon wafers", *Appl. Phys. Lett.,* vol. 110, no. 19, p. 191604, 2017.
[http://dx.doi.org/10.1063/1.4983380]

[41] J. Zhang, S.T. Lee, and B. Sun, "Effect of series and shunt resistance on organic-inorganic hybrid solar cells performance", *Electrochim. Acta,* vol. 146, pp. 845-849, 2014.
[http://dx.doi.org/10.1016/j.electacta.2014.08.065]

[42] S.P. Phang, and D. Macdonald, "Effect of boron codoping and phosphorus concentration on phosphorus diffusion gettering", *IEEE J. Photovolt.,* vol. 4, no. 1, pp. 64-69, 2014.
[http://dx.doi.org/10.1109/JPHOTOV.2013.2281740]

[43] H. Choi, C. Nahm, J. Kim, C. Kim, S. Kang, T. Hwang, and B. Park, "Review paper: Toward highly efficient quantum-dot- and dye-sensitized solar cells", *Curr. Appl. Phys.,* vol. 13, pp. S2-S13, 2013.
[http://dx.doi.org/10.1016/j.cap.2013.01.023]

[44] M.A. Green, "Accuracy of analytical expressions for solar cell fill factors", *Solar Cells,* vol. 7, no. 3, pp. 337-340, 1982.
[http://dx.doi.org/10.1016/0379-6787(82)90057-6]

[45] R.A. Sinton, and A. Cuevas, "A quasi-steady-state open-circuit voltage method for solar cell characterization", *16th Eur. Photovolt. Sol. Energy Conf.,* pp. 1-4, 2000.

[46] A. Cuevas and R. A. Sinton, "IIb-4 - Characterisation and Diagnosis of Silicon Wafers and Devices," T. Markvart and L. B. T.-P. H. of P. Castañer, Eds. Amsterdam: Elsevier Science, 2003, pp. 227–252.

[47] L. Duan, and A. Uddin, "Progress in stability of organic solar cells", *Adv. Sci. (Weinh.),* vol. 7, no. 11, p. 1903259, 2020.
[http://dx.doi.org/10.1002/advs.201903259] [PMID: 32537401]

[48]   Z. Lin, J. Chang, C. Zhang, D. Chen, J. Wu, and Y. Hao, "Enhanced performance and stability of polymer solar cells by *in situ* formed AlO $_x$ passivation and doping", *J. Phys. Chem. C,* vol. 121, no. 19, pp. 10275-10281, 2017.
[http://dx.doi.org/10.1021/acs.jpcc.6b12459]

[49]   N.G. Park, "Perovskite solar cells: an emerging photovoltaic technology", *Mater. Today,* vol. 18, no. 2, pp. 65-72, 2015.
[http://dx.doi.org/10.1016/j.mattod.2014.07.007]

[50]   H. Li, and W. Zhang, "Perovskite tandem solar cells: from fundamentals to commercial deployment", *Chem. Rev.,* vol. 120, no. 18, pp. 9835-9950, 2020.
[http://dx.doi.org/10.1021/acs.chemrev.9b00780] [PMID: 32786417]

[51]   A. Al-Ashouri, "Monolithic perovskite/silicon tandem solar cell with >29% efficiency by enhanced hole extraction", *Science,* vol. 370, no. 6522, pp. 1300-1309, 2020.
[http://dx.doi.org/10.1126/science.abd4016]

[52]   R. Wang, T. Huang, J. Xue, J. Tong, K. Zhu, and Y. Yang, "Prospects for metal halide perovskite-based tandem solar cells", *Nat. Photonics,* vol. 15, no. 6, pp. 411-425, 2021.
[http://dx.doi.org/10.1038/s41566-021-00809-8]

[53]   A.D. Vos, "Detailed balance limit of the efficiency of tandem solar cells", *J. Phys. D Appl. Phys.,* vol. 13, no. 5, pp. 839-846, 1980.
[http://dx.doi.org/10.1088/0022-3727/13/5/018]

[54]   A. Martí, and G.L. Araújo, "Limiting efficiencies for photovoltaic energy conversion in multigap systems", *Sol. Energy Mater. Sol. Cells,* vol. 43, no. 2, pp. 203-222, 1996.
[http://dx.doi.org/10.1016/0927-0248(96)00015-3]

[55]   S. Albrecht, and B. Rech, "Perovskite solar cells: On top of commercial photovoltaics", *Nat. Energy,* vol. 2, no. 1, p. 16196, 2017.
[http://dx.doi.org/10.1038/nenergy.2016.196]

[56]   A. Rai, P. Phogat, and R. Shreya, Jha, and S. Singh, "Microwave Assisted Zinc Sulphide Quantum Dots for Energy Device Applications", *MATEC Web Conf.,* vol. 393, 2024.
[http://dx.doi.org/10.1051/matecconf/202439301011]

[57]   P. Shreya, Phogat, S. Singh, and R. Jha, "Reduction mechanism of hydrothermally synthesized wide band gap ZnWO4 nanorods for HER application", *MATEC Web Conf.,* vol. 393, 2024.
[http://dx.doi.org/10.1051/matecconf/202439301004]

[58]   M. Jošt, L. Kegelmann, L. Korte, and S. Albrecht, "Monolithic perovskite tandem solar cells: a review of the present status and advanced characterization methods toward 30% efficiency", *Adv. Energy Mater.,* vol. 10, no. 26, p. 1904102, 2020.
[http://dx.doi.org/10.1002/aenm.201904102]

[59]   C. Nolden, J. Barnes, and J. Nicholls, "Community energy business model evolution: A review of solar photovoltaic developments in England", *Renew. Sustain. Energy Rev.,* vol. 122, p. 109722, 2020.
[http://dx.doi.org/10.1016/j.rser.2020.109722]

[60]   A. Fontaine, "Debating the sustainability of solar energy: Examining resource construction processes for local photovoltaic projects in France", *Energy Res. Soc. Sci.,* vol. 69, p. 101725, 2020.
[http://dx.doi.org/10.1016/j.erss.2020.101725]

[61]   M.M. Hasan, S. Hossain, M. Mofijur, Z. Kabir, I.A. Badruddin, T.M. Yunus Khan, and E. Jassim, "Harnessing solar power: A review of photovoltaic innovations, solar thermal systems, and the dawn of energy storage solutions", *Energies,* vol. 16, no. 18, p. 6456, 2023.
[http://dx.doi.org/10.3390/en16186456]

[62]   B. Zhu, L. Fan, N. Mushtaq, R. Raza, M. Sajid, Y. Wu, W. Lin, J-S. Kim, P.D. Lund, and S. Yun, "Semiconductor electrochemistry for clean energy conversion and storage", *Electrochemical Energy Reviews,* vol. 4, no. 4, pp. 757-792, 2021.

[http://dx.doi.org/10.1007/s41918-021-00112-8]

[63]   J. Lv, J. Xie, A.G.A. Mohamed, X. Zhang, and Y. Wang, "Photoelectrochemical energy storage materials: design principles and functional devices towards direct solar to electrochemical energy storage", *Chem. Soc. Rev.,* vol. 51, no. 4, pp. 1511-1528, 2022.
       [http://dx.doi.org/10.1039/D1CS00859E] [PMID: 35137737]

[64]   P. Forouzandeh, V. Kumaravel, and S.C. Pillai, "Electrode materials for supercapacitors: a review of recent advances", *Catalysts,* vol. 10, no. 9, p. 969, 2020.
       [http://dx.doi.org/10.3390/catal10090969]

[65]   T. Saga, "Advances in crystalline silicon solar cell technology for industrial mass production", *NPG Asia Mater.,* vol. 2, no. 3, pp. 96-102, 2010.
       [http://dx.doi.org/10.1038/asiamat.2010.82]

[66]   R. Pandey, A. Khanna, K. Singh, S.K. Patel, H. Singh, and J. Madan, "Device simulations: Toward the design of >13% efficient PbS colloidal quantum dot solar cell", *Sol. Energy,* vol. 207, pp. 893-902, 2020.
       [http://dx.doi.org/10.1016/j.solener.2020.06.099]

[67]   I. Dharmadasa, and A. Alam, "How to achieve efficiencies beyond 22.1% for CdTe-based thin-film solar cells", *Energies,* vol. 15, no. 24, p. 9510, 2022.
       [http://dx.doi.org/10.3390/en15249510]

[68]   F. Sahli, J. Werner, B.A. Kamino, M. Bräuninger, R. Monnard, B. Paviet-Salomon, L. Barraud, L. Ding, J.J. Diaz Leon, D. Sacchetto, G. Cattaneo, M. Despeisse, M. Boccard, S. Nicolay, Q. Jeangros, B. Niesen, and C. Ballif, "Fully textured monolithic perovskite/silicon tandem solar cells with 25.2% power conversion efficiency", *Nat. Mater.,* vol. 17, no. 9, pp. 820-826, 2018.
       [http://dx.doi.org/10.1038/s41563-018-0115-4] [PMID: 29891887]

[69]   M. Nakamura, K. Yamaguchi, Y. Kimoto, Y. Yasaki, T. Kato, and H. Sugimoto, "Cd-Free Cu(In,Ga)(Se,S) 2 thin-film solar cell with record efficiency of 23.35%", *IEEE J. Photovolt.,* vol. 9, no. 6, pp. 1863-1867, 2019.
       [http://dx.doi.org/10.1109/JPHOTOV.2019.2937218]

[70]   Q. Ma, Z. Jia, L. Meng, J. Zhang, H. Zhang, W. Huang, J. Yuan, F. Gao, Y. Wan, Z. Zhang, and Y. Li, "Promoting charge separation resulting in ternary organic solar cells efficiency over 17.5%", *Nano Energy,* vol. 78, p. 105272, 2020.
       [http://dx.doi.org/10.1016/j.nanoen.2020.105272]

[71]   J.J. Yoo, G. Seo, M.R. Chua, T.G. Park, Y. Lu, F. Rotermund, Y.K. Kim, C.S. Moon, N.J. Jeon, J.P. Correa-Baena, V. Bulović, S.S. Shin, M.G. Bawendi, and J. Seo, "Efficient perovskite solar cells *via* improved carrier management", *Nature,* vol. 590, no. 7847, pp. 587-593, 2021.
       [http://dx.doi.org/10.1038/s41586-021-03285-w] [PMID: 33627807]

[72]   L. Schmidt-Mende, U. Bach, R. Humphry-Baker, T. Horiuchi, H. Miura, S. Ito, S. Uchida, and M. Grätzel, "Organic dye for highly efficient solid-state dye-sensitized solar cells", *Adv. Mater.,* vol. 17, no. 7, pp. 813-815, 2005.
       [http://dx.doi.org/10.1002/adma.200401410]

[73]   J. Carrillo, A. Guerrero, S. Rahimnejad, O. Almora, I. Zarazua, E. Mas-Marza, J. Bisquert, and G. Garcia-Belmonte, "Ionic reactivity at contacts and aging of methylammonium lead triiodide perovskite solar cells", *Adv. Energy Mater.,* vol. 6, no. 9, p. 1502246, 2016.
       [http://dx.doi.org/10.1002/aenm.201502246]

[74]   B. Hallam, A. Sugianto, L. Mai, G.Q. Xu, C. Chan, M. Abbott, S. Wenham, A. Uruena, E. Cornagliotti, and M. Aleman, "Hydrogen passivation of laser-induced defects for laser-doped silicon solar cells", *IEEE J. Photovolt.,* vol. 4, no. 6, pp. 1413-1420, 2014.
       [http://dx.doi.org/10.1109/JPHOTOV.2014.2347804]

[75]   A. Descoeudres, Z.C. Holman, L. Barraud, S. Morel, S. De Wolf, and C. Ballif, ">21% Efficient Silicon Heterojunction Solar Cells on n- and p-Type Wafers Compared", *IEEE J. Photovolt.,* vol. 3,

no. 1, pp. 83-89, 2013.
[http://dx.doi.org/10.1109/JPHOTOV.2012.2209407]

[76] J.F. Geisz, R.M. France, K.L. Schulte, M.A. Steiner, A.G. Norman, H.L. Guthrey, M.R. Young, T. Song, and T. Moriarty, "Six-junction III–V solar cells with 47.1% conversion efficiency under 143 Suns concentration", *Nat. Energy,* vol. 5, no. 4, pp. 326-335, 2020.
[http://dx.doi.org/10.1038/s41560-020-0598-5]

[77] S.P. Philipps, "Present Status in the Development of III–V Multi-Junction Solar Cells BT - Next Generation of Photovoltaics: New Concepts", A. B. Cristóbal López, A. Martí Vega, and A. Luque López, Eds. Berlin, Heidelberg: Springer Berlin Heidelberg, 2012, pp. 1–21.

[78] H. Cao, Z. Zhang, M. Zhang, A. Gu, H. Yu, H. Ban, Q. Sun, Y. Shen, X-L. Zhang, J. Zhu, and M. Wang, "The effect of defects in tin-based perovskites and their photovoltaic devices", *Materials Today Physics,* vol. 21, p. 100513, 2021.
[http://dx.doi.org/10.1016/j.mtphys.2021.100513]

[79] H. Sung, N. Ahn, M.S. Jang, J-K. Lee, H. Yoon, N-G. Park, and M. Choi, "Transparent conductive oxide-free graphene-based perovskite solar cells with over 17% efficiency", *Adv. Energy Mater.,* vol. 6, no. 3, p. 1501873, 2016.
[http://dx.doi.org/10.1002/aenm.201501873]

[80] Z.A. Ansari, T.J. Singh, S.M. Islam, S. Singh, P. Mahala, A. Khan, and K.J. Singh, "Photovoltaic solar cells based on graphene/gallium arsenide Schottky junction", *Optik (Stuttg.),* vol. 182, pp. 500-506, 2019.
[http://dx.doi.org/10.1016/j.ijleo.2019.01.078]

[81] L. Barraud, Z.C. Holman, N. Badel, P. Reiss, A. Descoeudres, C. Battaglia, S. De Wolf, and C. Ballif, "Hydrogen-doped indium oxide/indium tin oxide bilayers for high-efficiency silicon heterojunction solar cells", *Sol. Energy Mater. Sol. Cells,* vol. 115, pp. 151-156, 2013.
[http://dx.doi.org/10.1016/j.solmat.2013.03.024]

[82] L. Han, A. Islam, H. Chen, C. Malapaka, B. Chiranjeevi, S. Zhang, X. Yang, and M. Yanagida, "High-efficiency dye-sensitized solar cell with a novel co-adsorbent", *Energy Environ. Sci.,* vol. 5, no. 3, pp. 6057-6060, 2012.
[http://dx.doi.org/10.1039/c2ee03418b]

[83] D. Zielke, A. Pazidis, F. Werner, and J. Schmidt, "Organic-silicon heterojunction solar cells on n-type silicon wafers: The BackPEDOT concept", *Sol. Energy Mater. Sol. Cells,* vol. 131, pp. 110-116, 2014.
[http://dx.doi.org/10.1016/j.solmat.2014.05.022]

[84] R. Pandey, S. Bhattarai, K. Sharma, J. Madan, A.K. Al-Mousoi, M.K.A. Mohammed, and M.K. Hossain, "Halide composition engineered a non-toxic perovskite–silicon tandem solar cell with 30.7% conversion efficiency", *ACS Appl. Electron. Mater.,* vol. 5, no. 10, pp. 5303-5315, 2023.
[http://dx.doi.org/10.1021/acsaelm.2c01574]

[85] S.M. Yakout, "Spintronics: future technology for new data storage and communication devices", *J. Supercond. Nov. Magn.,* vol. 33, no. 9, pp. 2557-2580, 2020.
[http://dx.doi.org/10.1007/s10948-020-05545-8]

[86] X. Huang, D. Ji, H. Fuchs, W. Hu, and T. Li, "Recent progress in organic phototransistors: semiconductor materials, device structures and optoelectronic applications", *ChemPhotoChem,* vol. 4, no. 1, pp. 9-38, 2020.
[http://dx.doi.org/10.1002/cptc.201900198]

[87] R. Khatri, and N.K. Puri, "Electrochemical study of hydrothermally synthesised reduced MoS2 layered nanosheets", *Vacuum,* vol. 175, p. 109250, 2020.
[http://dx.doi.org/10.1016/j.vacuum.2020.109250]

[88] R. Jha, M. Bhushan, and R. Bhardwaj, *Studies on Synthesis and Various Characteristics of Green Materials for Energy Conversion Applications.,* Springer International Publishing, 2020.
[http://dx.doi.org/10.1007/978-3-030-50108-2_1]

[89] Y. Wang, and A. Hu, "Carbon quantum dots: synthesis, properties and applications", *J. Mater. Chem. C Mater. Opt. Electron. Devices,* vol. 2, no. 34, pp. 6921-6939, 2014.
[http://dx.doi.org/10.1039/C4TC00988F]

[90] R. Bhardwaj, R. Jha, M. Bhushan, and R. Sharma, "Comparative study of the electrochemical properties of mesoporous 1-D and 3-D nano- structured rhombohedral nickel sulfide in alkaline electrolytes", *J. Phys. Chem. Solids,* vol. 144, no. April, p. 109503, 2020.
[http://dx.doi.org/10.1016/j.jpcs.2020.109503]

[91] F. Amirian, M. Molaei, M. Karimipour, and A.R. Bahador, "A new and simple UV-assisted approach for synthesis of water soluble ZnS and transition metals doped ZnS nanoparticles (NPs) and investigating optical and photocatalyst properties", *J. Lumin.,* vol. 196, pp. 174-180, 2018.
[http://dx.doi.org/10.1016/j.jlumin.2017.12.005]

[92] P. Sharma, and V. Kumar, "Current technology of supercapacitors: a review", *J. Electron. Mater.,* vol. 49, no. 6, pp. 3520-3532, 2020.
[http://dx.doi.org/10.1007/s11664-020-07992-4]

[93] W. Raza, F. Ali, N. Raza, Y. Luo, K-H. Kim, J. Yang, S. Kumar, A. Mehmood, and E.E. Kwon, "Recent advancements in supercapacitor technology", *Nano Energy,* vol. 52, pp. 441-473, 2018.
[http://dx.doi.org/10.1016/j.nanoen.2018.08.013]

[94] E. Raymundo-Piñero, F. Leroux, and F. Béguin, "A high-performance carbon for supercapacitors obtained by carbonization of a seaweed biopolymer", *Adv. Mater.,* vol. 18, no. 14, pp. 1877-1882, 2006.
[http://dx.doi.org/10.1002/adma.200501905]

[95] Q. Cheng, J. Tang, N. Shinya, and L.C. Qin, "Co(OH) 2 nanosheet-decorated graphene–CNT composite for supercapacitors of high energy density", *Sci. Technol. Adv. Mater.,* vol. 15, no. 1, p. 014206, 2014.
[http://dx.doi.org/10.1088/1468-6996/15/1/014206] [PMID: 27877633]

[96] M. Seredych, D. Hulicova-Jurcakova, G.Q. Lu, and T.J. Bandosz, "Surface functional groups of carbons and the effects of their chemical character, density and accessibility to ions on electrochemical performance", *Carbon,* vol. 46, no. 11, pp. 1475-1488, 2008.
[http://dx.doi.org/10.1016/j.carbon.2008.06.027]

[97] A.G. Pandolfo, and A.F. Hollenkamp, "Carbon properties and their role in supercapacitors", *J. Power Sources,* vol. 157, no. 1, pp. 11-27, 2006.
[http://dx.doi.org/10.1016/j.jpowsour.2006.02.065]

[98] L. Lai, H. Yang, L. Wang, B.K. Teh, J. Zhong, H. Chou, L. Chen, W. Chen, Z. Shen, R.S. Ruoff, and J. Lin, "Preparation of supercapacitor electrodes through selection of graphene surface functionalities", *ACS Nano,* vol. 6, no. 7, pp. 5941-5951, 2012.
[http://dx.doi.org/10.1021/nn3008096] [PMID: 22632101]

[99] P. Simon, and Y. Gogotsi, "Materials for electrochemical capacitors", *Nat. Mater.,* vol. 7, no. 11, pp. 845-854, 2008.
[http://dx.doi.org/10.1038/nmat2297] [PMID: 18956000]

[100] X. Lu, G. Li, and Y. Tong, "A review of negative electrode materials for electrochemical supercapacitors", *Sci. China Technol. Sci.,* vol. 58, no. 11, pp. 1799-1808, 2015.
[http://dx.doi.org/10.1007/s11431-015-5931-z]

[101] V.V.N. Obreja, "Supercapacitors specialities - Materials review", *AIP Conf. Proc.,* vol. 1597, no. 1, pp. 98-120, 2014.
[http://dx.doi.org/10.1063/1.4878482]

[102] A. González, E. Goikolea, J.A. Barrena, and R. Mysyk, "Review on supercapacitors: Technologies and materials", *Renew. Sustain. Energy Rev.,* vol. 58, pp. 1189-1206, 2016.
[http://dx.doi.org/10.1016/j.rser.2015.12.249]

[103] R. Kötz, and M. Carlen, "Principles and applications of electrochemical capacitors", *Electrochim. Acta,* vol. 45, no. 15-16, pp. 2483-2498, 2000.
[http://dx.doi.org/10.1016/S0013-4686(00)00354-6]

[104] W.G. Pell, and B.E. Conway, "Voltammetry at a de Levie brush electrode as a model for electrochemical supercapacitor behaviour", *J. Electroanal. Chem. (Lausanne),* vol. 500, no. 1-2, pp. 121-133, 2001.
[http://dx.doi.org/10.1016/S0022-0728(00)00423-X]

[105] G.M.A. Girard, M. Hilder, H. Zhu, D. Nucciarone, K. Whitbread, S. Zavorine, M. Moser, M. Forsyth, D.R. MacFarlane, and P.C. Howlett, "Electrochemical and physicochemical properties of small phosphonium cation ionic liquid electrolytes with high lithium salt content", *Phys. Chem. Chem. Phys.,* vol. 17, no. 14, pp. 8706-8713, 2015.
[http://dx.doi.org/10.1039/C5CP00205B] [PMID: 25820549]

[106] H. Zheng, H. Zhang, Y. Fu, T. Abe, and Z. Ogumi, "Temperature effects on the electrochemical behavior of spinel LiMn(2)O(4) in quaternary ammonium-based ionic liquid electrolyte", *J. Phys. Chem. B,* vol. 109, no. 28, pp. 13676-13684, 2005.
[http://dx.doi.org/10.1021/jp051238i] [PMID: 16852714]

[107] N. Nandihalli, "Thermoelectric films and periodic structures and spin Seebeck effect systems: facets of performance optimization", *Mater. Today Energy,* vol. 25, p. 100965, 2022.
[http://dx.doi.org/10.1016/j.mtener.2022.100965]

[108] S. Ayachi, X. He, and H.J. Yoon, "Solar Thermoelectricity for Power Generation", *Adv. Energy Mater.,* vol. 13, no. 28, p. 2300937, 2023.
[http://dx.doi.org/10.1002/aenm.202300937]

[109] H. Zhou, F. Matoba, R. Matsuno, Y. Wakayama, and T. Yamada, "Direct conversion of phase-transition entropy into electrochemical thermopower and the peltier effect", *Adv. Mater.,* vol. 35, no. 36, p. 2303341, 2023.
[http://dx.doi.org/10.1002/adma.202303341] [PMID: 37315308]

[110] L. Yang, Z.G. Chen, M.S. Dargusch, and J. Zou, "High performance thermoelectric materials: progress and their applications", *Adv. Energy Mater.,* vol. 8, no. 6, p. 1701797, 2018.
[http://dx.doi.org/10.1002/aenm.201701797]

[111] S. Duan, N. Man, J. Xu, Q. Wu, G. Liu, X. Tan, H. Shao, K. Guo, X. Yang, and J. Jiang, "Thermoelectric (Bi,Sb)2Te3–Ge0.5Mn0.5Te composites with excellent mechanical properties", *J. Mater. Chem. A Mater. Energy Sustain.,* vol. 7, no. 15, pp. 9241-9246, 2019.
[http://dx.doi.org/10.1039/C9TA01962F]

[112] Y.S. Wang, L.L. Huang, D. Li, J. Zhang, and X.Y. Qin, "Enhanced thermoelectric performance of Bi 0.4 Sb 1.6 Te 3 based composites with CuInTe 2 inclusions", *J. Alloys Compd.,* vol. 758, pp. 72-77, 2018.
[http://dx.doi.org/10.1016/j.jallcom.2018.05.035]

[113] Y.S. Wang, L.L. Huang, C. Zhu, J. Zhang, D. Li, H.X. Xin, M.H. Danish, and X.Y. Qin, "Simultaneously enhanced power factor and phonon scattering in Bi0.4Sb1.6Te3 alloy doped with germanium", *Scr. Mater.,* vol. 154, pp. 118-122, 2018.
[http://dx.doi.org/10.1016/j.scriptamat.2018.05.026]

[114] W.T. Chiu, C.L. Chen, and Y.Y. Chen, "A strategy to optimize the thermoelectric performance in a spark plasma sintering process", *Sci. Rep.,* vol. 6, no. 1, p. 23143, 2016.
[http://dx.doi.org/10.1038/srep23143] [PMID: 26975209]

[115] F. Hao, P. Qiu, Q. Song, H. Chen, P. Lu, D. Ren, X. Shi, and L. Chen, "Roles of Cu in the enhanced thermoelectric properties in Bi0.5Sb1.5Te3", *Materials (Basel),* vol. 10, no. 3, p. 251, 2017.
[http://dx.doi.org/10.3390/ma10030251] [PMID: 28772610]

[116] X. Guo, J. Qin, X. Jia, H. Ma, and H. Jia, "Quaternary thermoelectric materials: Synthesis,

microstructure and thermoelectric properties of the (Bi,Sb)2(Te,Se)3 alloys", *J. Alloys Compd.,* vol. 705, pp. 363-368, 2017.
[http://dx.doi.org/10.1016/j.jallcom.2017.02.182]

[117] F. Serrano-Sánchez, M. Gharsallah, N.M. Nemes, N. Biskup, M. Varela, J.L. Martínez, M.T. Fernández-Díaz, and J.A. Alonso, "Enhanced figure of merit in nanostructured (Bi,Sb)2Te3 with optimized composition, prepared by a straightforward arc-melting procedure", *Sci. Rep.,* vol. 7, no. 1, p. 6277, 2017.
[http://dx.doi.org/10.1038/s41598-017-05428-4] [PMID: 28740227]

[118] N. Gao, B. Zhu, X. Wang, Y. Yu, and F. Zu, "Simultaneous optimization of Seebeck, electrical and thermal conductivity in free-solidified Bi0.4Sb1.6Te3 alloy *via* liquid-state manipulation", *J. Mater. Sci.,* vol. 53, no. 12, pp. 9107-9116, 2018.
[http://dx.doi.org/10.1007/s10853-018-2209-4]

[119] S. Il Kim, "Dense dislocation arrays embedded in grain boundaries for high-performance bulk thermoelectrics", *Science,* vol. 348, no. 6230, pp. 109-114, 2015.
[http://dx.doi.org/10.1126/science.aaa4166]

[120] Z. Fan, H. Wang, Y. Wu, X.J. Liu, and Z.P. Lu, "Thermoelectric high-entropy alloys with low lattice thermal conductivity", *RSC Advances,* vol. 6, no. 57, pp. 52164-52170, 2016.
[http://dx.doi.org/10.1039/C5RA28088E]

[121] Y. Zheng, C. Liu, L. Miao, H. Lin, J. Gao, X. Wang, J. Chen, S. Wu, X. Li, and H. Cai, "Cost effective synthesis of p-type Zn-doped MgAgSb by planetary ball-milling with enhanced thermoelectric properties", *RSC Advances,* vol. 8, no. 62, pp. 35353-35359, 2018.
[http://dx.doi.org/10.1039/C8RA06765A] [PMID: 35547930]

[122] X. Chen, W. Dai, T. Wu, W. Luo, J. Yang, W. Jiang, and L. Wang, "Thin film thermoelectric materials: classification, characterization, and potential for wearable applications", *Coatings,* vol. 8, no. 7, p. 244, 2018.
[http://dx.doi.org/10.3390/coatings8070244]

[123] Z. Fan, P. Li, D. Du, and J. Ouyang, "Significantly enhanced thermoelectric properties of pedot:pss films through sequential post-treatments with common acids and bases", *Adv. Energy Mater.,* vol. 7, no. 8, p. 1602116, 2017.
[http://dx.doi.org/10.1002/aenm.201602116]

[124] I.H. Jung, C.T. Hong, U.H. Lee, Y.H. Kang, K.S. Jang, and S.Y. Cho, "High thermoelectric power factor of a diketopyrrolopyrrole-based low bandgap polymer *via* Finely Tuned Doping Engineering", *Sci. Rep.,* vol. 7, no. 1, p. 44704, 2017.
[http://dx.doi.org/10.1038/srep44704] [PMID: 28317929]

[125] L. Zhang, T. Goto, I. Imae, Y. Sakurai, and Y. Harima, "Thermoelectric properties of PEDOT films prepared by electrochemical polymerization", *J. Polym. Sci., B, Polym. Phys.,* vol. 55, no. 6, pp. 524-531, 2017.
[http://dx.doi.org/10.1002/polb.24299]

[126] S.N. Patel, A.M. Glaudell, D. Kiefer, and M.L. Chabinyc, "Increasing the thermoelectric power factor of a semiconducting polymer by doping from the vapor phase", *ACS Macro Lett.,* vol. 5, no. 3, pp. 268-272, 2016.
[http://dx.doi.org/10.1021/acsmacrolett.5b00887] [PMID: 35614719]

[127] E. Lim, K.A. Peterson, G.M. Su, and M.L. Chabinyc, "Thermoelectric Properties of Poly(3-hexylthiophene) (P3HT) Doped with 2,3,5,6-Tetrafluoro-7,7,8,8-tetracyanoquinodimethane (F 4 TCNQ) by Vapor-Phase Infiltration", *Chem. Mater.,* vol. 30, no. 3, pp. 998-1010, 2018.
[http://dx.doi.org/10.1021/acs.chemmater.7b04849]

[128] H. Li, M.E. DeCoster, R.M. Ireland, J. Song, P.E. Hopkins, and H.E. Katz, "Modification of the Poly(bisdodecylquaterthiophene) Structure for High and Predominantly Nonionic Conductivity with Matched Dopants", *J. Am. Chem. Soc.,* vol. 139, no. 32, pp. 11149-11157, 2017.

[http://dx.doi.org/10.1021/jacs.7b05300] [PMID: 28737034]

[129] C. Pan, L. Wang, T. Liu, X. Zhou, T. Wan, S. Wang, Z. Chen, C. Gao, and L. Wang, "Polar side chain effects on the thermoelectric properties of benzo[1,2-b:4,5-b′]dithiophene-based conjugated polymers", *Macromol. Rapid Commun.,* vol. 40, no. 12, p. 1900082, 2019.
[http://dx.doi.org/10.1002/marc.201900082] [PMID: 30942939]

[130] Y. Kang, Q. Zhang, C. Fan, W. Hu, C. Chen, L. Zhang, F. Yu, Y. Tian, and B. Xu, "High pressure synthesis and thermoelectric properties of polycrystalline Bi2Se3", *J. Alloys Compd.,* vol. 700, pp. 223-227, 2017.
[http://dx.doi.org/10.1016/j.jallcom.2017.01.062]

[131] A. Nozariasbmarz, J.S. Krasinski, and D. Vashaee, "N-type bismuth telluride nanocomposite materials optimization for thermoelectric generators in wearable applications", *Materials (Basel),* vol. 12, no. 9, p. 1529, 2019.
[http://dx.doi.org/10.3390/ma12091529] [PMID: 31083307]

[132] B. Zhu, Z.Y. Huang, X.Y. Wang, Y. Yu, N. Gao, and F.Q. Zu, "Enhanced thermoelectric properties of n-type direction solidified Bi2Te2.7Se0.3 alloys by manipulating its liquid state", *Scr. Mater.,* vol. 146, pp. 192-195, 2018.
[http://dx.doi.org/10.1016/j.scriptamat.2017.11.045]

[133] X. Wang, H. Wang, B. Xiang, H. Shang, B. Zhu, Y. Yu, H. Jin, R. Zhao, Z. Huang, L. Liu, F. Zu, and Z. Chen, "Attaining reduced lattice thermal conductivity and enhanced electrical conductivity in as-sintered pure n-type Bi2Te3 alloy", *J. Mater. Sci.,* vol. 54, no. 6, pp. 4788-4797, 2019.
[http://dx.doi.org/10.1007/s10853-018-3172-9]

[134] M. Hong, T.C. Chasapis, Z.G. Chen, L. Yang, M.G. Kanatzidis, G.J. Snyder, and J. Zou, "n -Type Bi 2 Te 3–x Se x nanoplates with enhanced thermoelectric efficiency driven by wide-frequency phonon scatterings and synergistic carrier scatterings", *ACS Nano,* vol. 10, no. 4, pp. 4719-4727, 2016.
[http://dx.doi.org/10.1021/acsnano.6b01156] [PMID: 27058746]

[135] H. Jang, Y.J. Park, X. Chen, T. Das, M.S. Kim, and J.H. Ahn, "Graphene-based flexible and stretchable electronics", *Adv. Mater.,* vol. 28, no. 22, pp. 4184-4202, 2016.
[http://dx.doi.org/10.1002/adma.201504245] [PMID: 26728114]

[136] D. Madan, X. Zhao, R.M. Ireland, D. Xiao, and H.E. Katz, "Conductivity and power factor enhancement of n-type semiconducting polymers using sodium silica gel dopant", *APL Mater.,* vol. 5, no. 8, p. 086106, 2017.
[http://dx.doi.org/10.1063/1.4990139]

[137] D.S. Montgomery, C.A. Hewitt, R. Barbalace, T. Jones, and D.L. Carroll, "Spray doping method to create a low-profile high-density carbon nanotube thermoelectric generator", *Carbon,* vol. 96, pp. 778-781, 2016.
[http://dx.doi.org/10.1016/j.carbon.2015.09.029]

[138] W. Zhou, Q. Fan, Q. Zhang, L. Cai, K. Li, X. Gu, F. Yang, N. Zhang, Y. Wang, H. Liu, W. Zhou, and S. Xie, "High-performance and compact-designed flexible thermoelectric modules enabled by a reticulate carbon nanotube architecture", *Nat. Commun.,* vol. 8, no. 1, p. 14886, 2017.
[http://dx.doi.org/10.1038/ncomms14886] [PMID: 28337987]

[139] S. Horike, T. Fukushima, T. Saito, T. Kuchimura, Y. Koshiba, M. Morimoto, and K. Ishida, "Highly stable n-type thermoelectric materials fabricated *via* electron doping into inkjet-printed carbon nanotubes using oxygen-abundant simple polymers", *Mol. Syst. Des. Eng.,* vol. 2, no. 5, pp. 616-623, 2017.
[http://dx.doi.org/10.1039/C7ME00063D]

[140] S. Park, and S. Jayaraman, "Smart textiles: wearable electronic systems", *MRS Bull.,* vol. 28, no. 8, pp. 585-591, 2003.
[http://dx.doi.org/10.1557/mrs2003.170]

[141] C. Gonçalves, A. Ferreira da Silva, J. Gomes, and R. Simoes, "Wearable e-textile technologies: a

review on sensors, actuators and control elements", *Inventions (Basel),* vol. 3, no. 1, p. 14, 2018.
[http://dx.doi.org/10.3390/inventions3010014]

[142] R. Ranjan, P. Kumar, and N. Kumar, "Demonstration and performance assessment of dopant free tfet including lattice heating and temperature effects", *Silicon,* pp. 1-8, 2024.
[http://dx.doi.org/10.1007/s12633-024-03008-6]

[143] S. Neethirajan, "Recent advances in wearable sensors for animal health management", *Sens. Biosensing Res.,* vol. 12, pp. 15-29, 2017.
[http://dx.doi.org/10.1016/j.sbsr.2016.11.004]

[144] R. Pailes-Friedman, "Electronics and fabrics: the development of garment-based wearables", *Adv. Mater. Technol.,* vol. 3, no. 10, p. 1700307, 2018.
[http://dx.doi.org/10.1002/admt.201700307]

[145] P. Singh, A. Raman, D.S. Yadav, N. Kumar, A. Dixit, and M.D.H.R. Ansari, "Ultra thin finger-like source region-based tfet: temperature sensor", *IEEE Sens. Lett.,* vol. 8, no. 5, pp. 1-4, 2024.
[http://dx.doi.org/10.1109/LSENS.2024.3390689]

[146] A.K. Dąbrowska, G.M. Rotaru, S. Derler, F. Spano, M. Camenzind, S. Annaheim, R. Stämpfli, M. Schmid, and R.M. Rossi, "Materials used to simulate physical properties of human skin", *Skin Res. Technol.,* vol. 22, no. 1, pp. 3-14, 2016.
[http://dx.doi.org/10.1111/srt.12235] [PMID: 26096898]

[147] J. Huang, D. Li, M. Zhao, P. Lv, L. Lucia, and Q. Wei, "Highly stretchable and bio-based sensors for sensitive strain detection of angular displacements", *Cellulose,* vol. 26, no. 5, pp. 3401-3413, 2019.
[http://dx.doi.org/10.1007/s10570-019-02313-3]

[148] N. Kumar, A. Dixit, A. Rezaei, T. Dutta, C.P. García, and V. Georgiev, "Insights into the Ultra-Steep Subthreshold Slope Gate-all-around Feedback-FET for memory and sensing applications", *2023 IEEE Nanotechnology Materials and Devices Conference (NMDC),* IEEE., pp. 617-620, 2023.
[http://dx.doi.org/10.1109/NMDC57951.2023.10343913]

[149] P. Singh, and D.S. Yadav, "Assessment of temperature and ITCs on single gate L-shaped tunnel FET for low power high frequency application", *Engineering Research Express,* vol. 6, no. 1, p. 015319, 2024.
[http://dx.doi.org/10.1088/2631-8695/ad32b0]

[150] G.H. Büscher, R. Kõiva, C. Schürmann, R. Haschke, and H.J. Ritter, "Flexible and stretchable fabric-based tactile sensor", *Robot. Auton. Syst.,* vol. 63, pp. 244-252, 2015.
[http://dx.doi.org/10.1016/j.robot.2014.09.007]

[151] J. Dahiya, P. Phogat, A. Hooda, and S. Khasa, "Investigations of Praseodymium doped LiF-Zn--Bi2O3-B2O3 glass matrix for photonic applications", *AIP Conf. Proc.,* vol. 2995, no. 1, p. 020065, 2024.
[http://dx.doi.org/10.1063/5.0178197]

[152] N. Kumar, C.P. García, A. Dixit, A. Rezaei, and V. Georgiev, "Charge dynamics of amino acids fingerprints and the effect of density on FinFET-based Electrolyte-gated sensor", *Solid-State Electron.,* vol. 210, p. 108789, 2023.
[http://dx.doi.org/10.1016/j.sse.2023.108789]

[153] P. Singh, and D.S. Yadav, "Impact of work function variation for enhanced electrostatic control with suppressed ambipolar behavior for dual gate L-TFET", *Curr. Appl. Phys.,* vol. 44, pp. 90-101, 2022.
[http://dx.doi.org/10.1016/j.cap.2022.09.014]

[154] V. Galstyan, A. Ponzoni, I. Kholmanov, M.M. Natile, E. Comini, and G. Sberveglieri, "Highly sensitive and selective detection of dimethylamine through Nb-doping of TiO2 nanotubes for potential use in seafood quality control", *Sens. Actuators B Chem.,* vol. 303, p. 127217, 2020.
[http://dx.doi.org/10.1016/j.snb.2019.127217]

[155] N. Kumar, R. Dhar, C. Pascual Garcia, and V.P. Georgiev, "A novel computational framework for

simulations of bio-field effect transistors", *ECS Trans.*, vol. 111, no. 1, pp. 249-260, 2023.
[http://dx.doi.org/10.1149/11101.0249ecst]

[156]   F. Falsafi, "Sm-doped cobalt ferrite nanoparticles: A novel sensing material for conductometric hydrogen leak sensor", In: *Ceram. Int.* vol. 43. , 2017, no. 1, Part B, pp. 1029-1037.
[http://dx.doi.org/10.1016/j.ceramint.2016.10.035]

[157]   J.M. Tulliani, and P. Bonville, "Influence of the dopants on the electrical resistance of hematite-based humidity sensors", *Ceram. Int.*, vol. 31, no. 4, pp. 507-514, 2005.
[http://dx.doi.org/10.1016/j.ceramint.2004.06.015]

[158]   J.M. Tulliani, C. Baroni, L. Zavattaro, and C. Grignani, "Strontium-doped hematite as a possible humidity sensing material for soil water content determination", *Sensors (Basel)*, vol. 13, no. 9, pp. 12070-12092, 2013.
[http://dx.doi.org/10.3390/s130912070] [PMID: 24025555]

[159]   K. Galatsis, L. Cukrov, W. Wlodarski, P. McCormick, K. Kalantar-zadeh, E. Comini, and G. Sberveglieri, "p- and n-type Fe-doped SnO_2 gas sensors fabricated by the mechanochemical processing technique", *Sens. Actuators B Chem.*, vol. 93, no. 1-3, pp. 562-565, 2003.
[http://dx.doi.org/10.1016/S0925-4005(03)00233-8]

[160]   N.S. Ramgir, Y.K. Hwang, S.H. Jhung, H-K. Kim, J-S. Hwang, I.S. Mulla, and J-S. Chang, "CO sensor derived from mesostructured Au-doped SnO_2 thin film", *Appl. Surf. Sci.*, vol. 252, no. 12, pp. 4298-4305, 2006.
[http://dx.doi.org/10.1016/j.apsusc.2005.07.015]

[161]   S. Wang, Y. Zhao, J. Huang, Y. Wang, F. Kong, S. Wu, S. Zhang, and W. Huang, "Preparation and CO gas-sensing behavior of Au-doped SnO_2 sensors", *Vacuum*, vol. 81, no. 3, pp. 394-397, 2006.
[http://dx.doi.org/10.1016/j.vacuum.2006.05.004]

[162]   P. Singh, and D.S. Yadav, "Performance analysis of ITCs on analog/RF, linearity and reliability performance metrics of tunnel FET with ultra-thin source region", *Appl. Phys., A Mater. Sci. Process.*, vol. 128, no. 7, p. 612, 2022.
[http://dx.doi.org/10.1007/s00339-022-05741-4]

[163]   Y. Shen, T. Yamazaki, Z. Liu, D. Meng, T. Kikuta, N. Nakatani, M. Saito, and M. Mori, "Microstructure and H2 gas sensing properties of undoped and Pd-doped SnO_2 nanowires", *Sens. Actuators B Chem.*, vol. 135, no. 2, pp. 524-529, 2009.
[http://dx.doi.org/10.1016/j.snb.2008.09.010]

[164]   J.K. Choi, I-S. Hwang, S-J. Kim, J-S. Park, S-S. Park, U. Jeong, Y.C. Kang, and J-H. Lee, "Design of selective gas sensors using electrospun Pd-doped SnO_2 hollow nanofibers", *Sens. Actuators B Chem.*, vol. 150, no. 1, pp. 191-199, 2010.
[http://dx.doi.org/10.1016/j.snb.2010.07.013]

[165]   Y.C. Lee, H. Huang, O.K. Tan, and M.S. Tse, "Semiconductor gas sensor based on Pd-doped SnO_2 nanorod thin films", *Sens. Actuators B Chem.*, vol. 132, no. 1, pp. 239-242, 2008.
[http://dx.doi.org/10.1016/j.snb.2008.01.028]

[166]   P. Singh, and D. S. Yadav, "Design and investigation of f-shaped tunnel fet with enhanced analog/rf parameters", *Silicon*, pp. 1-16, 2021.

[167]   K.Y. Dong, J-K. Choi, I-S. Hwang, J-W. Lee, B.H. Kang, D-J. Ham, J-H. Lee, and B-K. Ju, "Enhanced H2S sensing characteristics of Pt doped SnO_2 nanofibers sensors with micro heater", *Sens. Actuators B Chem.*, vol. 157, no. 1, pp. 154-161, 2011.
[http://dx.doi.org/10.1016/j.snb.2011.03.043]

[168]   P. Ivanov, E. Llobet, X. Vilanova, J. Brezmes, J. Hubalek, and X. Correig, "Development of high sensitivity ethanol gas sensors based on Pt-doped SnO_2 surfaces", *Sens. Actuators B Chem.*, vol. 99, no. 2-3, pp. 201-206, 2004.
[http://dx.doi.org/10.1016/j.snb.2003.11.012]

[169] P. Singh, and D.S. Yadav, "Impactful study of f-shaped tunnel fet", *Silicon,* vol. 14, no. 10, pp. 5359-5365, 2022.
[http://dx.doi.org/10.1007/s12633-021-01319-6]

[170] J. Kappler, N. Bârsan, U. Weimar, A. Dièguez, J.L. Alay, A. Romano-Rodriguez, J.R. Morante, and W. Göpel, "Correlation between XPS, Raman and TEM measurements and the gas sensitivity of Pt and Pd doped SnO 2 based gas sensors", *Fresenius J. Anal. Chem.,* vol. 361, no. 2, pp. 110-114, 1998.
[http://dx.doi.org/10.1007/s002160050844]

[171] L. Zhu, and W. Zeng, "Room-temperature gas sensing of ZnO-based gas sensor: A review", *Sens. Actuators A Phys.,* vol. 267, pp. 242-261, 2017.
[http://dx.doi.org/10.1016/j.sna.2017.10.021]

[172] A. Moumen, N. Kaur, N. Poli, D. Zappa, and E. Comini, "One dimensional ZnO nanostructures: growth and chemical sensing performances", *Nanomaterials (Basel),* vol. 10, no. 10, p. 1940, 2020.
[http://dx.doi.org/10.3390/nano10101940] [PMID: 33003427]

[173] R. Kumar, O. Al-Dossary, G. Kumar, and A. Umar, "Zinc oxide nanostructures for NO2 gas–sensor applications: A review", *Nano-Micro Lett.,* vol. 7, no. 2, pp. 97-120, 2015.
[http://dx.doi.org/10.1007/s40820-014-0023-3] [PMID: 30464961]

[174] D.P. Volanti, A.A. Felix, M.O. Orlandi, G. Whitfield, D-J. Yang, E. Longo, H.L. Tuller, and J.A. Varela, "The role of hierarchical morphologies in the superior gas sensing performance of CuO-Based chemiresistors", *Adv. Funct. Mater.,* vol. 23, no. 14, pp. 1759-1766, 2013.
[http://dx.doi.org/10.1002/adfm.201202332]

[175] C. Yang, X. Su, J. Wang, X. Cao, S. Wang, and L. Zhang, "Facile microwave-assisted hydrothermal synthesis of varied-shaped CuO nanoparticles and their gas sensing properties", *Sens. Actuators B Chem.,* vol. 185, pp. 159-165, 2013.
[http://dx.doi.org/10.1016/j.snb.2013.04.100]

[176] P. Singh, A. Raman, and N. Kumar, "Spectroscopic and simulation analysis of facile PEDOT: PSS layer deposition-silicon for perovskite solar cell", *Silicon,* vol. 12, no. 8, pp. 1769-1777, 2020.
[http://dx.doi.org/10.1007/s12633-019-00284-5]

[177] J.W. Yoon, H.J. Kim, H.M. Jeong, and J.H. Lee, "Gas sensing characteristics of p-type Cr2O3 and Co3O4 nanofibers depending on inter-particle connectivity", *Sens. Actuators B Chem.,* vol. 202, pp. 263-271, 2014.
[http://dx.doi.org/10.1016/j.snb.2014.05.081]

[178] C. Wang, X. Cui, J. Liu, X. Zhou, X. Cheng, P. Sun, X. Hu, X. Li, J. Zheng, and G. Lu, "Design of superior ethanol gas sensor based on al-doped nio nanorod-flowers", *ACS Sens.,* vol. 1, no. 2, pp. 131-136, 2016.
[http://dx.doi.org/10.1021/acssensors.5b00123]

[179] N. Kaur, E. Comini, D. Zappa, N. Poli, and G. Sberveglieri, "Nickel oxide nanowires: vapor liquid solid synthesis and integration into a gas sensing device", *Nanotechnology,* vol. 27, no. 20, p. 205701, 2016.
[http://dx.doi.org/10.1088/0957-4484/27/20/205701] [PMID: 27053627]

[180] B. Liu, H. Yang, H. Zhao, L. An, L. Zhang, R. Shi, L. Wang, L. Bao, and Y. Chen, "Synthesis and enhanced gas-sensing properties of ultralong NiO nanowires assembled with NiO nanocrystals", *Sens. Actuators B Chem.,* vol. 156, no. 1, pp. 251-262, 2011.
[http://dx.doi.org/10.1016/j.snb.2011.04.028]

[181] P.C. Chou, H-I. Chen, I-P. Liu, C-C. Chen, J-K. Liou, K-S. Hsu, and W-C. Liu, "On the ammonia gas sensing performance of a rf sputtered nio thin-film sensor", *IEEE Sens. J.,* vol. 15, no. 7, pp. 3711-3715, 2015.
[http://dx.doi.org/10.1109/JSEN.2015.2391286]

[182] T.P. Mokoena, H.C. Swart, and D.E. Motaung, "A review on recent progress of p-type nickel oxide

based gas sensors: Future perspectives", *J. Alloys Compd.,* vol. 805, pp. 267-294, 2019.
[http://dx.doi.org/10.1016/j.jallcom.2019.06.329]

[183] C. Cantalini, "Cr2O3, WO3 single and Cr/W binary oxide prepared by physical methods for gas sensing applications", *J. Eur. Ceram. Soc.,* vol. 24, no. 6, pp. 1421-1424, 2004.
[http://dx.doi.org/10.1016/S0955-2219(03)00442-4]

[184] T.H. Kim, J.W. Yoon, Y.C. Kang, F. Abdel-Hady, A.A. Wazzan, and J.H. Lee, "A strategy for ultrasensitive and selective detection of methylamine using p-type Cr2O3: Morphological design of sensing materials, control of charge carrier concentrations, and configurational tuning of Au catalysts", *Sens. Actuators B Chem.,* vol. 240, pp. 1049-1057, 2017.
[http://dx.doi.org/10.1016/j.snb.2016.09.098]

[185] J. Cao, Y. Xu, L. Sui, X. Zhang, S. Gao, X. Cheng, H. Zhao, and L. Huo, "Highly selective low-temperature triethylamine sensor based on Ag/Cr2O3 mesoporous microspheres", *Sens. Actuators B Chem.,* vol. 220, pp. 910-918, 2015.
[http://dx.doi.org/10.1016/j.snb.2015.06.023]

[186] H. Ma, Y. Xu, Z. Rong, X. Cheng, S. Gao, X. Zhang, H. Zhao, and L. Huo, "Highly toluene sensing performance based on monodispersed Cr2O3 porous microspheres", *Sens. Actuators B Chem.,* vol. 174, pp. 325-331, 2012.
[http://dx.doi.org/10.1016/j.snb.2012.08.073]

[187] M. Reza Gholipour, C.T. Dinh, F. Béland, and T.O. Do, "Nanocomposite heterojunctions as sunlight-driven photocatalysts for hydrogen production from water splitting", *Nanoscale,* vol. 7, no. 18, pp. 8187-8208, 2015.
[http://dx.doi.org/10.1039/C4NR07224C] [PMID: 25804291]

[188] M. Carmo, D.L. Fritz, J. Mergel, and D. Stolten, "A comprehensive review on PEM water electrolysis", *Int. J. Hydrogen Energy,* vol. 38, no. 12, pp. 4901-4934, 2013.
[http://dx.doi.org/10.1016/j.ijhydene.2013.01.151]

[189] J. Yu, and A. Kudo, "Effects of structural variation on the photocatalytic performance of hydrothermally synthesized BiVO 4", *Adv. Funct. Mater.,* vol. 16, no. 16, pp. 2163-2169, 2006.
[http://dx.doi.org/10.1002/adfm.200500799]

[190] F. Tran, and P. Blaha, "Accurate band gaps of semiconductors and insulators with a semilocal exchange-correlation potential", *Phys. Rev. Lett.,* vol. 102, no. 22, p. 226401, 2009.
[http://dx.doi.org/10.1103/PhysRevLett.102.226401] [PMID: 19658882]

[191] T. Yasuda, M. Kato, M. Ichimura, and T. Hatayama, "SiC photoelectrodes for a self-driven water-splitting cell", *Appl. Phys. Lett.,* vol. 101, no. 5, p. 053902, 2012.
[http://dx.doi.org/10.1063/1.4740079]

[192] A. Kushwaha, and M. Aslam, "ZnS shielded ZnO nanowire photoanodes for efficient water splitting", *Electrochim. Acta,* vol. 130, pp. 222-231, 2014.
[http://dx.doi.org/10.1016/j.electacta.2014.03.008]

[193] V. Cristino, S. Caramori, R. Argazzi, L. Meda, G.L. Marra, and C.A. Bignozzi, "Efficient photoelectrochemical water splitting by anodically grown WO3 electrodes", *Langmuir,* vol. 27, no. 11, pp. 7276-7284, 2011.
[http://dx.doi.org/10.1021/la200595x] [PMID: 21542603]

[194] Y. Wei, L. Ke, J. Kong, H. Liu, Z. Jiao, X. Lu, H. Du, and X.W. Sun, "Enhanced photoelectrochemical water-splitting effect with a bent ZnO nanorod photoanode decorated with Ag nanoparticles", *Nanotechnology,* vol. 23, no. 23, p. 235401, 2012.
[http://dx.doi.org/10.1088/0957-4484/23/23/235401] [PMID: 22609803]

[195] R. Abe, "Recent progress on photocatalytic and photoelectrochemical water splitting under visible light irradiation", *J. Photochem. Photobiol. Photochem. Rev.,* vol. 11, no. 4, pp. 179-209, 2010.
[http://dx.doi.org/10.1016/j.jphotochemrev.2011.02.003]

[196] N. Serpone, and A.V. Emeline, "Semiconductor photocatalysis — past, present, and future outlook", *J. Phys. Chem. Lett.,* vol. 3, no. 5, pp. 673-677, 2012.
[http://dx.doi.org/10.1021/jz300071j] [PMID: 26286164]

[197] M. Silva, I. Oliveira-Inocêncio, R.C. Martins, R. Quinta-Ferreira, M. Gmurek, A. Nogueira, and S. Castro-Silva, "Optimization of heterogeneous photosensitized oxidation for winery effluent treatment", *Water,* vol. 15, no. 13, p. 2340, 2023.
[http://dx.doi.org/10.3390/w15132340]

[198] Y.J. Yuan, H.W. Lu, Z.T. Yu, and Z.G. Zou, "Noble-metal-free molybdenum disulfide cocatalyst for photocatalytic hydrogen production", *ChemSusChem,* vol. 8, no. 24, pp. 4113-4127, 2015.
[http://dx.doi.org/10.1002/cssc.201501203] [PMID: 26586523]

[199] A. Fujishima, and K. Honda, "Electrochemical photolysis of water at a semiconductor electrode", *Nature,* vol. 238, no. 5358, pp. 37-38, 1972.
[http://dx.doi.org/10.1038/238037a0] [PMID: 12635268]

[200] G.G. Bessegato, T.T. Guaraldo, J.F. de Brito, M.F. Brugnera, and M.V.B. Zanoni, "Achievements and trends in photoelectrocatalysis: from environmental to energy applications", *Electrocatalysis,* vol. 6, no. 5, pp. 415-441, 2015.
[http://dx.doi.org/10.1007/s12678-015-0259-9]

[201] A. Kudo, and Y. Miseki, "Heterogeneous photocatalyst materials for water splitting", *Chem. Soc. Rev.,* vol. 38, no. 1, pp. 253-278, 2009.
[http://dx.doi.org/10.1039/B800489G] [PMID: 19088977]

[202] Y. Ma, X. Wang, Y. Jia, X. Chen, H. Han, and C. Li, "Titanium dioxide-based nanomaterials for photocatalytic fuel generations", *Chem. Rev.,* vol. 114, no. 19, pp. 9987-10043, 2014.
[http://dx.doi.org/10.1021/cr500008u] [PMID: 25098384]

Semiconductor Nanoscale Devices, 2025, 155-172

Measurement Techniques for Determining the Thermal Conductivity of Bulk Samples and Thin Films

Simrandeep Kour[1], Rikky Sharma[1], Sameena Sulthana[2] and Rupam Mukherjee[2,*]

[1] *Department of Physics, Lovely Professional University, Phagwara, Punjab, 144001, India*

[2] *Department of Physics, Presidency University, Bangalore, Karnataka, 560064, India*

Abstract: Thermal conductivity is one class of basic transport properties of materials that characterizes the flow of heat through it. Over recent years, the transformation of smart materials of atomically thin layers to small-size bulk samples has further made it difficult to determine the thermal conductivity more accurately due to the second law of thermodynamics, which prevents full control over heat flux during measurement. Heat flux and small temperature gradients are the two most important parameters that need to be considered while measuring the thermal properties of small dimensional samples. The difficulty in thermal measurements is associated with the thermal anchoring and controlling the heat loss that takes place due to conduction, convection, and radiation processes. In addition, controlled temperature difference coupled with high-speed data acquisition allows to study the thermal properties in a more extensive way. In this chapter, some of the reliable and effective techniques are mentioned that can help us to measure thermal conductivity with the lowest possible error. The importance of maintaining a high vacuum, choosing a proper heat source, and selecting heat sinks with desirable electrical outlets is also discussed here. Moreover, depending on the nature and dimension of the samples, different measuring techniques need to be used to extract the thermal conductivity of samples accurately. In general, understanding these properties is significant for predicting the performance of electronic materials in real-world applications, such as heat exchangers, evaporators, thermocouples, refrigerators, gas turbine engine applications, automotive parts, and biomedical parts. Further, these properties can be helpful in analyzing carbon nanotubes, selecting suitable ceramic coatings, assessing polymers, *etc.*

Keywords: Bulk sample, Heat flux, Heat sink, Radiation, Thin film, Temperature gradient.

* **Corresponding author Rupam Mukherjee:** Department of Physics, Presidency University, Bangalore, Karnataka, 560064, India; E-mail: rupam.mukherjee@presidencyuniversity.in

Ashish Raman, Prabhat Singh, Naveen Kumar & Ravi Ranjan (Eds.)

INTRODUCTION

Thermal conductivity is an important parameter in the preparation of devices using solid-state materials where temperature and thermal flux are of primary concern. However, the characterization of the thermal conductivity of bulk or thin film samples always remains an uphill task as the second law of thermodynamics denies full control over heat flux during measurement. For instance, the interaction of the sample with its surroundings leads to continuous heat exchange, which facilitates unavoidable power loss owing to erroneous thermal measurements. To date, various measuring techniques have been adopted to determine the thermal conductivity of various dimensional systems. This chapter provides an overview of thermal measuring techniques and the possible errors corresponding to each type of measurement. The discussion can help the researchers to combat unwanted errors while performing these experiments.

Depending on the nature, geometry, and size limitations of samples, the measurement technique for thermal conductivity could be sample-specific. For characterizing the thermal properties of bulk materials, measurement techniques, like the steady state method (absolute and comparative technique), transient dynamic technique, and laser flash analysis, are widely accepted [1 - 4]. Each of these techniques has its own merits and limitations depending on the sample specimen. However, the same thermal measurement techniques may not be supported by thin films whose thickness ranges from a few nanometers to hundreds of micrometers grown on different substrates. A new approach corresponding to thin films, like 3ω and the transient thermoreflectance technique, has proven to be more flexible, formidable, and reliable [5, 6].

Thermal conductivity is a property of the materials to carry heat along the temperature gradient. The measurement of the thermal properties of solid-state materials has been carried out since 1950. Over the period, new materials have emerged with dimensions ranging from bulk to a single atomic layer [7 - 13]. This development has also urged scientific engineers to redesign and modify transport measuring techniques, which is crucial and tricky at the same time. These days, commercial instruments are extensively used to measure the thermal conductivity of samples [14 - 16]. However, to modify the measurements, students and researchers should have a clear picture of how these experimental techniques are carried out. How to control heat loss through conduction, convection, and radiation processes; what effort a sample preparation needs, and how to measure temperature gradient are some of the questions that need to be addressed in this context. Moreover, controlled temperature difference coupled with high-speed data acquisition allows to study the thermal properties in a more extensive way. Hopefully, this short review of measurement techniques will benefit both the readers and the scientific community in understanding the fundamentals of science and possible sources of error in thermal conductivity measurements.

STEADY-STATE METHOD FOR BULK SAMPLES

Absolute Technique

Thermal characterization of materials can pose various types of problems depending on which measurement technique one is using. According to the fundamental concept of heat flow, thermal conductivity κ can be extracted if a temperature gradient ΔT across a sample is measured when a certain amount of heating power is applied. In mathematical terms, it is represented as:

$$\kappa = \dot{Q} \ l/(A\Delta T) \tag{1}$$

Where, $\dot{Q}$ is the thermal power flowing across the sample, l is the length in between the thermocouples, and A is the cross-sectional area of a sample. Now at a fixed base temperature, it is clear from the above equation that κ becomes a slope if a graph is plotted between $\dot{Q}$ and ΔT.

The absolute technique is usually good for samples that are rectangular, flat, or cylindrical in shape of thickness 1-2 mm [2, 3]. In this technique, the sample whose thermal conductivity needs to be determined is kept in between the heat source and a heat sink, as depicted in Fig. (**1**).

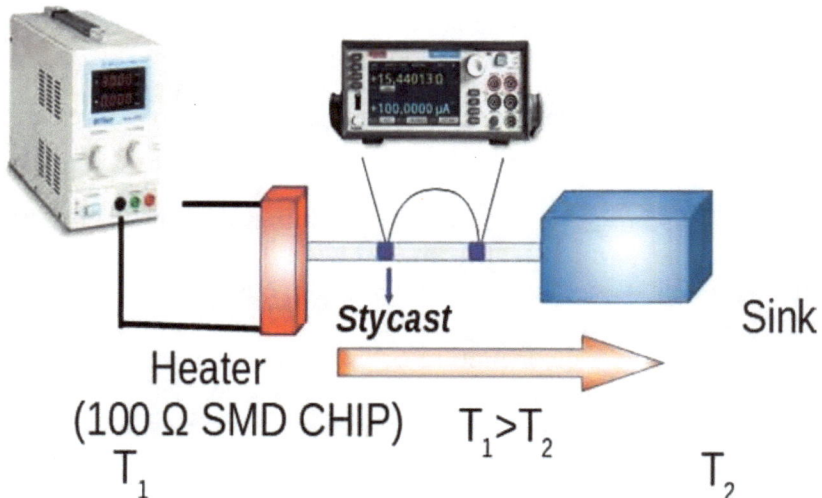

Fig. (1). An experimental setup where the temperature gradient is established across the sample kept in between heater and sink maintained at temperature T_1 and T_2, respectively.

A 100 Ω SMD chip provides a power (P) of I^2R due to Joule heating acting as an input heat source. A gold-coated copper plate used as a heat sink is connected to the other end of the sample. A differential-type thermocouple is glued to the sample with a stycast and electrically connected to the nano-voltmeter, as shown in the figure. The details of the differential thermocouple are discussed later in this chapter.

However, the determination of $\dot{Q}$ is not an easy task as a small fraction of input power is lost due to unavoidable thermal processes, like radiation, conduction, and convection [17]. Mathematically, this means the net power flowing through the specimen is obtained as:

$$\dot{Q} = P - P_{loss} \tag{2}$$

Where, P_{loss} is the parasitic heat loss due to the above-mentioned thermal losses through surrounding gases or the connection leads linking the heater and the sample. The major challenge here is to know the actual heat power (Q) flowing through the sample and the temperature gradient (ΔT) accurately. The two major sources of errors in calculating thermal conductivity are the uncertainty in dimensional measurement and the parasitic heat loss through multi-components of the system. These uncertainties cannot be removed completely although an appropriate experimental design of the setup could minimize the error and the thermal losses [1]. Conduction, convection, and radiation are usually more difficult to quantify and have serious adverse effects, like fluctuation of ΔT with time, non-uniform heating/cooling processes, and low signal-to-noise ratio.

Extensive efforts can be taken to circumvent the thermal issues. The process of radiation can be minimized if two layers of radiation shroud with gold or silver coating are provided around the sample. According to Stephen-Boltzmann, law radiant power loss is proportional to the cube power of the absolute temperature of the sample, *i.e.*, $P_{loss} = A\sigma^2 c(4T^3\Delta T)$, where σ is the Stephen-Boltzmann constant (5.67×10^{-8} W m^{-2} K^{-4}), A is the cross-sectional area, and c is the emissivity of the sample. If the ΔT across the sample varies more than $\geq 2\%$, then radiation loss is often seen in the thermal conductivity data in the form of radiation tail, which usually begins above 150 K. Thus, supportive care should be taken with the input power (P) of the heater such that ΔT remains less than 2-3 K, thus reducing the radiation tail substantially. The linear variation of input power with a temperature gradient (ΔT) across the thermocouple is shown in Fig. (**2**). Moreover, this T^3 tail can be reduced if the geometrical factor A/L (ratio of sample cross-sectional area and the corresponding length) is increased to make the thermal conductance of the sample much higher than the errors associated with the radiation thermal conductance after subtraction. This error is about ± 1 m W

K^{-1}. Further error minimization is possible by considering proper thermal anchoring, which involves good thermal contact and thermal grounding of the thermocouples with the cold sink [8]. For this, a copper heat sink needs to be used to cool the cold junction in order to lower temperatures and vary the heater power to maintain a time-invariant hot temperature junction across it.

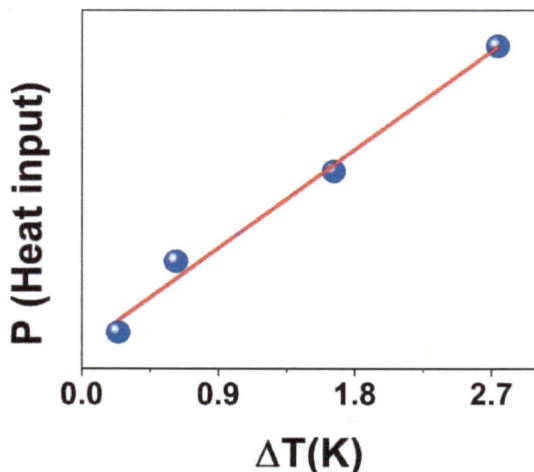

Fig. (2). Linear dependence of input heater power as a function of the generated temperature gradient across the sample.

Besides, to reduce the heat propagation through the thermocouple wires, it is preferable to use thin wires of thickness < 0.001 inch. In addition to bare contacts, such as thermocouple wires to samples or samples to sinks, thermal interface materials, like highly thermal-conducting thermal greases or thermal adhesives, are commonly used to reduce the thermal contact resistance. Also, to improve the thermal conductance through the sample, the effect of convection and radiation can be lowered by performing the experiment in a high vacuum surrounding ≈ 10^{-6} Torr.

A systematic positive error may also be introduced while calculating thermal conductivity by assuming a linear temperature profile from the heater input [15]. The error involved can be estimated as:

$$Error \quad \% \quad = \quad [(\Delta T/L \quad - \quad dT/dx|hj \quad)/(\Delta T/L)].100\% \quad \textbf{(3)}$$

Where, Δ T is the temperature difference across the sample of length L and $dT/dx|_{hj}$ is the temperature gradient at the hot junction [18]. Radiative heat transfer between the sample and the surrounding follows an inhomogeneous differential equation, given as:

$$A\frac{d^2T}{dx^2} - Ac\sigma(T^4 - T_s^4) = 0 \qquad\qquad (4)$$

Where, T and T_S are the temperature of the sample and the surrounding within the radiation shield, respectively. Therefore, $dT/dx|_{hj}$ can be calculated by solving the above equation using boundary value condition, which is at x = 0, T = $T_{cold\,junction}$ and at x = L, T = T_{hj} .

COMPARATIVE TECHNIQUE

In contrast to the absolute method where a testing sample is stuck in between the heater and sink, the comparative technique involves a known standard sample connected in series with the sample under testing. This technique yields better results than the former one as long as the steady heater power gives a repeatable nature of thermal conductivity to the known standard sample [2, 19 - 22]. But at the same time, as this comparative technique involves more thermal contact junctions compared to the absolute technique, it leads to more sources of potential error that need to be taken into account. Fig. (3) depicts the thermal arrangement involved in the comparative technique, carried out in a closed-cycle cryostat.

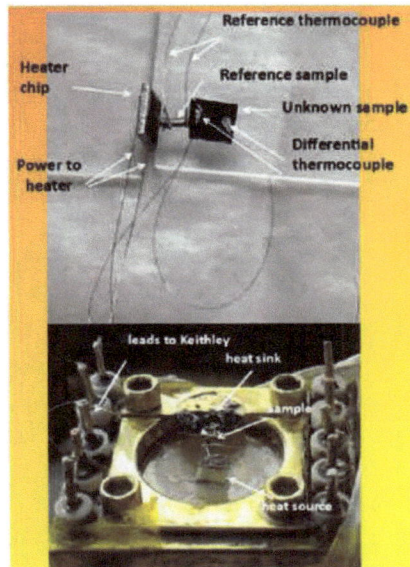

Fig. (3). Experimental setup for comparative technique where reference sample and sample under test connected in series are together kept between the heater and the sink. The sample stage along with the setup is also shown in the lower half of the figure.

In this method, since the heat flow is analogous to a series electrical circuit, it implies that the power through the standard sample is equal to the power through the unknown test sample [23]. This means that if K_S and K_T are the thermal conductivity of standard and test samples, respectively, then:

$$K_T \quad = \quad K_S \quad (A_S \quad \Delta T_S L_T \quad /A_T \quad \Delta T \quad L_S \quad) \quad \textbf{(5)}$$

Where, A stands for the cross-sectional area of the samples, ΔT represents the temperature difference across the samples, and L is the length of the samples [3, 21]. The suffixes S and T stand for standard and test samples, respectively. As the number of thermal junctions increases, the determining temperature gradient ΔT becomes more critical. To circumvent this problem, a differential-type thermocouple is used in the setup, as shown in Fig. (3). This differential arrangement of thermocouple provides better accuracy for the measurement of ΔT as it offers direct measurement of ΔT across the sample [24]. "T" type-based differential thermocouple, which is used in the temperature range 4 K < T < 350 K, has two-point junctions made up of copper and constantan. If T_H and T_C ($T_H > T_C$) are the temperatures at points A and B, respectively, while S_{Cu} and S_{Con} are the respective Seebeck coefficients for copper and constantan wires, then the thermal voltage V_{th} generated due to ΔT (T_H - T_C) is given as:

$$\Delta T = Vth/(SCu - SCon) \tag{6}$$

S_{Cu} - S_{Con} can be extracted from the NIST ITS-90 database for thermocouples [25]. The thermocouple on standard and test samples can easily be glued with silver paste or good thermal conductors, like GE varnish or stycast. A gentle effort should be given to keep the thermocouple tips electrically insulating from the sample. Moreover, a sample thickness of less than 2-3 mm is preferable for thermal measurement as it can avoid developing a vertical temperature gradient, which can lead to error in V_{th} [2, 15]. Despite all these efforts, the radiation loss can still be detrimental to the temperature-dependent thermal conductivity data of the samples. But, at the same time, we know that the total thermal conductivity of a sample is actually a summation of two parts, lattice (κ_L) and electronic (κ_E) thermal conductivity. To correct the radiation loss, the lattice thermal conductivity can be extracted by subtracting the electronic part of thermal conductivity from the total thermal conductivity (κ), i.e., $\kappa_L = \kappa - \kappa_E$ [5]. κ_E can be determined from Wiedemann-Franz law, which is given as $\kappa_E = L_o T/\rho$, where L_o is Lorentz number, T is the surrounding temperature, and ρ is the resistivity of the sample. The thermal conductivity of samples usually has a low-temperature phonon peak above which they exhibit a characteristics temperature dependence, i.e., $\kappa_L \approx 1/T$ behavior due to Umklapp phonon scattering [26, 27]. Reports of $1/T^{0.5}$ independent of temperature for other kinds of materials have also been acknowledged [8 - 11]. If the radiation loss is significantly low, then the mathematical fitting (1/T dependence) of κ_L *versus* T above the low-temperature peak would be reasonably good. On the other hand, if the radiation loss is not well controlled, then the lattice thermal conductivity would deviate from the theoretical fitting. These two distinguished cases are well depicted in Fig. (**4**). The left-hand side panel (a) shows 1/T fitting (red dots) above the low-temperature peak to 300 K, agreeing well with the theoretical dependence, whereas the deviation of κ_L from 1/T dependence above 140 K is shown in the right-hand side panel (b). In the latter case, the fitting should be applied till 140 K and then the κ_L can be extrapolated to 300 K, thus negating the effect of the radiation tail.

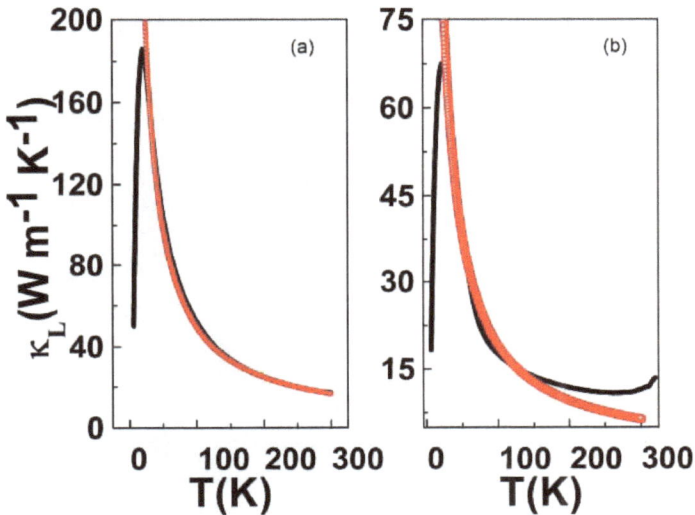

Fig. (4). Temperature-dependent thermal conductivity at low atmospheric pressure. (**a**) The red line shows 1/T fitting from low-temperature peak to 300 K. (**b**) The red line fitting shows deviation from 1/T dependence above 150 K, indicating the onset of radiation tail.

PARALLEL THERMAL CONDUCTANCE TECHNIQUE

The size of the samples often puts a limitation on the use of the above-mentioned techniques to determine the thermal properties of interesting materials. For instance, imagine taking a 1-3 mm size sample and putting electrical contact pads on the surface. If the total area of the contact electrode is greater than 20% of the total surface area of the sample, then a considerable error or thermal loss can be added to the thermal conductivity measurement. The parallel mode of thermal conductance technique has been recently developed by Tritt *et al.* for a small size needle shape single crystal ($2.0\times0.05 \times 0.1$ mm^3) [2, 28]. To support heaters and temperature sensors for thermal measurement, a new non-contact experimental design can be established as a variation of the steady-state process. Fig. (**5**) demonstrates the new sample stage, which allows holding the sample between the heat source and the sink. In this technique, the primary effort should be given to determine the thermal conductance of the sample stage alone before extracting the thermal conductivity of the sample. This should be because it will allow to quantify the background thermal conductance or the radiation loss associated with the stage and the attached thermocouples. After this process, the sample would be appended to the stage and once again the thermal conductance of the system could be determined. Thus, by subtracting these values, the thermal conductance of the test sample can be determined. The thermal conductivity of the sample is determined by multiplying the ratio L/A of the sample, where L is the sample length and A is the cross-sectional area. However, the most important challenge in this method is to determine the dimension of the samples with high accuracy and the repeatability of the thermal conductance of the holder with respect to time and temperature.

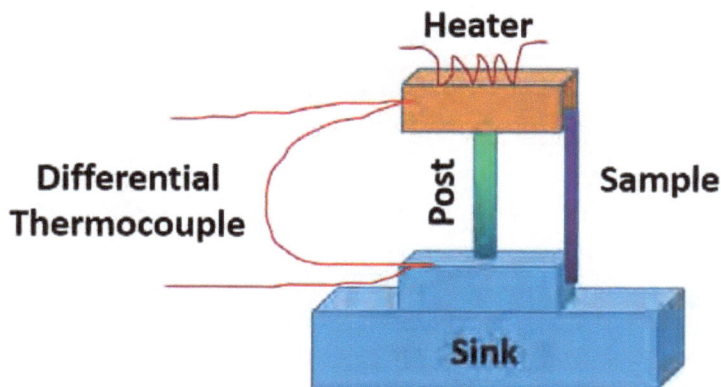

Fig. (5). Parallel thermal conductance setup, where heat is passed through the sample and post connected in parallel. A differential thermocouple is attached to the heater and the sink to measure the temperature difference.

TRANSIENT METHOD FOR BULK SAMPLES

The Pulse-power Method

The pulse power method also known as the Maldonado technique is used for measuring thermoelectric properties; it has an additional advantage over all other techniques discussed so far [29 - 32]. The experimental design in Fig. (**6**) shows the test sample to be attached in between the heater and the sink. Unlike the previous methods, the Maldonado technique does not require maintaining a steady-state temperature difference across the sample between the measurements. This is achieved by allowing the base temperature of the cold sink to drift slowly, while the heater on the top side is fed with a pulsed square wave current I(t), owing to a small temperature gradient $\Delta T \approx 0.3 - 0.5K$. This means that the time needed to attain thermal equilibrium as needed in steady-state techniques can be completely ignored. Moreover, the experimental errors, for example, the time to capture the thermal voltage signal and various offset drifts that arise from time delay can be substantially minimized [29]. The net heat produced by time-dependent current is manifested by the heater's Joule heating and the thermal conductance through the sample. If Q is the rate of change of heat, C is the specific heat of the heater at temperature T_h, R is the electrical resistance of the same heater, K is the sample thermal conductance, and T_c is the bath temperature, then the heat balance equation can be written as [30]:

$$\frac{dQ}{dt} = C(Th)\frac{dTh}{dt} = I^2(t)R(Th) - K(Th - Tc) \qquad (7)$$

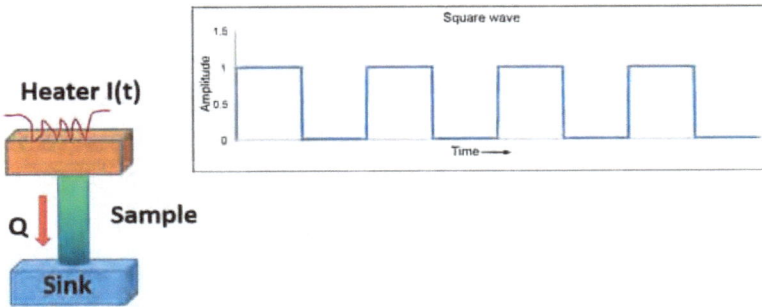

Fig. (6). A square wavefront current with a specific time period is passed through the heater, which produces a periodic heat flux. This heat flux is then passed through a sample in between the heater and a sink.

Regarding this technique, there are two vital points that need to be taken care of: (a) as K is dependent on temperature change, the temperature difference $(T_h - T_c)$ should be small compared to the mean temperature of sample $\frac{Th+Tc}{}$, so that K can be just a function of mean temperature. (b) The drift of T_c should be slow compared to the time period of square wave current (2τ) such that adiabatic approximation can be employed. Taking advantage of these two points, it can become much easier to convert the nonlinear equation 7 to a linear equation by replacing T_c with T_h. The solution of the above equation has a sawtooth form, as shown in Fig. (7) [3]. In Fig. (7), the two lines are drawn corresponding to the maximum and minimum of the periodic oscillation, and the separation between the lines is represented by ΔT_{pp}. This yields a final solution for thermal conductance, given as [33]:

$$K = RI^2\ tanh(K\tau/2C)/(\Delta Tpp) \qquad (8)$$

Numerical iteration is finally used to solve the thermal conductance K in equation 8 as all other parameters have been found to be a function of temperature. The measurement accuracy for this technique has been found to be better than 5%, as reported by Maldonado [3].

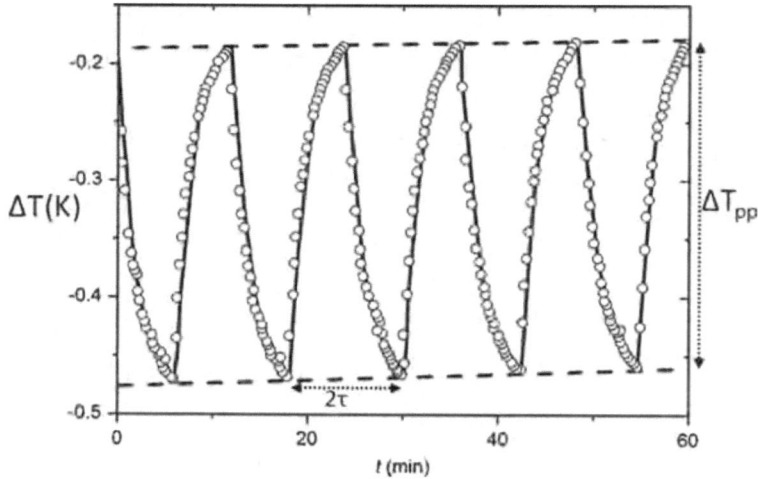

Fig. (7). The variation of temperature gradient with time is recorded for a sample. 2τ is the time period of the square wave and the periodic oscillation between maxima and minima is given by ΔT_{pp}.

The Laser Flash Technique

The laser flash technique has been introduced by Parker *et al.* as another non-contact and non-destructive technique to measure the thermal conductivity of samples with certain geometrical limitations [32 - 36]. This method involves a short laser pulse (< 1ms width) as a heating source on the front side of the test sample and an IR (Infrared Radiation) detector to monitor the increase in temperature on the back side of the test sample. To prevent the laser from flashing directly to IR, planar shape samples, like thin discs ≈ 2-inch diameter or plates, are recommendable. Moreover, both the front and back surfaces of the test sample are coated with thin layers of graphite network for better heat conduction. Here, the thermal diffusivity (α) is determined from a time-dependent temperature change profile, as shown in Fig. (**8**). If d is the sample thickness, C is the thermal heat capacity, and the density of the sample (ρ) is known, then the thermal conductivity (K) in the adiabatic condition can be extracted using the following equation [36]:

$$\alpha = 0.138 d^2 / t_{1/2} \rightarrow \alpha = K / (\rho C) \tag{9}$$

where, $t_{1/2}$ is the time taken (Fig. **8**) to reach one-half of the maximum temperature at the back side of the sample. Heat transfer and the temperature rise are both directly proportional to thermal diffusivity. This demands the sample to be highly emissive to maximize the heat flow from the front to the back side of the sample.

The laser flash technique can determine the thermal conductivity of a sample over a wide range of temperatures from 77 K to 2300 K with relative measurement uncertainties in the 3-5% range. However, the drawback of this method is that specific heat and the density of the sample are to be measured using separate techniques, which may result in the addition of uncertainties leading to significant errors. In addition, the IR detector should be fast enough to detect the signal emitting from the sample while the precision of temperature calibration should be within the limit of $\pm$ 0.2 K.

3ω Technique for Thin Films

Till now, all the experimental techniques that we have discussed so far are good for determining the thermal conductivity of mostly solid bulk samples. Other kinds of materials in the form of thin films grown on various substrates have also been studied widely throughout the scientific community. Due to the small thickness of the thin films, which usually ranges from a few nanometers to tens of micrometers, the measurement of thermal conductivity in these samples becomes a little tricky. A new technique called 3ω is now widely used to determine the thermal conductivity of both thin film and bulk materials in a wide temperature range from 77 K to 900 K [37 - 39]. This method was first introduced by O. Corbino in the year 1910 to determine the thermal diffusivity of metallic filaments for use in light bulbs [40]. Later, in the year 1990, Cahill utilized the 3ω technique to determine the thermal conductivity of a dielectric film. The schematic diagram of this method is shown in Fig. (**9**). The working principle of the 3ω technique has the additional advantage of being time-independent as it modulates the electric signal in a specific frequency domain. A thin metallic strip of aluminum or gold is deposited on top of an active substrate or the sample thin film on the substrate stack. The metallic strip, as shown in the figure, serves as both an electrical heater and a temperature sensor.

An alternating current (AC) with angular frequency ω is passed across the metallic strip. The current is expressed as $I = I_o \cos(\omega t)$, where I_o is the current amplitude. The Joule heat with power $P(t) = I^2(t)R$ generated by the current diffuses into the specimen. This power can also be written as [41]:

$$P(t) = \frac{1}{2}I_o^2 R_o(1 + cos(2\omega t)) \qquad (10)$$

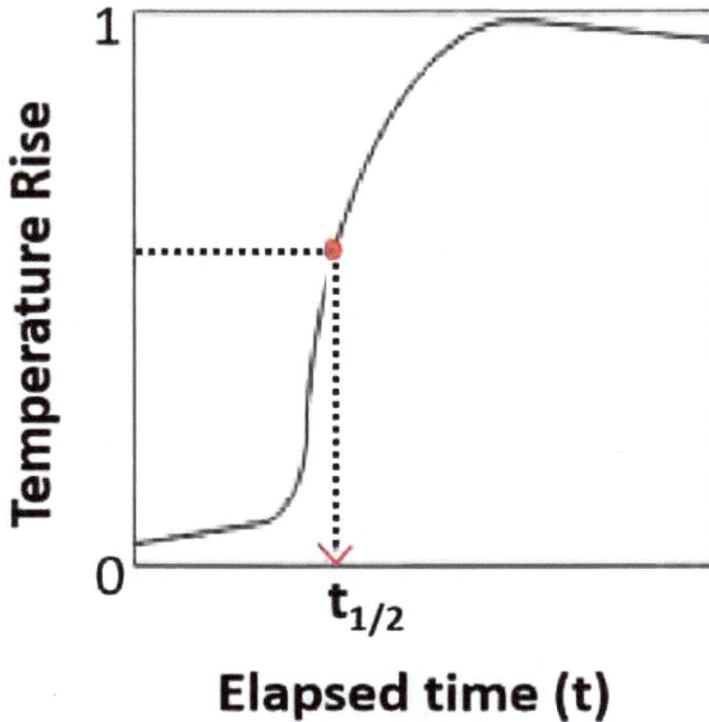

Fig. (8). With the elapsed time, the temperature rise of the back of the test sample is plotted. $t_{1/2}$ corresponds to the time taken to reach one-half of the maximum temperature (red dot) at the back side of the sample.

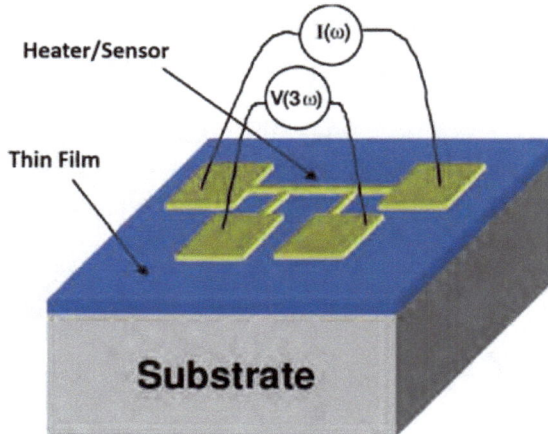

Fig. (9). The schematic setup for the 3ω technique. An alternating current of frequency 1ω is passed through the metallic strip shown in green and a voltage drop of 3ω is measured in a four-point contact configuration with the expression provided below.

$V(t) = I(t).R(t) = I_O R_O(1+\alpha TDC)Cos(\omega t) + \frac{1}{2}(\alpha T2\omega R_O\ I_OCos(\omega t+\varphi)) + \frac{1}{2}(\alpha T2\omega R_O$ **(11)**

$I_OCos(3\omega t+\varphi))$

which is basically a summation of the DC component () and AC component, which depends on the second harmonic oscillation 2ω. Such heating, which is proportional to 2ω, facilitates the temperature change of the heater sensor at 2ω frequency too. So the temperature-dependent heater resistance can now be written as $R(t) = R_o(1 + \alpha\Delta T)$, where R_o is the initial resistance of the heater, α is the temperature coefficient of resistance of the metallic heater, and $\Delta T = T_{DC} + T_{2\omega}(Cos(2\omega t +\varphi))$, with φ being phase difference between temperature oscillation and the excitation current. Consequently, the resistance of the heater gets modified at the second harmonic (2ω) frequency. Following Ohm's law, the potential drop (V) across the heater/sensor is given as [42] or it can be written as $V(t) = V_{DC} + V_{1\omega} + V_{3\omega}$. Since the voltage at 3ω is directly proportional to the temperature oscillation at 2ω, the third term in the above equation carries the information about the thermal dynamics of the sample. However, as the amplitude of $V_{3\omega}$ is weak compared to $V_{1\omega}$, the voltage signal at 3ω is detectable only by use of a lock-in amplifier. The temperature amplitude of the heater is then finally determined as:

$$T_{2\omega} = 2V_{3\omega}/(\alpha V_{1\omega})$$ **(12)**

The 3ω technique can measure the thermal conductivity in both in-plane and cross-plane directions. However, the coupling between the width of the metallic strip and the thickness of the thin film determines the measurement sensitivity. Another advantage of the 3ω method is that the radiation heat loss-assisted errors are greatly reduced below 2% even at high temperatures.

CONCLUSION

In summary, this chapter has reviewed measurement techniques, like steady-state and transient methods, laser flash, and 3ω methods for determining the thermal conductivity of both bulk samples and thin film samples. Possible experimental errors and uncertainties related to each kind of measurement technique have been discussed in detail. Thermal measurement methods have been found to be very sensitive to thermal losses, like conduction, convection, and radiation. Much care should be given while designing experimental apparatus, like the thickness of the wire leads, vacuum pressure, and the analysis of the data. Radiation tail is often seen in experimental data, which appears mostly after 150 K. This creates a major drift in data from actual temperature dependence (1/T) of thermal conductivity. Appropriate fitting and subtraction of the radiation effect are needed to determine the thermal conductivity of the sample at room temperature (300 K). In addition, almost all techniques are sensitive to sample geometry. Hence, precise determination of sample dimensions can pose a serious limiting issue. However, measuring thermal conductivity with less than 5% error is acceptable but challenging.

KEY POINTS

1. Several measurement techniques for determining thermal conductivity are discussed in the chapter. Each technique has its advantages and limitations, and the choice depends on factors, such as sample geometry, material properties, required resolution, and available instrumentation.
2. Precise temperature control is critical for accurate measurements, especially in techniques involving temperature differentials. Determining the temperature gradient using differential thermocouples is discussed here.
3. Calibration with reference materials of known thermal conductivity is essential to validate the measurement setup and ensure accuracy. The steady-state method, which is a comparative method, is discussed in the chapter.
4. To determine the thermal conductivity of solid bulk samples, the use of measuring techniques, like the absolute method, comparative method, and parallel thermal conductance method, is discussed. Time-dependent measurement techniques, like a transient method for bulk samples and laser flash technique, are also discussed.
5. For thin film samples, 3 Omega method is discussed, which utilizes a thin metal film as both a heater and a thermometer. A high-frequency AC current is passed through the film, and the voltage response is measured. Thermal conductivity is inferred from the phase lag between the current and the voltage.
6. Sample preparation should be tuned according to the need of each measurement technique, as well discussed in the chapter. Moreover, the samples must be prepared carefully to ensure uniformity and eliminate any artifacts that may affect the measurements.

ACKNOWLEDGEMENTS

The authors would like to acknowledge the support provided by the Central Instrumentation Facility, Lovely Professional University Punjab, India, and Presidency University Bangalore India for providing experimental and characterization facilities. They would also like to thank their research scholars for their extended support in performing the thermal measurements.

REFERENCES

[1] L. Gerald, "Kapitza resistance", In: *In: Reviews of Modern Physics* vol. 41. , 1969, no. 1, p. 48.

[2] M Terry, *Thermal conductivity: theory, properties, and applications* Springer Science & Business Media, 2005.

[3] D. Zhao, X. Qian, X. Gu, S.A. Jajja, and R. Yang, "Measurement techniques for thermal conductivity and interfacial thermal conductance of bulk and thin film materials", *J. Electron. Packag.,* vol. 138, no. 4, p. 040802, 2016.
[http://dx.doi.org/10.1115/1.4034605]

[4] K.M.F. Shahil, and A.A. Balandin, "Thermal properties of graphene and multilayer graphene: Applications in thermal interface materials", *Solid State Commun.,* vol. 152, no. 15, pp. 1331-1340, 2012.
[http://dx.doi.org/10.1016/j.ssc.2012.04.034]

[5] D.G. Cahill, "Thermal conductivity measurement from 30 to 750 K: the 3ω method", *Rev. Sci. Instrum.,* vol. 61, no. 2, pp. 802-808, 1990.
[http://dx.doi.org/10.1063/1.1141498]

[6] N. Yüksel, "The review of some commonly used methods and techniques to measure the thermal conductivity of insulation materials", In: *Insulation materials in context of sustainabil- ity.* IntechOpen, 2016.
[http://dx.doi.org/10.5772/64157]

[7] C.H. Lui, Z. Ye, C. Ji, K.C. Chiu, C.T. Chou, T.I. Andersen, C. Means-Shively, H. Anderson, J.M. Wu, T. Kidd, Y.H. Lee, and R. He, "Observation of interlayer phonon modes in van der Waals heterostructures", *Phys. Rev. B Condens. Matter Mater. Phys.,* vol. 91, no. 16, p. 165403, 2015.
[http://dx.doi.org/10.1103/PhysRevB.91.165403]

[8] Y. Wang, N. Xu, D. Li, and J. Zhu, "Thermal properties of two dimensional layered materials", *Adv. Funct. Mater.,* vol. 27, no. 19, p. 1604134, 2017.
[http://dx.doi.org/10.1002/adfm.201604134]

[9] V. Goyal, D. Teweldebrhan, and A.A. Balandin, "Mechanically-exfoliated stacks of thin films of Bi2Te3 topological insulators with enhanced thermoelectric performance", *Appl. Phys. Lett.,* vol. 97, no. 13, p. 133117, 2010.
[http://dx.doi.org/10.1063/1.3494529]

[10] B. Sharma, R. Sharma, S. Kour, M.D. Sharma, O. Amin, A.R. Maity, and R. Mukherjee, "Fractional exponents of electrical and thermal conductivity of vanadium intercalated layered 2H-NbS2 bulk crystal", *Indian J. Phys. Proc. Indian Assoc. Cultiv. Sci.,* vol. 96, no. 5, pp. 1335-1339, 2022.
[http://dx.doi.org/10.1007/s12648-021-02045-w]

[11] N. Kumari, M. Kalyan, S. Ghosh, A.R. Maity, and R. Mukherjee, "Possible negative correlation between electrical and thermal conductivity in p-doped WSe 2 single crystal", *Mater. Res. Express,* vol. 8, no. 4, p. 045902, 2021.
[http://dx.doi.org/10.1088/2053-1591/abf682]

[12] A.A. Balandin, S. Ghosh, W. Bao, I. Calizo, D. Teweldebrhan, F. Miao, and C.N. Lau, "Superior thermal conductivity of single-layer graphene", *Nano Lett.,* vol. 8, no. 3, pp. 902-907, 2008.
[http://dx.doi.org/10.1021/nl0731872] [PMID: 18284217]

[13] S.M. Lee, D.G. Cahill, and R. Venkatasubramanian, "Thermal conductivity of Si–Ge superlattices", *Appl. Phys. Lett.,* vol. 70, no. 22, pp. 2957-2959, 1997.
[http://dx.doi.org/10.1063/1.118755]

[14] M.N. Luckyanova, J. Garg, K. Esfarjani, A. Jandl, M.T. Bulsara, A.J. Schmidt, A.J. Minnich, S. Chen, M.S. Dresselhaus, Z. Ren, E.A. Fitzgerald, and G. Chen, "Coherent phonon heat conduction in superlattices", *Science,* vol. 338, no. 6109, pp. 936-939, 2012.
[http://dx.doi.org/10.1126/science.1225549] [PMID: 23161996]

[15]   D. Kraemer, and G. Chen, "A simple differential steady-state method to measure the thermal conductivity of solid bulk materials with high accuracy", *Rev. Sci. Instrum.,* vol. 85, no. 2, p. 025108, 2014.
[http://dx.doi.org/10.1063/1.4865111] [PMID: 24593397]

[16]   Y.K. Koh, S.L. Singer, W. Kim, J.M.O. Zide, H. Lu, D.G. Cahill, A. Majumdar, and A.C. Gossard, "Comparison of the 3ω method and time-domain thermoreflectance for measurements of the cross-plane thermal conductivity of epitaxial semiconductors", *J. Appl. Phys.,* vol. 105, no. 5, p. 054303, 2009.
[http://dx.doi.org/10.1063/1.3078808]

[17]   W. Zhao, Y. Yang, Z. Bao, D. Yan, and Z. Zhu, "Methods for measuring the effective thermal conductivity of metal hydride beds: A review", *Int. J. Hydrogen Energy,* vol. 45, no. 11, pp. 6680-6700, 2020.
[http://dx.doi.org/10.1016/j.ijhydene.2019.12.185]

[18]   C.G.S. Pillai, and A.M. George, "An improved comparative thermal conductivity apparatus for measurements at high temperatures", *Int. J. Thermophys.,* vol. 12, no. 3, pp. 563-576, 1991.
[http://dx.doi.org/10.1007/BF00502369]

[19]   A. Tiwari, K. Boussois, B. Nait-Ali, D.S. Smith, and P. Blanchart, "Anisotropic thermal conductivity of thin polycrystalline oxide samples", *AIP Adv.,* vol. 3, no. 11, p. 112129, 2013.
[http://dx.doi.org/10.1063/1.4836555]

[20]   T.C. Harman, J.H. Cahn, and M.J. Logan, "Measurement of thermal conductivity by utilization of the Peltier effect", *J. Appl. Phys.,* vol. 30, no. 9, pp. 1351-1359, 1959.
[http://dx.doi.org/10.1063/1.1735334]

[21]   A. Beck, "A steady state method for the rapid measurement of the thermal conductivity of rocks", *J. Sci. Instrum.,* vol. 34, no. 5, pp. 186-189, 1957.
[http://dx.doi.org/10.1088/0950-7671/34/5/304]

[22]   N.S. Rasor, and J.D. McClelland, "Thermal property measurements at very high temperatures", *Rev. Sci. Instrum.,* vol. 31, no. 6, pp. 595-604, 1960.
[http://dx.doi.org/10.1063/1.1931263]

[23]   M.H. Sehhat, A. Mahdianikhotbesara, and F. Yadegari, "Ali Mahdianikhotbesara, and Farzad Yadegari. "Experimental validation of conductive heat transfer theory: thermal resistivity and system effects"", *Computational Research Progress In Applied Science & Engineering,* vol. 7, no. 4, pp. 1-6, 2021.
[http://dx.doi.org/10.52547/crpase.7.4.2415]

[24]   K.C. Sloneker, "High-resolution differential thermocouple measurements using an im- proved noise cancellation and magnetic amplification technique", In: *In: AIP Conference Proceedings.* vol. 684. American Institute of Physics, 2003, no. 1, pp. 997-1002.

[25]   L. Weston, "Calibration of cryogenic resistance thermometers between 0.65 K and 165 K on the international temperature scale of 1990", In: *In: NIST Special Publication* vol. 250. American Institute of Physics, 2015, p. 91.

[26]   Y. Guo, S. Zhou, Y. Bai, and J. Zhao, "Tunable thermal conductivity of silicene by Germanium doping", *J. Supercond. Nov. Magn.,* vol. 29, no. 3, pp. 717-720, 2016.
[http://dx.doi.org/10.1007/s10948-015-3305-1]

[27]   D.L. Nika, E.P. Pokatilov, A.S. Askerov, and A.A. Balandin, "Phonon thermal conduction in graphene: Role of Umklapp and edge roughness scattering", *Phys. Rev. B Condens. Matter Mater. Phys.,* vol. 79, no. 15, p. 155413, 2009.
[http://dx.doi.org/10.1103/PhysRevB.79.155413]

[28]   M. Bartosz, "Zawilski, Roy T Littleton IV, and Terry M Tritt. "Description of the parallel ther- mal conductance technique for the measurement of the thermal conductivity of small diameter samples"",

Rev. Sci. Instrum., vol. 72, no. 3, pp. 1770-1774, 2001.
[http://dx.doi.org/10.1063/1.1347980]

[29] O. Maldonado, "Pulse method for simultaneous measurement of electric thermopower and heat conductivity at low temperatures", *Cryogenics,* vol. 32, no. 10, pp. 908-912, 1992.
[http://dx.doi.org/10.1016/0011-2275(92)90358-H]

[30] V.E. Borisenko, V.V. Gribkovskii, V.A. Labunov, and S.G. Yudin, "Pulsed heating of semiconductors", *Phys. Status Solidi, A Appl. Res.,* vol. 86, no. 2, pp. 573-583, 1984.
[http://dx.doi.org/10.1002/pssa.2210860214]

[31] N.M. Kharalkar, L.J. Hayes, and J.W. Valvano, "Pulse-power integrated-decay technique for the measurement of thermal conductivity", *Meas. Sci. Technol.,* vol. 19, no. 7, p. 075104, 2008.
[http://dx.doi.org/10.1088/0957-0233/19/7/075104]

[32] S.E. Gustafsson, "Transient plane source techniques for thermal conductivity and thermal diffusivity measurements of solid materials", *Rev. Sci. Instrum.,* vol. 62, no. 3, pp. 797-804, 1991.
[http://dx.doi.org/10.1063/1.1142087]

[33] C.R.B. Lister, "The pulse-probe method of conductivity measurement", *Geophys. J. Int.,* vol. 57, no. 2, pp. 451-461, 1979.
[http://dx.doi.org/10.1111/j.1365-246X.1979.tb04788.x]

[34] A.J. Griffin Jr, F.R. Brotzen, and P.J. Loos, "Effect of thickness on the transverse thermal conductivity of thin dielectric films", *J. Appl. Phys.,* vol. 75, no. 8, pp. 3761-3764, 1994.
[http://dx.doi.org/10.1063/1.356049]

[35] T.W. Wojtatowicz, and K. Rozniakowski, "On the application of the laser flash method for different materials", In: *Thermal Conductivity 20.* Springer, 1990, pp. 367-375.

[36] W.J. Parker, R.J. Jenkins, C.P. Butler, and G.L. Abbott, "Flash method of determining thermal diffusivity, heat capacity, and thermal conductivity", *J. Appl. Phys.,* vol. 32, no. 9, pp. 1679-1684, 1961.
[http://dx.doi.org/10.1063/1.1728417]

[37] J. Kopp, and G.A. Slack, "Thermal contact problems in low temperature thermocouple thermometry", *Cryogenics,* vol. 11, no. 1, pp. 22-25, 1971.
[http://dx.doi.org/10.1016/0011-2275(71)90005-1]

[38] T.Y. Choi, D. Poulikakos, J. Tharian, and U. Sennhauser, "Measurement of the thermal conductivity of individual carbon nanotubes by the four-point three-omega method", *Nano Lett.,* vol. 6, no. 8, pp. 1589-1593, 2006.
[http://dx.doi.org/10.1021/nl060331v] [PMID: 16895340]

[39] W. Jaber, and P.O. Chapuis, "Non-idealities in the 3 ω method for thermal characterization in the low- and high-frequency regimes", *AIP Adv.,* vol. 8, no. 4, p. 045111, 2018.
[http://dx.doi.org/10.1063/1.5027396]

[40] F. Volklein, and T. Starz, "Thermal conductivity of thin films-experimental methods and theoretical interpretation", *In: XVI ICT'97. Proceedings ICT'97. 16th International Conference on Thermoelectrics (Cat. No. 97TH8291). IEEE,* 1997pp. 711-718
[http://dx.doi.org/10.1109/ICT.1997.667630]

[41] A.A. Guermoudi, P.Y. Cresson, A. Ouldabbes, G. Boussatour, and T. Lasri, "Thermal conductivity and interfacial effect of parylene C thin film using the 3-omega method", *J. Therm. Anal. Calorim.,* vol. 145, no. 1, pp. 1-12, 2021.
[http://dx.doi.org/10.1007/s10973-020-09612-z]

[42] Wassim Jaber and Pierre-Olivier Chapuis, "Non-idealities in the 3ω method for thermal characterization in the low-and high-frequency regimes", In: *In: AIP Advances* vol. 8.4. , 2018, p. 04511.

CHAPTER 6

Structural Analysis of Feedback Field Effect Transistor and its Applications

Simranjit Singh[1,*], Ashish Raman[1], Ravi Ranjan[2] and Prabhat Singh[1]

[1] *Dr. B. R. Ambedkar National Institute of Technology, Jalandhar, Punjab, India*

[2] *Tyndall National Institute, Lee Maltings Complex Dyke Parade, Cork, Cork, Ireland*

Abstract: This book chapter provides a comprehensive overview of the Feedback Field-effect Transistor (FBFET), detailing its structure, working principle, and diverse applications. The chapter explores the unique characteristics of FBFETs, including using positive feedback phenomena to enhance current flow, leading to a high on/off current ratio and exceptional subthreshold swing. Subthreshold Swing (SS) is an important parameter in evaluating the performance of a Field-effect Transistor (FET), such as a Feedback Field-Effect Transistor (FBFET). It indicates how efficiently the transistor can switch between the off state and the on state. Essentially, SS measures the sharpness of the transition from the off current (leakage current) to the on current (drive current) in an FET. Additionally, the chapter discusses the different types of device architectures and the operational theory of the device, highlighting its potential as a memory device due to hysteresis effects. This chapter provides a valuable resource for grasping the innovative design and versatile applications of FBFET technology. The optimal steep switching property of the alternative switching technology, *i.e.*, the Feedback Field-effect Transistor (FBFET), has drawn attention. Utilizing the positive feedback phenomena, there is a significant increase in the overall quantity of holes and electrons contributing to drain current. FBFETs have a high on/off ratio of current (~10 10) and a great subthreshold swing (~00 millivolt/decade at 300 Kelvin) due to the positive feedback phenomena. Until the operation starts, the power utilization of the turn-off and on states is very small.

Keywords: Feedback field-effect transistor, Positive feedback, Steep-switching, Subthreshold swing.

INTRODUCTION

Over the span of the preceding six decades, Metal Oxide Semiconductor Field-effect Transistors (MOSFETs) have gone through size reduction to achieve enhanced performance, high density, and cost-effectiveness [1, 2]. While scaling

---

* **Corresponding author Simranjit Singh:** Dr. B. R. Ambedkar National Institute of Technology, Jalandhar, Punjab, India; E-mail: simranjits.ec.22@nitj.ac.in

Ashish Raman, Prabhat Singh, Naveen Kumar & Ravi Ranjan (Eds.)

has brought numerous advantages, it has also introduced challenges as devices have become smaller. Issues, such as increased operational temperature and power consumption, have emerged significantly [2]. Additionally, the current leakage has risen to the limit that it may exceed dynamic power utilization. Many approaches have been suggested to mitigate power usage and leakage current. Still, progress has been hindered by the MOSFET's theoretical subthreshold swing limit, which stands at around 060 mV/decade at 300 Kelvin [2]. To tackle this, numerous novel devices for steep-switching have been developed to enhance the transistor's subthreshold swing. Some novel devices can be classified as follows: the Negative Capacitance FET (NCFET) utilizes the negative effect of the capacitance of a ferroelectric layer to achieve a steep slope [3 - 7]. In contrast, the phase FET incorporates extra components to exploit the effects of phase transition substances, enabling flexible resistivity switching [8]. The Nano-electro mechanical (NEM) relay operates by mechanically disconnecting and connecting channels [9]. The impact Ionization MOS (IMOS) relies on impact ionization from a high electric field [10 - 12], while the Tunnel FET (TFET) works on the principle of Band-to-band Tunneling (BTBT) [13 - 15].

Positive feedback Field Effect Transistors (FETs) have garnered particular interest. Feedback FETs (FBFETs) exhibit high on-/off current ratios ($\sim 10^{10}$) and great subthreshold swing (~ 00 mV/decade 300 kelvin) [16, 17]. Due to these factors, a large number of researchers have shown interest in Field-band FETs (FBFETs) and proposed a variety of FBFET types [18-20], including Z^2-FETs, which means that subthreshold slope is zero and impact ionization is null [21 - 27], and Z^3-FETs zero gate, where subthreshold slope is zero and null impact ionization takes place [28 - 30]. The operating mechanism, types of structures, and transfer properties of FBFETs are all summed up in this chapter.

The enhanced characteristics of FBFETs over standard FETs can be applied to neuromorphic devices in particular or in a circuit of inverters. One of these characteristics concerns a distinct saturation area that changes with gate voltage (Vg). Power provided to a device in both on and off states can be meager, especially at threshold voltage (Vth), and power utilization is small both at the time of ongoing operating state and also in case of off-state.

FBFETs have shown promise as memory devices, particularly in the context of Dynamic Random-access Memory (DRAM) without capacitors. While the concept of 1T-DRAM was introduced around twenty years ago [31], it has largely remained within the realm of academic research. Challenges, such as low reliability, high consumption of power, and suitability issues with standard technologies of processes, like A2RAM [32 - 38] and meta-stable DRAM [32 - 34], have hindered its commercial viability. In contrast, the long-established 1-

transistor 1-capacitor DRAM has managed to address development challenges more effectively. Despite efforts, the 1T-DRAM has not yet proven itself as a feasible replacement for conventional DRAM due to the difficulty in using the scaled capacitor in the bit-cell to match the access transistor [39, 40]. Numerous research efforts have explored the substitution of conventional DRAM with a new type that incorporates FBFETs, showing significant promise [41 - 45]. Recent research has explored the use of FBFETs in DRAM design, leveraging the positive feedback mechanism to create hysteresis through the assemblage in the potential well of charge carriers. This unique approach could potentially lead to the development of a new and promising 1T-DRAM architecture that overcomes current challenges faced by traditional DRAM technologies [46-50].

Creating a new SRAM that is static random access memory using unit cells of the FBFET has been suggested. High-speed memory operation is made possible for the SRAM unit cells by a unique capability brought about by the positive feedback operation. Moreover, a straightforward device topology employing FBFETs addresses the large cell region, which has been identified as a constraint of the traditional SRAM. The FBFET-based SRAM unit cell ensures both efficient switching characteristics and low power consumption due to its small switching current [49 - 51]. This design allows for high-density memory performance with a compact area of $8F^2$. Additionally, the SRAM unit cell, utilizing positive feedback, demonstrates impressive operational performance in terms of retention duration, write speed, and read speed [52 - 62]. These findings show how promising FBFET is for applications using next-generation DRAM/SRAM memory.

Principle of Positive Feedback

Feedback involves sending back a portion or the entire output to the input. In positive feedback, initially, input rises after a single operation, and this enhanced input state continues into the next cycle. This leads to a sustained positive amplification of regenerative cycles. Similar to thyristors, FBFETs have two potential barriers positioned near the channel adjoining the drain and source, as expressed in Fig. (**1**), obstructing the stream of electrons and holes. Conversely, representing the carriers' lowest energy state, a potential wall is simultaneously created along with the potential barrier. Positive feedback occurs as the potential barrier decreases with the supply of a gate voltage. This reduction in the potential barrier enables charge carriers to move to the drain from the source. Electrons or holes may become caught in the potential wall within the region of the channel [63 - 65], impacting the energy band due to the charges generated by carriers confined within the potential wall.

When electrons uplift the band of energy in the nearby region, the barrier of the potential wall on the other side decreases. On the contrary, holes elevate the energy band in an opposing manner. The existence of trapped holes close to the potential wall can cause the barrier to break down. The reduced barrier allows for the continuous flow of additional carriers to sustain the ongoing process, as the barriers have been decreased in height.

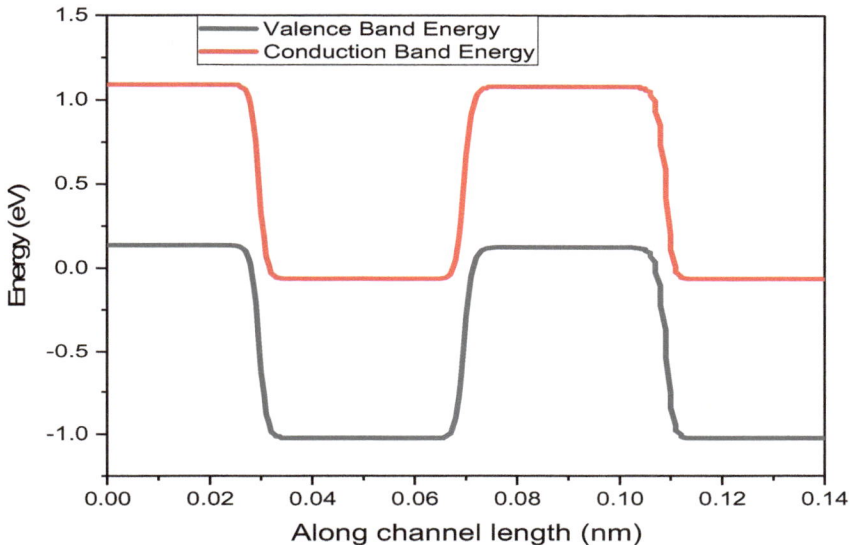

Fig. (1). Structure of energy band of a FBFET.

STRUCTURE OF FBFET

In 2008, the concept of FBFETs was introduced, presenting a new architecture akin to a P-I-N diode operating under the condition of forward bias, as illustrated in Fig. (**2**). This new FET design features two advantageous potential barriers with positive feedback. Gate-sidewall spacers on the FBFET serve to capture charges around the channel area near the source and drain. P-I-N diodes can function as FBFETs due to their ability to navigate barriers hindering the stream of electrons and holes effectively.

FBFETs offer a solution to the drawbacks associated with traditional MOSFETs. These devices can achieve a peak on/off ratio of current of around 10^7 and exhibit sharp switching characteristics (subthreshold swing approximately 2 mV/decade at 300 kelvin). Depending on the gate voltage applied, FBFETs can operate as either n-type or p-type. However, challenges arise due to trapped charge issues. For example, instability can occur when the spacer is stimulated during device operation. It is essential to streamline the process of storing carriers in the spacer linked to the region of source and region of drain.

Fig. (2). A cross-sectional representation of P-I-N type feedback field-effect transistor.

Several research studies have focused on improving the electrical characteristics of FBFETs through structural modifications [23 - 32, 37 - 42]. A typical approach involves redefining the formation of potential barriers caused by charges trapped in spacers. Two main methods have been under scrutiny: initially, barriers of potential produced by the intersections of silicon regions with inconsistent doping levels can replace these trapped charges. Depending on the gate's place concerning the source or drain, the FET can function as either n-type or p-type. As a result, a device design modification was implemented, integrating dual gates in proximity to the source and drain zones, enabling them to serve dual functions as both n-type and p-type transistors concurrently [63 - 65]. An alternative method includes utilizing energy that is electrical as a potential barrier, accomplished by incorporating a second gate near the source (S) and drain (D) sites. This setup enables the device to function as either an n-type or p-type FET contingent on the gate voltage. Nonetheless, a limitation of this method is the energy usage associated with the gate voltage required to establish the barrier. The following are presented the types of structures of feedback field effect transistors [66-81].

Recessed SOI FBFET

In the depiction presented in Fig. (3), the configuration of an n-type Recessed Silicon on Insulator (SOI) FBFET [81] is showcased, with the relevant device parameters summarized in Table 1. For the n-type Recessed SOI FBFET design, a heavily doped p-n-p-n structure is examined, contrasting the n-p-n-p structure with heavy doping for the p-type Recessed SOI FBFET. The gate oxide composition, depicted in Fig. (3), comprises a HfO_2/SiO_2 stack. While FBFET devices notably excel in OFF current performance compared to traditional counterparts, their ON current exhibits a modest decline. To address this, implementing the recessed source and drain (Re-S/D) technique diminishes source and drain resistance by embedding these regions further into the layer of Buried Oxide (BOX).

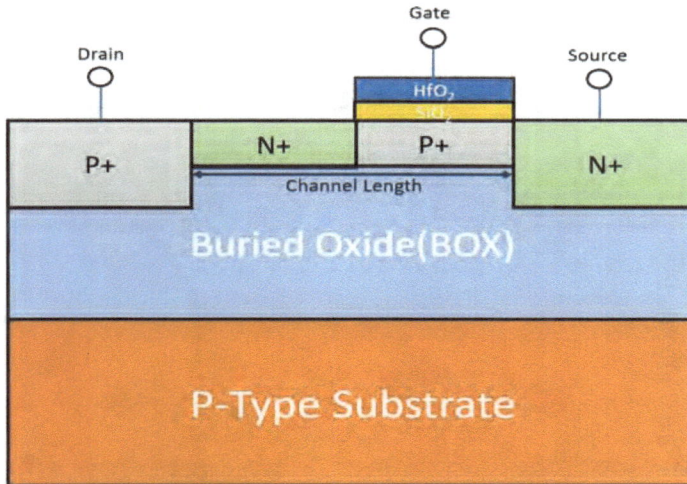

Fig. (3). n-type recessed SOI FBFET structure.

Table 1. Optimized parameters for recessed SOI FBFET.

Parameters	Values	Description
L_{ch}	80 nm	Channel length
t_{ox}	3 nm	The thickness of gate oxide (HfO_2)
$t_{Re-S/D}$	50 nm	Thickness of recessed source/drain
ϕ	4.6 eV	Metal gate work function
$N_{D/S}$	1×10^{20} cm^{-3}	Impurity concentration in channel
N_{ch}	1×10^{20} cm^{-3}	Impurity concentration source/drain

The graphical representation of I_d-V_{gs} characteristics of both recessed S/D and recessed type SOI FBFETs, encompassing variations in n- and p-type configurations, is detailed in Fig. (4). Manipulating V_{gs} through forward and reverse sweeps enables the observation of hysteresis behavior. A rise in the recessed S/D thickness results in a decrease in the memory window of recessed SOI FBFET when collated with standard SOI FBFETs. This decrease in the threshold voltage during the reverse sweep for the recessed S/D configuration with a greater thickness is attributed to a reduced gathering of charge carriers within the potential well, which is a departure from the characteristics seen in conventional SOI FBFETs.

The fine-tuning of the window of memory using Re S/D thickness underscores the applicability of the recessed SOI FBFET architecture for the design of logic and memory circuits [82-86]. Furthermore, an increase in recessed S/D thickness correlates with a rise in drain current, and all Id-Vgs characteristics showcase a subthreshold swing approaching zero, indicative of enhanced operational efficiency.

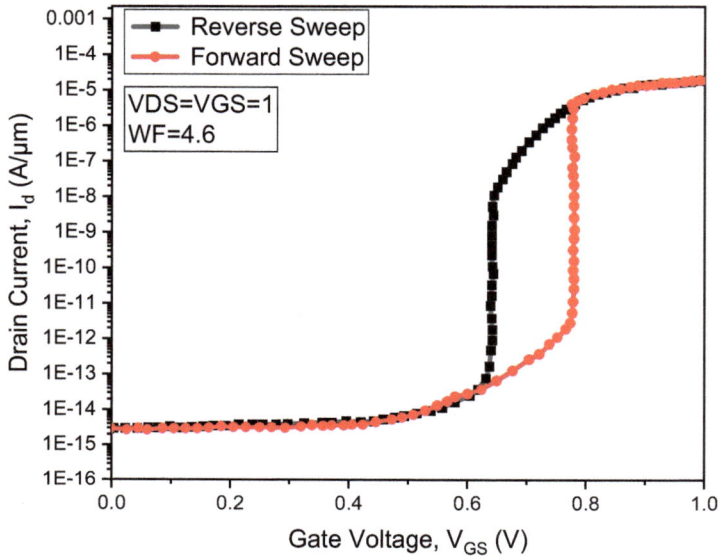

Fig. (4). I_d-V_{GS} graph of recessed SOI FBFET structure.

Vertical FBFET

Fig. (**5**) depicts the conceptual vertical FBFET device structure, consisting of four key regions: source (S), channel 1 (Ch1), channel 2 (Ch2), and drain (D) [82]. The parameters used are as outlined in Table **2**. The device layering includes various components, such as a 200-nanometer substrate layer, a 200-nanometer layer of Buried Oxide (BOX), a p+ type doping drain layer, an n+ channel two regions, p+ channel one region doping, and an n+ type source region (PNPN). Each of these regions is doped with a doping concentration of 1×10^{20} cm^{-3}. To optimize performance, an ohmic contact extends towards the p+ silicon drain region through a downward path, effectively reducing the resistance of drain contact. Moreover, a standard double gate is symmetrically positioned on channel one above the 1 nm thick gate oxide layer, completing the device's layout.

For the comprehensive evaluation of the vertical FBFET's functionality and operational efficacy, the widely used Sentaurus TCAD tool was engaged. A hydrodynamic transport model was meticulously applied to ensure precise simulation output [87, 88]. In order to examine the consequence of charges trapped at the interface, a uniform fixed charge density over the interface was taken into account. Based on earlier studies, the interface charge density at the Si/Al_2O_3 interface ranged from -1×10^{12} cm^{-2} to 1×10^{12} cm^{-2}. Transient simulation analysis was conducted to extract DC characteristics for effective handling of convergence challenges. Fig. (**6**) displays the transfer characteristics of the N-type vertical FBFET at $V_{DS}=1V$. The selection of a metal work function of 4.6eV was instrumental in setting the threshold voltage at 0.46V. Calculations indicated an ON-state drive current of 3.06×10^{-5} A/µm, accompanied by an outstanding ON/OFF ratio of current of 10^{11}. The mean Subthreshold Swing (SS) for the vertical FBFET of N-type stood at 8.77 mV/dec. Notably, a minimum SS point of 0.03mV/dec was observed during the vertical FBFET assessments.

Table 2. Optimized parameters for vertical FBFET.

Parameters	Values	Description
L_{ch}	70 nm	Channel length
t_{ox}	2 nm	The width of gate oxide (HfO_2)
L_{BOX}	210 nm	Length of BOX
$L_{S/D}$	40 nm	Length of source/drain
L_{Sub}	200 nm	Substrate length
ϕ	4.6	Metal gate work function
$N_{D/S}$	1×10^{20} cm^{-3}	Impurity concentration in the channel
N_{ch}	1×10^{20} cm^{-3}	Impurity concentration source/drain

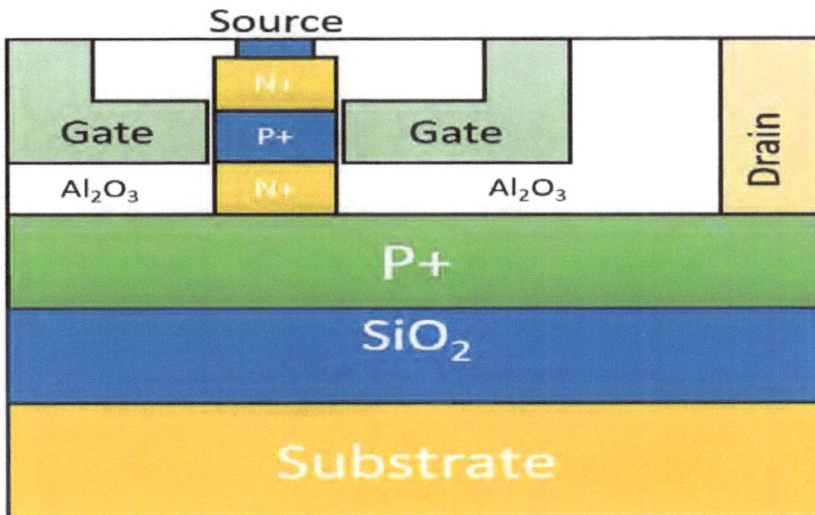

Fig. (5). Schematic structure of vertical FBFET.

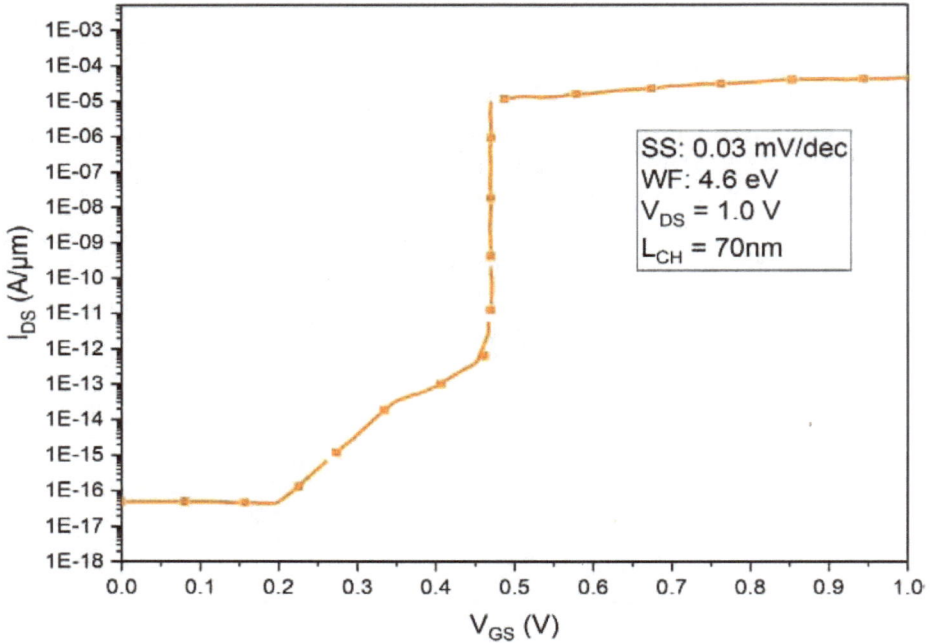

Fig. (6). I_d-V_{GS} graph of vertical FBFET.

Si-SiGe Heterostructure FBFET

The top view in Fig. (**7**) showcases an N-type HDGFBFET, comprising extremely doped N+-P+-N+-P+ regions: the drain sections, channel-1 region, channel-2 region, and the source [83]. These regions play a crucial role in establishing an S-shaped diagram of the energy band within the device and facilitating a positive feedback mechanism under specific bias settings. The use of Si material in the source and channel-1 areas enhances the overall device performance, while $Si_{1-x}Ge_x$ material is integrated into channel-2 and drain regions for optimization. Positioned near the source end on channel one, the gate electrode layer operates the n-type FBFET device. The gate electrodes, composed of polysilicon with a broadness of 1 nanometer and a Work Function (WF) of 4.65eV, are instrumental in device's functionality.

Table **3** represents the comparative analysis of the various FBFETs. Fig. (**8**) showcases a notable I_{ON} current of 2.55×10^{-4} A/µm, an exceptional I_{ON}/I_{OFF} of three times 10^{11}, and a minimal SS of approximately 0.8 mV/decade. The fine-tuning of various parameters, such as the work function of gate metal, doping concentrations in distinct regions, the molar fraction of Ge, and channel length, contributes to achieving a reduced V_{TH} of 0.42V. A commendable SS of 0.8

mV/decade at 300K is effectively attained. Similar to the SOI FBFET, the Si-SiGe heterostructure device operates across equilibrium (V_{DS}=0V, V_{GS}=0V), off (V_{GS}=0V V_{DS}=1.0V), and on (V_{GS}=1.0 V, V_{DS}=1.0V) states. The integration of $Si_{1-x}Ge_x$ in the Ch-2 and drain regions is deliberated, emphasizing a reduction in the bandgap (Eg) across all scenarios. This adjustment results in heightened hole involvement in the feedback mechanism by encountering fewer potential barriers.

Table 3. Comparison of parameters of feedback FET devices.

Structure [Ref.]	SS (mV/dec)	I_{ON}/I_{OFF}	Memory Window (V)	Year
SOI,gatesidewallSpacers [14]	~2	10^7	<0.1	2008
SiNWFET P+N+I-Si-N+ [21]	0.005	10^{11}	2.02	2016
Z^2 FET Dual gate [22]	~1	10^{10}	NA	2016
Nitride Spacers single gate [25]	3.79	10^9	NA	2016
FinFET P+N+P+N+ [24]	<1	10^9	<0.1	2018
SiNWFET P+I-Si-N+ [33]	1.36	10^5	0.2	2019
Stacked SOI FBFET [15]	<1	10^7	NA	2020
Vertical FBFET [16]	0.03	10^{11}	1.1	2023
Si-SiGe hetero-structure FBFET	0.8	2.3×10^{11}	4.99	2023

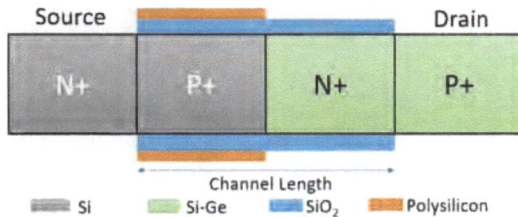

Fig. (7). Schematic structure of Si-SiGe heterostructure FBFET.

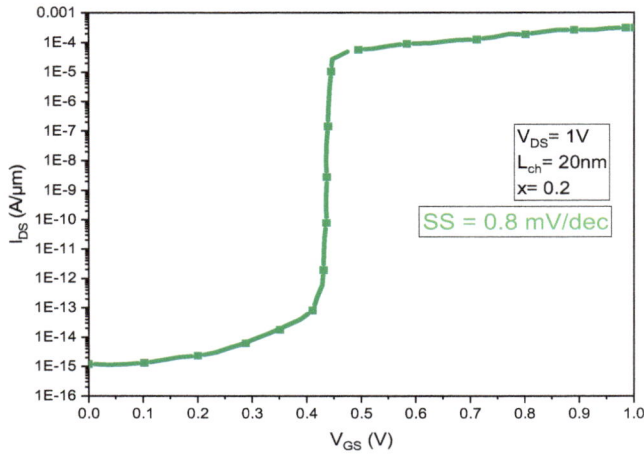

Fig. (8). Schematic structure of Si-SiGe heterostructure FBFET.

Characteristics of FBFET

At 300 K, the conceptual restriction of SS is 60 mV/decade [2]. With innovative gadget designs and materials, it is possible to exceed the theoretical limit of SS. To get around this restriction, several steep-switching devices have been investigated [3 - 16]. Among these, gadgets with remarkable capabilities that rely on the mechanism of positive feedback have been proposed and shown [19 - 30]. FBFETs exhibit high on-/off current ratios ($\sim 10^{10}$) and great subthreshold slope (~ 00 mV/decade at 300 K) [16, 17]. As temperatures or VDS values increase, the on-/off current ratio also increases. The device demonstrates characteristics that are more compatible with FDSOI structures when the channel depth decreases. An escalation in the dielectric constant of the insulator layer brings an increment in the threshold voltage (V_{TH}), while an increment in the gate metal work function results in a decrease in the V_{TH}.

The length of the channel and V_{TH} have restrictions determined by the voltage at the gate terminal and gate length. By situating the gate near the source/drain, N-/p-type FBFETs can be formed. As the length of the channel exceeds five hundred nanometers, voltage V_{on} increases, current I_{on} diminishes, and rapid switching diminishes due to the rapid replenishment of charge carriers in minority. Additionally, as the length of the channel expands, hysteresis disappears [70].

Numerous modeling works have revealed that the FBFET channel's performance improves with its length. Reducing the channel length enhances the switching properties, and unlike current MOSFETs, the exceptional features of the FBFET are reflected in the advantages of scaling a device [75]. On the other hand, the SS value and on/off ratio of current are low in the absence of saturation and can be obtained as the channel length grows. FBFETs do not have the best switching properties for channel lengths over 100 nm. However, at 300 K, no SS value exceeds 60 mV/decade.

Nevertheless, contrary to the better properties achieved on a ~10 nm scale, current research has demonstrated that the positive feedback mechanism could be more functional. The cause is related to the function of potential barriers, which are essential to FBFET operation. The potential barrier expands with respect to the extent of the channel. When the length of the channel exceeds 40 nanometers, more energy and time are needed to break through the potential barrier. On the other hand, short channel effects are caused when the potential barrier gets too narrow at more minor scales and is predicted to obstruct the carrier flow. It is vital to reflect on the solutions to overcome the short-channel effects, like in traditional MOSFETs, to control this restriction [65].

FBFET APPLICATIONS

Logic Device

The pullup and pulldown mechanisms must operate effectively for an inverter design to work well. Since the device needs to be utilized with a forward biasing, the drain to source V_{DS} should have a larger magnitude than the FBFET's intrinsic potential [71]. This can be optimized by combining several bias combinations, but it may require much work. However, an inverter featuring a combined FBFET has the following benefits: FBFETs have a big current on/off ratio of currents, steep switching, and a distinct region of saturation, particularly beyond V_{TH}. Stated differently, the gate voltage (V_g) determines how much current is given to the operational and non-operational regions, allowing for minimal utilization of power during operation and off-state. Digital integrated circuits make use of the FBFET's steep-switching feature.

An illustration of this concept can be seen in an inverter where an FBFET regulates the switching process. In a standard circuit of inverter, as the input voltage rises, the capacitor charge depletes gradually over time [71, 72]. Conversely, if the FBFET is tasked with supervising the input supply voltage for the inverter's operation, the voltage reserved in the capacitor experiences rapid fluctuations at the FBFET's V_{TH}. Additionally, the voltage supply demonstrates consistent on/off states and enters a saturation region beyond V_{TH}.

Memory Cells

The traditional 6-transistor static RAM (6T-SRAM) delivers peak execution but at the cost of lower density [73 - 78], whereas Dynamic Random-access Memory

(DRAM) offers higher density, but with relatively lower performance. FBFETs present a solution for the next generation of SRAMs and DRAMs, addressing each individual's technical limitations. Extensive research has been conducted to showcase the competitive features within the DRAM domain, with a particular focus on 1T-DRAM, which is capacitor-less and was initially proposed around two decades ago [31]. This 1T-DRAM utilizes side issues, such as hysteresis, a phenomenon viewed by some studies as detrimental, and thus, efforts have been made to eliminate it. By storing parasitic carriers in the body to elevate the potential, leading to a lowered V_{th} and enabling a significant flow of current, the "1" state is achieved. Conversely, removing carriers from the body results in reduced current flow in the "0" state.

Currently, the 1T-DRAM as a possible next-generation option has remained stagnant. Despite the significant structural benefits of 1T-DRAM, its ability to fully enter the market must be proven before it can entirely replace conventional DRAM. Unfortunately, for a number of the suggested models, fatal flaws, including poor dependability, excessive consumption of power, or non-compatibility with standard process technologies, have surfaced [35 - 40]. Most notably, conventional DRAM is still thriving and faces no significant issues. The traditional 1C-1T DRAM memory cell is currently approaching its scalability limit. The inability of the cell capacitor to scale with the transistor is one of its main issues [39, 40].

Additionally, a novel SRAM cell design utilizing FBFETs has been put forward. These SRAM cells enable high-speed memory operations due to the sharp-switching characteristic facilitated by positive feedback operation. The issue of high cell area is a drawback in traditional SRAM designs. The compact $8F^2$ area of the cell allows for the development of highly dense integrated memory cells, and the reduced switching current significantly enhances the switching characteristics, facilitating low-power memory functionality. The impressive operational performance of SRAM is highlighted by the SRAM bit-cell, showcasing fast read and write speeds, along with effective retention periods affected by charges trapped in the channel area near the source and drain region [49]. These outcomes emphasize the significant capability of FBFET in DRAM and SRAM for future memory applications.

CONCLUSION

This chapter has delved into the exploration of various device structures stemming from the initial proposal of the FBFET. It has elucidated the positive feedback mechanism integral to their operation. Thanks to this phenomenon, FBFETs exhibit remarkable subthreshold swing (~0 millivolt/decade at 300 K) and peak on/off current ratios ($\sim 10^{10}$), positioning them as promising candidates for memory cells of the next generation with hysteresis. Notably, FBFETs find application in 1T-1C DRAM to create capacitor-less DRAM and in 6T-SRAM to form 2T-SRAM, offering solutions that surpass conventional device limitations. Additionally, their integration into neuromorphic circuits showcases significantly enhanced performance, effectively addressing power consumption and spatial requirements. These findings substantiate the potential for replacing existing devices with next-generation alternatives characterized by ultra-low power consumption, high performance, and increased density.

KEY POINTS

1. The chapter facilitates an understanding of the importance of the Feedback Field-effect Transistor (FBFET) in low-power applications.
2. It analyzes the different types of structures of feedback field effect transistors.
3. It further describes the working principle of the feedback field effect transistor.
4. It also provides an insight into the characteristics of FBFETs and their various low-power applications.

ACKNOWLEDGEMENTS

The authors would like to thank the Department of Electronics and Communication Engineering, Dr. B.R Ambedkar National Institute of Technology, Jalandhar, Punjab, for providing valuable support to carry out this study.

REFERENCES

[1] P.K. Chatterjee, and R.R. Doering, "The future of microelectronics", *Proc. IEEE,* vol. 86, no. 1, pp. 176-183, 1998.
[http://dx.doi.org/10.1109/5.658769]

[2] T. Sakurai, "Perspectives of low-power VLSI's", *IEICE Trans. Electron.,* vol. 87, no. 4, pp. 429-436, 2004.

[3] A.I. Khan, K. Chatterjee, B. Wang, S. Drapcho, L. You, C. Serrao, S.R. Bakaul, R. Ramesh, and S. Salahuddin, "Negative capacitance in a ferroelectric capacitor", *Nat. Mater.,* vol. 14, no. 2, pp. 182-186, 2015.
[http://dx.doi.org/10.1038/nmat4148] [PMID: 25502099]

[4] E. Ko, J.W. Lee, and C. Shin, "Negative capacitance FinFET with sub-20-mV/decade subthreshold slope and minimal hysteresis of 0.48 V", *IEEE Electron Device Lett.,* vol. 38, no. 4, pp. 418-421, 2017.
[http://dx.doi.org/10.1109/LED.2017.2672967]

[5] E. Ko, J. Shin, and C. Shin, "Steep switching devices for low power applications: negative differential capacitance/resistance field effect transistors", *Nano Converg.,* vol. 5, no. 1, p. 2, 2018.
[http://dx.doi.org/10.1186/s40580-018-0135-4] [PMID: 29399434]

[6] J. Jo, W.Y. Choi, J.D. Park, J.W. Shim, H.Y. Yu, and C. Shin, "Negative capacitance in organic/ferroelectric capacitor to implement steep switching MOS devices", *Nano Lett.,* vol. 15, no. 7,

pp. 4553-4556, 2015.
[http://dx.doi.org/10.1021/acs.nanolett.5b01130] [PMID: 26103511]

[7]   E. Ko, and C. Shin, "Effective drive current in steep slope FinFET (vs. conventional FinFET)", *Appl. Phys. Lett.,* vol. 111, no. 15, p. 152105, 2017.
[http://dx.doi.org/10.1063/1.4998508]

[8]   N. Shukla, A.V. Thathachary, A. Agrawal, H. Paik, A. Aziz, D.G. Schlom, S.K. Gupta, R. Engel-Herbert, and S. Datta, "A steep-slope transistor based on abrupt electronic phase transition", *Nat. Commun.,* vol. 6, no. 1, p. 7812, 2015.
[http://dx.doi.org/10.1038/ncomms8812] [PMID: 26249212]

[9]   T-J. Liu, D. Markovic, V. Stojanovic, and E. Alon, "The relay reborn", *IEEE Spectr.,* vol. 49, no. 4, pp. 40-43, 2012.
[http://dx.doi.org/10.1109/MSPEC.2012.6172808]

[10]   R.H. Dennard, F.H. Gaensslen, H-N. Yu, V.L. Rideout, E. Bassous, and A.R. LeBlanc, "Design of ion-implanted MOSFET's with very small physical dimensions", *IEEE J. Solid-State Circuits,* vol. 9, no. 5, pp. 256-268, 1974.
[http://dx.doi.org/10.1109/JSSC.1974.1050511]

[11]   K. Gopalakrishnan, P.B. Griffin, and J.D. Plummer, "Impact ionization MOS (I-MOS)-Part I: device and circuit simulations", *IEEE Trans. Electron Dev.,* vol. 52, no. 1, pp. 69-76, 2005.
[http://dx.doi.org/10.1109/TED.2004.841344]

[12]   K. Gopalakrishnan, P.B. Griffin, and J.D. Plummer, "I-MOS: A novel semiconductor device with a subthreshold slope lower than kT/q", *International Electron Devices Meeting,* pp. 289-292, 2002.
[http://dx.doi.org/10.1109/IEDM.2002.1175835]

[13]   Byung-Gook Park, "Tunneling field-effect transistors (TFETs) with subthreshold swing (SS) less than 60 mV/dec", *IEEE Electron Device Lett.,* vol. 28, no. 8, pp. 743-745, 2007.
[http://dx.doi.org/10.1109/LED.2007.901273]

[14]   K. Choe, and C. Shin, "Adjusting the operating voltage of an nanoelectromechanical relay using negative capacitance", *IEEE Trans. Electron Dev.,* vol. 64, no. 12, pp. 5270-5273, 2017.
[http://dx.doi.org/10.1109/TED.2017.2756676]

[15]   N. Cui, R. Liang, J. Wang, W. Zhou, and J. Xu, "A PNPN tunnel field-effect transistor with high- k gate and low- k fringe dielectrics", *J. Semicond.,* vol. 33, no. 8, p. 084004, 2012.
[http://dx.doi.org/10.1088/1674-4926/33/8/084004]

[16]   J. Wan, C.L. Royer, A. Zaslavsky, and S. Cristoloveanu, "A systematic study of the sharp-switching Z2-FET device: From mechanism to modeling and compact memory applications", *Solid-State Electron.,* vol. 90, pp. 2-11, 2013.
[http://dx.doi.org/10.1016/j.sse.2013.02.060]

[17]   H. El Dirani, P. Fonteneau, Y. Solaro, C-A. Legrand, D. Marin-Cudraz, P. Ferrari, and S. Cristoloveanu, "Sharp-switching band-modulation back-gated devices in advanced FDSOI technology", *Solid-State Electron.,* vol. 128, pp. 180-186, 2017.
[http://dx.doi.org/10.1016/j.sse.2016.10.008]

[18]   S.M. Joe, H-J. Kang, N. Choi, M. Kang, B-G. Park, and J-H. Lee, "Diode-type NAND flash memory cell string having super-steep switching slope based on positive feedback", *IEEE Trans. Electron Dev.,* vol. 63, no. 4, pp. 1533-1538, 2016.
[http://dx.doi.org/10.1109/TED.2016.2533019]

[19]   K.B. Choi, S.Y. Woo, W.M. Kang, S. Lee, C.H. Kim, J.H. Bae, S. Lim, and J.H. Lee, "A split-gate positive feedback device with an integrate-and-fire capability for a high-density low-power neuron circuit", *Front. Neurosci.,* vol. 12, p. 704, 2018.
[http://dx.doi.org/10.3389/fnins.2018.00704] [PMID: 30356702]

[20]   Choi, N. Y., Joe, S. M., Park, B. G., & Lee, J. H. "Design consideration of diode-type NAND flash

memory cell string having super-steep switching slope", *IEEE Journal of the Electron Devices Society,* vol. 4, no. 5, pp. 328-334, 2016.

[21] S. Cristoloveanu, J. Wan, C. Le Royer, and A. Zaslavsky, "(Invited) Innovative Sharp Switching Devices", *ECS Trans.,* vol. 54, no. 1, pp. 65-75, 2013.
[http://dx.doi.org/10.1149/05401.0065ecst]

[22] J. Wan, S. Cristoloveanu, C. Le Royer, and A. Zaslavsky, "A feedback silicon-on-insulator steep switching device with gate-controlled carrier injection", *Solid-State Electron.,* vol. 76, pp. 109-111, 2012.
[http://dx.doi.org/10.1016/j.sse.2012.05.061]

[23] S. Cristoloveanu, J. Wan, C. Le Royer, and A. Zaslavsky, "Sharp switching SOI devices", *ECS Trans.,* vol. 53, no. 5, pp. 3-13, 2013.
[http://dx.doi.org/10.1149/05305.0003ecst]

[24] H. El Dirani, "Sharp-switching Z2-FET device in 14 nm FDSOI technology", *in 2015 45th European Solid State Device Research Conference (ESSDERC),* pp. 364-367, 2015.

[25] H.E. Dirani, Y. Solaro, P. Fonteneau, P. Ferrari, and S. Cristoloveanu, "Properties and mechanisms of Z2-FET at variable temperature", *Solid-State Electron.,* vol. 115, pp. 201-206, 2016.
[http://dx.doi.org/10.1016/j.sse.2015.08.015]

[26] S. Navarro, C. Marquez, K.H. Lee, C. Navarro, M. Parihar, H. Park, P. Galy, M. Bawedin, Y.T. Kim, S. Cristoloveanu, and F. Gamiz, "Investigation of thin gate-stack Z2-FET devices as capacitor-less memory cells", *Solid-State Electron.,* vol. 159, pp. 12-18, 2019.
[http://dx.doi.org/10.1016/j.sse.2019.03.040]

[27] C. Marquez, "Temperature and gate leakage influence on the Z2-FET memory operation", *in ESSDERC 2019-49th European Solid-State Device Research Conference (ESSDERC),* pp. 154-157, 2019.
[http://dx.doi.org/10.1109/ESSDERC.2019.8901803]

[28] H. El Dirani, "A sharp-switching gateless device (Z3-FET) in advanced FDSOI technology", *2016 Joint International EUROSOI Workshop and International Conference on Ultimate Integration on Silicon (EUROSOI-ULIS),* pp. 204-207, 2016.
[http://dx.doi.org/10.1109/ULIS.2016.7440070]

[29] Y. Solaro, P. Fonteneau, C.A. Legrand, C. Fenouillet-Beranger, P. Ferrari, and S. Cristoloveanu, "A sharp-switching device with free surface and buried gates based on band modulation and feedback mechanisms", *Solid-State Electron.,* vol. 116, pp. 8-11, 2016.
[http://dx.doi.org/10.1016/j.sse.2015.10.010]

[30] H. El Dirani, "Novel FDSOI band-modulation device: Z 2-FET with dual ground planes", *in 2016 46th European Solid-State Device Research Conference (ESSDERC). IEEE,* 2016.
[http://dx.doi.org/10.1109/ESSDERC.2016.7599623]

[31] H-J. Wann, and C. Hu, "A capacitorless DRAM cell on SOI substrate", *Proceedings of IEEE International Electron Devices Meeting,* IEEE, pp. 169-172, 1993.

[32] J. Lacord, "MSDRAM, A2RAM and Z 2-FET performance benchmark for 1T-DRAM applications", *2018 International Conference on Simulation of Semiconductor Processes and Devices (SISPAD),* IEEE, 2018.
[http://dx.doi.org/10.1109/SISPAD.2018.8551674]

[33] M. Bawedin, S. Cristoloveanu, and D. Flandre, "A capacitorless 1T-DRAM on SOI based on dynamic coupling and double-gate operation", *IEEE Electron Device Lett.,* vol. 29, no. 7, pp. 795-798, 2008.
[http://dx.doi.org/10.1109/LED.2008.2000601]

[34] A. Hubert, M. Bawedin, S. Cristoloveanu, and T. Ernst, "Dimensional effects and scalability of Meta-Stable Dip (MSD) memory effect for 1T-DRAM SOI MOSFETs", *Solid-State Electron.,* vol. 53, no. 12, pp. 1280-1286, 2009.

[http://dx.doi.org/10.1016/j.sse.2009.09.020]

[35]   N. Rodriguez, S. Cristoloveanu, and F. Gamiz, "Novel capacitorless 1T-DRAM cell for 22-nm node compatible with bulk and SOI substrates", *IEEE Trans. Electron Dev.,* vol. 58, no. 8, pp. 2371-2377, 2011.
[http://dx.doi.org/10.1109/TED.2011.2147788]

[36]   N. Rodriguez, C. Navarro, F. Gamiz, F. Andrieu, O. Faynot, and S. Cristoloveanu, "Experimental demonstration of capacitorless A2RAM cells on silicon-on-insulator", *IEEE Electron Device Lett.,* vol. 33, no. 12, pp. 1717-1719, 2012.
[http://dx.doi.org/10.1109/LED.2012.2221074]

[37]   K.W. Song, J-Y. Kim, J-M. Yoon, S. Kim, H. Kim, H-W. Chung, H. Kim, K. Kim, H-W. Park, H.C. Kang, N-K. Tak, D. Park, W-S. Kim, Y-T. Lee, Y.C. Oh, G-Y. Jin, J. Yoo, D. Park, K. Oh, C. Kim, and Y-H. Jun, "A 31 ns Random Cycle VCAT-Based 4F^2 DRAM With Manufacturability and Enhanced Cell Efficiency", *IEEE J. Solid-State Circuits,* vol. 45, no. 4, pp. 880-888, 2010.
[http://dx.doi.org/10.1109/JSSC.2010.2040229]

[38]   W. Mueller, "Challenges for the DRAM cell scaling to 40nm", *IEEE International Electron Devices Meeting,* pp. 727-730, 2005.
[http://dx.doi.org/10.1109/IEDM.2005.1609344]

[39]   S. Eminente, S. Cristoloveanu, R. Clerc, A. Ohata, and G. Ghibaudo, "Ultra-thin fully-depleted SOI MOSFETs: Special charge properties and coupling effects", *Solid-State Electron.,* vol. 51, no. 2, pp. 239-244, 2007.
[http://dx.doi.org/10.1016/j.sse.2007.01.016]

[40]   S. Cristoloveanu, S. Athanasiou, M. Bawedin, and P. Galy, "Evidence of supercoupling effect in ultrathin silicon layers using a four-gate MOSFET", *IEEE Electron Device Lett.,* vol. 38, no. 2, pp. 157-159, 2017.
[http://dx.doi.org/10.1109/LED.2016.2637563]

[41]   J. Wan, "Z 2-FET used as 1-transistor high-speed DRAM", *in 2012 Proceedings of the European Solid-State Device Research Conference (ESSDERC). IEEE,* 2012.
[http://dx.doi.org/10.1109/ESSDERC.2012.6343338]

[42]   J. Wan, C. Le Royer, A. Zaslavsky, and S. Cristoloveanu, "Progress in Z2-FET 1T-DRAM: Retention time, writing modes, selective array operation, and dual bit storage", *Solid-State Electron.,* vol. 84, pp. 147-154, 2013.
[http://dx.doi.org/10.1016/j.sse.2013.02.010]

[43]   S. Cristoloveanu, K.H. Lee, M.S. Parihar, H. El Dirani, J. Lacord, S. Martinie, C. Le Royer, J-C. Barbe, X. Mescot, P. Fonteneau, P. Galy, F. Gamiz, C. Navarro, B. Cheng, M. Duan, F. Adamu-Lema, A. Asenov, Y. Taur, Y. Xu, Y-T. Kim, J. Wan, and M. Bawedin, "A review of the Z 2 -FET 1T-DRAM memory: Operation mechanisms and key parameters", *Solid-State Electron.,* vol. 143, pp. 10-19, 2018.
[http://dx.doi.org/10.1016/j.sse.2017.11.012]

[44]   C. Navarro, C. Marquez, S. Navarro, C. Lozano, S. Kwon, Y-T. Kim, and F. Gamiz, "Simulation Perspectives of Sub-1V Single-Supply Z 2 -FET 1T-DRAM Cells for Low-Power", *IEEE Access,* vol. 7, pp. 40279-40284, 2019.
[http://dx.doi.org/10.1109/ACCESS.2019.2907151]

[45]   H. Kang, J. Cho, Y. Kim, D. Lim, S. Woo, K. Cho, and S. Kim, "Nonvolatile and volatile memory characteristics of a silicon nanowire feedback field-effect transistor with a nitride charge-storage layer", *IEEE Trans. Electron Dev.,* vol. 66, no. 8, pp. 3342-3348, 2019.
[http://dx.doi.org/10.1109/TED.2019.2924961]

[46]   J. Wan, A. Zaslavsky, C. Le Royer, and S. Cristoloveanu, "Novel bipolar-enhanced tunneling FET with simulated high on-current", *IEEE Electron Device Lett.,* vol. 34, no. 1, pp. 24-26, 2013.
[http://dx.doi.org/10.1109/LED.2012.2228159]

[47]   J. Wan, ""Z 2-FET: A zero-slope switching device with gate-controlled hysteresis," in Proceedings of Technical Program of 2012 VLSI Technology", *System and Application. IEEE,* no. Jun, 2012.
[http://dx.doi.org/10.1109/VLSI-TSA.2012.6210667]

[48]   K.H. Lee, "Sharp switching, hysteresis-free characteristics of Z 2-FET for fast logic applications", *in 2018 48th European Solid-State Device Research Conference (ESSDERC). IEEE,* 2018.
[http://dx.doi.org/10.1109/ESSDERC.2018.8486915]

[49]   J. Cho, D. Lim, S. Woo, K. Cho, and S. Kim, "Static random access memory characteristics of single-gated feedback field-effect transistors", *IEEE Trans. Electron Dev.,* vol. 66, no. 1, pp. 413-419, 2019.
[http://dx.doi.org/10.1109/TED.2018.2881965]

[50]   M. Duan, C. Navarro, B. Cheng, F. Adamu-Lema, X. Wang, V.P. Georgiev, F. Gamiz, C. Millar, and A. Asenov, "Thorough understanding of retention time of Z2FET memory operation", *IEEE Trans. Electron Dev.,* vol. 66, no. 1, pp. 383-388, 2019.
[http://dx.doi.org/10.1109/TED.2018.2877977]

[51]   M.W. Kwon, K. Park, M-H. Baek, J. Lee, and B-G. Park, "A low-energy high-density capacitor-less I&F neuron circuit using feedback FET co-integrated with CMOS", *IEEE J. Electron Devices Soc.,* vol. 7, pp. 1080-1084, 2019.
[http://dx.doi.org/10.1109/JEDS.2019.2941917]

[52]   M.-H. Oh, "A new device characteristic model generation by machine learning", *in 2019 Electron Devices Technology and Manufacturing Conference (EDTM). IEEE,* 2019.
[http://dx.doi.org/10.1109/EDTM.2019.8731336]

[53]   Padilla, "Feedback FET: A novel transistor exhibiting steep switching behaviour at low bias voltages", *2008 IEEE International Electron Devices Meeting,* IEEE, pp. 1-4, 2008.
[http://dx.doi.org/10.1109/IEDM.2008.4796643]

[54]   W. Yeung, "Programming characteristics of the steep turn-on/off feedback FET (FBFET)", *in 2009 Symposium on VLSI Technology. IEEE,* pp. 36-37, 2009.

[55]   W. Yeung, *Steep on/off transistors for future low-power electronics.* University of California: Berkeley, 2014.

[56]   H. El Dirani, Y. Solaro, P. Fonteneau, C-A. Legrand, D. Marin-Cudraz, D. Golanski, P. Ferrari, and S. Cristoloveanu, "A band-modulation device in advanced FDSOI technology: Sharp switching characteristics", *Solid-State Electron.,* vol. 125, pp. 103-110, 2016.
[http://dx.doi.org/10.1016/j.sse.2016.07.018]

[57]   M.S. Parihar, K.H. Lee, H.J. Park, J. Lacord, S. Martinie, J-C. Barbé, Y. Xu, H. El Dirani, Y. Taur, S. Cristoloveanu, and M. Bawedin, "Insight into carrier lifetime impact on band-modulation devices", *Solid-State Electron.,* vol. 143, pp. 41-48, 2018.
[http://dx.doi.org/10.1016/j.sse.2017.12.007]

[58]   H. Kim, "Back biasing effects in a feedback steep switching device with charge spacer", *in 2016 IEEE Silicon Nanoelectronics Workshop (SNW). IEEE,,* 2016.
[http://dx.doi.org/10.1109/SNW.2016.7578037]

[59]   Y. Jeon, M. Kim, Y. Kim, and S. Kim, "Switching characteristics of nanowire feedback field-effect transistors with nanocrystal charge spacers on plastic substrates", *ACS Nano,* vol. 8, no. 4, pp. 3781-3787, 2014.
[http://dx.doi.org/10.1021/nn500494a] [PMID: 24635681]

[60]   Y. Kim, D. Lim, J. Cho, and S. Kim, "Feedback and tunneling operations of a p+ - i - n+ silicon nanowire field-effect transistor", *Nanotechnology,* vol. 29, no. 43, p. 435202, 2018.
[http://dx.doi.org/10.1088/1361-6528/aad9df] [PMID: 30102245]

[61]   D. Lim, and S. Kim, "Polarity control of carrier injection for nanowire feedback field-effect transistors", *Nano Res.,* vol. 12, no. 10, pp. 2509-2514, 2019.
[http://dx.doi.org/10.1007/s12274-019-2477-6]

[62]   D. Lim, and S. Kim, "Optically tunable feedback operation of silicon nanowire transistors", *Semicond. Sci. Technol.,* vol. 34, no. 11, p. 115014, 2019.
[http://dx.doi.org/10.1088/1361-6641/ab3586]

[63]   Y. Kim, and S. Kim, "Effect of substrates on the electrical characteristics of a silicon nanowire feedback field-effect transistor under bending stresses", *Semicond. Sci. Technol.,* vol. 33, no. 10, p. 105009, 2018.
[http://dx.doi.org/10.1088/1361-6641/aadfb5]

[64]   M. Kim, Y. Kim, D. Lim, S. Woo, K. Cho, and S. Kim, "Steep switching characteristics of single-gated feedback field-effect transistors", *Nanotechnology,* vol. 28, no. 5, p. 055205, 2017.
[http://dx.doi.org/10.1088/1361-6528/28/5/055205] [PMID: 28032609]

[65]   C. Lee, E. Ko, and C. Shin, "Steep slope silicon-on-insulator feedback field-effect transistor: Design and performance analysis", *IEEE Trans. Electron Dev.,* vol. 66, no. 1, pp. 286-291, 2019.
[http://dx.doi.org/10.1109/TED.2018.2879653]

[66]   S. Hwang, "'Si 1-x Ge x positive feedback field-effect transistor with steep subthreshold swing for low-voltage operation," JSTS", *Journal of Semiconductor Technology and Science,* vol. 17, no. 2, pp. 216-222, 2017.
[http://dx.doi.org/10.5573/JSTS.2017.17.2.216]

[67]   F. Raissii, "A brief analysis of the field effect diode and breakdown transistor", *IEEE Trans. Electron Dev.,* vol. 43, no. 2, pp. 362-365, 1996.
[http://dx.doi.org/10.1109/16.481742]

[68]   Y. Yang, "Scaling of the SOI field effect diode (FED) for memory application", *2009 International Semiconductor Device Research Symposium,* IEEE, pp. 1-2, 2009.
[http://dx.doi.org/10.1109/ISDRS.2009.5378045]

[69]   N. Manavizadeh, F. Raissi, E.A. Soleimani, M. Pourfath, and S. Selberherr, "Performance assessment of nanoscale field-effect diodes", *IEEE Trans. Electron Dev.,* vol. 58, no. 8, pp. 2378-2384, 2011.
[http://dx.doi.org/10.1109/TED.2011.2152844]

[70]   K. Lee, "The hysteresis characteristic of Feedback field-effect transistors with fluctuation of gate oxide and metal gate", *Journal of IKEEE,* vol. 22, no. 2, pp. 488-490, 2018.
[http://dx.doi.org/10.5370/JEIEE.2018.22.2.488]

[71]   W.C. Chen, H-T. Lue, Y-H. Hsiao, and C-Y. Lu, "A novel supersteep subthreshold slope dual-channel FET utilizing a gate-controlled thyristor mode-induced positive feedback current", *IEEE Trans. Electron Dev.,* vol. 64, no. 3, pp. 1336-1342, 2017.
[http://dx.doi.org/10.1109/TED.2017.2656903]

[72]   A. Tura, and J.C.S. Woo, "Performance comparison of silicon steep subthreshold FETs", *IEEE Trans. Electron Dev.,* vol. 57, no. 6, pp. 1362-1368, 2010.
[http://dx.doi.org/10.1109/TED.2010.2047066]

[73]   Kaczer, "Atomistic approach to variability of bias-temperature instability in circuit simulations", *2011 International Reliability Physics Symposium,* IEEE, p. 5B, 2011.
[http://dx.doi.org/10.1109/IRPS.2011.5784604]

[74]   D.J. Frank, R.H. Dennard, E. Nowak, P.M. Solomon, and Y. Taur, "Device scaling limits of Si MOSFETs and their application dependencies", *Proc. IEEE,* vol. 89, no. 3, pp. 259-288, 2001.
[http://dx.doi.org/10.1109/5.915374]

[75]   W. Haensch, "Silicon CMOS devices beyond scaling", *IBM Journal of Research and Development,* vol. 50, no. 4.5, pp. 339-361, 2006.
[http://dx.doi.org/10.1147/rd.504.0339]

[76]   A. Asenov, B. Cheng, Xingsheng Wang, A.R. Brown, C. Millar, C. Alexander, S.M. Amoroso, J.B. Kuang, and S.R. Nassif, "Variability aware simulation based design-technology cooptimization (DTCO) flow in 14 nm FinFET/SRAM cooptimization", *IEEE Trans. Electron Dev.,* vol. 62, no. 6, pp.

1682-1690, 2015.
[http://dx.doi.org/10.1109/TED.2014.2363117]

[77]   B. Cheng, A.R. Brown, and A. Asenov, "Impact of NBTI/PBTI on SRAM stability degradation", *IEEE Electron Device Lett.,* vol. 32, no. 6, pp. 740-742, 2011.
[http://dx.doi.org/10.1109/LED.2011.2136316]

[78]   M. Duan, "Hot carrier aging and its variation under use-bias: Kinetics, prediction, impact on Vdd and SRAM", *2015 IEEE International Electron Devices Meeting (IEDM),* IEEE, p. 27.1, 2015.
[http://dx.doi.org/10.1109/IEDM.2015.7409742]

[79]   C. Lee, J. Sung, and C. Shin, "Understanding of feedback field-effect transistor and its applications", *Appl. Sci. (Basel),* vol. 10, no. 9, p. 3070, 2020.
[http://dx.doi.org/10.3390/app10093070]

[80]   N. Kumar, "Insights into the Ultra-Steep Subthreshold Slope Gate-all-around Feedback-FET for memory and sensing applications", *in 2023 IEEE Nanotechnology Materials and Devices Conference (NMDC). IEEE,* 2023.
[http://dx.doi.org/10.1109/NMDC57951.2023.10343913]

[81]   S.K. Suddarsi, K.J. Dhanaraj, and G.K. Saramekala, "Investigation of switching and inverter characteristics of Recessed-Source/Drain (Re–S/D) Silicon-on-Insulator (SOI) Feedback Field Effect Transistor (FBFET)", *Microelectronics J.,* vol. 138, p. 105855, 2023.
[http://dx.doi.org/10.1016/j.mejo.2023.105855]

[82]   S.S. Katta, T. Kumari, S. Das, and P.K. Tiwari, "Design and performance assessment of a vertical feedback FET", *Microelectronics J.,* vol. 137, p. 105806, 2023.
[http://dx.doi.org/10.1016/j.mejo.2023.105806]

[83]   S. Das, S.S. Katta, P. Raj, J. Singh, and P.K. Tiwari, "Design and performance analysis of Si-SiGe heterostructure based double gate feedback FET", *Phys. Scr.,* vol. 99, no. 2, p. 025939, 2024.
[http://dx.doi.org/10.1088/1402-4896/ad1a31]

[84]   P. Singh, D.P. Samajdar, and D.S. Yadav, "A low power single gate l-shaped tfet for high frequency application", *in 2021 6th International Conference for Convergence in Technology (I2CT). IEEE,* pp. 85-90, 2021.
[http://dx.doi.org/10.1109/I2CT51068.2021.9418075]

[85]   P. Singh, and D.S. Yadav, "Impact of temperature on analog/RF, linearity and reliability performance metrics of tunnel FET with ultra-thin source region", *Appl. Phys., A Mater. Sci. Process.,* vol. 127, no. 9, p. 671, 2021.
[http://dx.doi.org/10.1007/s00339-021-04813-1]

[86]   Y. Thakur, B. Raj, and S.S. Gill, "Design and performance analysis of pentacene organic field effect transistor with high-K dielectric materials", *Optoelectron. Adv. Mater. Rapid Commun.,* vol. 17, no. 7-8, pp. 335-343, 2023.
[http://dx.doi.org/10.46390/oamm17.07.08.52]

[87]   P. Singh, and D.S. Yadav, "Impact of work function variation for enhanced electrostatic control with suppressed ambipolar behavior for dual gate L-TFET", *Curr. Appl. Phys.,* vol. 44, pp. 90-101, 2022.
[http://dx.doi.org/10.1016/j.cap.2022.09.014]

[88]   Y. Thakur, B. Raj, and B. Raj, "Design of pentacene thin-film transistor based hydrogen gas sensor with high-k dielectric materials for high sensitivity", *ECS J. Solid State Sci. Technol.,* vol. 13, no. 4, p. 047005, 2024.
[http://dx.doi.org/10.1149/2162-8777/ad3d86]

Semiconductor Nanoscale Devices, 2025, 193-212

CHAPTER 7

GaN-Based High Electron Mobility Transistor

Nipun Sharma[1,*], **Ashish Raman**[1] and **Ravi Ranjan**[2]

[1] *Dr. B. R. Ambedkar National Institute of Technology, Jalandhar, Punjab, India*

[2] *Tyndall National Institute, Lee Maltings Complex Dyke Parade, Cork, Cork, Ireland*

Abstract: A next-generation of highly efficient power devices is under development, utilizing wide bandgap semiconductors, such as GaN and SiC. These materials are gaining traction as attractive alternatives to silicon due to their superior properties. GaN, in particular, has garnered significant interest due to its excellent characteristics, such as a high electric field, saturation velocity, electron mobility, and thermal stability. GaN-based High Electron Mobility Transistors (HEMTs) exhibit superior performance, enabling operation at higher currents, voltages, temperatures, and frequencies. As a result, they are well-suited for the next wave of high-efficiency power converters, including applications in electric vehicles, phone chargers, renewable energy systems, and data centers. This chapter aims to provide an overview of the technological and scientific aspects of current GaN HEMT technology, including normally-on and normally-off. It starts by summarizing recent semiconductor market advancements and key application areas. A comparison between GaN and other materials is then presented, followed by a principle of HEMT and calculations of bound charge. The chapter also delves into normally-off GaN HEMT technology, focusing on aspects, like the recessed gate technique, p-GaN gate, and fluorine implantation. Additionally, reliability concerns associated with GaN HEMTs, such as low positive threshold voltage, 2DEG degradation, leakage current, and different degradation types, are examined. Finally, the study touches upon the use of different normally-off techniques in combination form to improve device parameters, such as threshold voltage and 2DEG concentration.

Keywords: AlGaN, Bandgap, Barrier layer, Channel layer, E-mode, GaN, HEMT, Normally-off, Normally-on, Piezoelectric, Polarization.

INTRODUCTION

The foundation of contemporary electronics is semiconductor materials, which allow for altering complex gadgets that drive our technological society. Among these materials, silicon and III-V semiconductors are notable for their unique qualities and wide range of uses. The industry workhorse for a long time has been

* **Corresponding author Nipun Sharma:** Dr. B. R. Ambedkar National Institute of Technology, Jalandhar, Punjab, India; E-mail: nipuns.ec.23@nitj.ac.in

Ashish Raman, Prabhat Singh, Naveen Kumar & Ravi Ranjan (Eds.)

silicon, a fundamental component of semiconductor technology. With time, silicon-based power devices have reached their maximum theoretical limits; design engineers face increasing current, voltage, and power ratings. This has led to the shift of the semiconductor industry towards comprehensive bandgap materials. Wide bandgap materials serve an essential function in the design and performance of HEMTs, offering distinct merits over traditional semiconductor materials, like silicon. Due to their unique properties, gallium nitride and silicon carbide are two prominent wide bandgap materials extensively used. One key advantage of wide bandgap materials is their ability to withstand higher electric fields, leading to higher breakdown voltages compared to silicon. This property enables HEMTs based on GaN to operate at elevated voltages, making them suitable for high-power applications.

Moreover, wide bandgap materials exhibit superior electron mobility, which is essential for high-frequency operation and fast switching speeds in HEMTs. The creation of a 2DEG at the junction boundary within these materials not only boosts electron transport efficiency, but also enhances overall device performance.

Wide bandgap materials also offer exceptional thermal stability, which is critical in maintaining the reliability and performance of devices in high-temperature conditions. This quality makes GaN HEMTs highly attractive for applications where effective heat dissipation is crucial, such as in electrical power control for electric vehicles and renewable energy systems. The properties of different semiconductors are shown in Table **1** [1]. GaN is more exceptional than the other III-V materials for developing heterostructures due to the large energy gap and in-built polarization field. Due to its suitable properties, it is extensively used as an essential component in radar, 5-G communication, and satellite communication [2 - 5].

Table 1. Properties of different semiconductor materials.

Properties	Silicon	Silicon Carbide	Indium Phosphide	Gallium Nitride
Bandgap (eV)	1.12	3.26	1.35	3.4
Carrier mobility (cm^2/V-s)	1500	900	4000	1500
Thermal conductivity (W/m.K)	150	400	80	200
Crystal structure	Diamond cubic	Hexagonal	Zincblende	Wurtzide

High-electron-mobility transistor is a semiconductor device used in high-intensity frequency and power applications. It is particularly known for its capability to operate at RF frequencies [6, 7]. Traditional HEMT is usually normally-on due to the generation of 2-DEG, which acts as the channel at the AlGaN/GaN boundary. GaN materials have piezoelectric properties, and piezoelectric polarization is the principle behind the formation of 2DEG in HEMT. The normally-off HEMT is preferred over the normally-on one because the normally-off HEMT is the ideal device for applications involving fast switching and significant power requirements [8].

WORKING PRINCIPLE

The basic structure of HEMT is created when a layer of higher bandgap (barrier layer) material is grown on a layer of lower bandgap (channel layer). Both the layers have different bandgaps and piezoelectric fields. We know that gallium nitride possesses piezoelectric properties. When two semiconductor layers of distinct lattice constants are brought into contact, the layers stretch or strain, forming a piezoelectric polarization field. A bound charge is generated at the heterointerface of GaN and AlGaN; if the bound charge at the interface is positive, negatively charged electrons tend to neutralize for the positive charge, leading to the generation of 2DEG below the channel layer. Fig. (**1**) depicts the bound charge generated at the AlGaN/GaN junction [1]. The electrons in 2DEG have enhanced mobility, so we use the term high electron mobility transistor. The 2DEG serves as the channel of HEMT, and the gate controls the electron movement amid the source and drain.

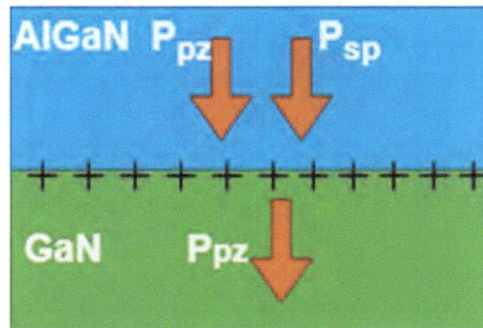

Fig. (1). Bound charge at the interface.

Bound Charge Calculation at the Interface

The bound charge formed at the junction of AlGaN/GaN is primary for the formation of an electron layer in HEMT. The barrier layer and channel layer of GaN HEMT are grown in the [0001] direction. The AlGaN layer on the gallium nitride layer stretches itself to match the properties of the gallium nitride layer. Due to this, a positive bound charge is formed [1]. The following equation can be used to calculate the bound charge:

$$\rho = -\nabla \cdot \mathbf{P} \tag{1}$$

Where, ρ denotes the charge density and $\mathbf{P}$ is the polarization field.

Integrating the above equation, we get:

$$\iiint \rho dV = -\iiint (\nabla \cdot \mathbf{P}) dV \tag{2}$$

$$\iiint \rho dV = Q_b \tag{3}$$

Here, Q_b denotes the bound charge at the junction.

By applying the divergence theorem, equation 3 reduces to:

$$\iiint (\nabla \cdot \mathbf{P}) dV = -\iint \mathbf{P} \cdot \mathbf{n} dS = -Q_b \tag{4}$$

The positive bound charge is given by:

$$\iint \mathbf{P} \cdot \mathbf{n} dS = Q_b \quad \blacktriangleright \quad -P_{AlGaN}S \quad + \quad P_{GaN}S = -Q_b \tag{5}$$

$$P_{AlGaN} \quad - \quad P_{GaN} = Q_b/S = \sigma$$

$P_{AlGaN}S$ and $P_{GaN}S$ represent the polarization field magnitudes of AlGaN and GaN layers.

Given that the GaN layer is in a relaxed state while the AlGaN layer is under tension:

$$P_{GaN} = P^{sp}_{GaN} \tag{6}$$

$$P_{AlGaN} = P^{sp}_{AlGaN} + P^{pz}_{AlGaN} \tag{7}$$

Where, P^{sp} represents the spontaneous polarization field and P^{pz} represents the piezo-electric polarization field. The spontaneous polarization of the barrier layer can be determined by calculating:

$$P_{AlGaN} = P^{sp}_{AlGaN} + P^{pz}_{AlGaN} \tag{8}$$

ALGaN/GaN HEMT STRUCTURE

The conventional structure of HEMT is composed of many layers. Fig. (**2**) shows the different layers of HEMT. It contains the barrier layer, channel layer, nucleation and buffer layers, and finally, the substrate. In between the AlGaN and GaN layers lies the layer of electrons (2DEG) that serves as the channel in the

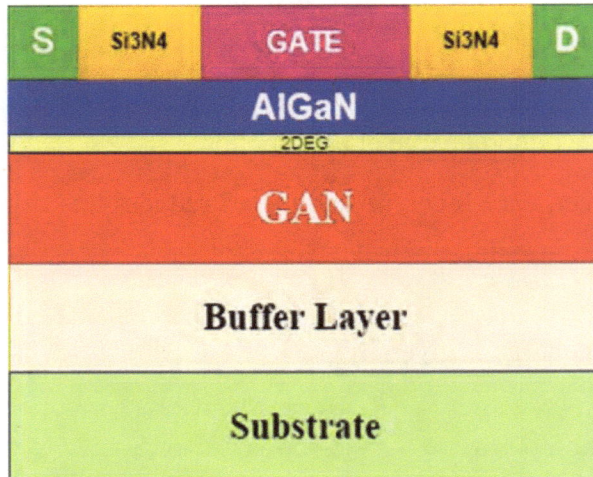

Fig. (2). Structure of normally-on HEMT.

HEMT structure. HEMT has various layers. The material and function of multiple layers are described as follows:

Substrate

The substrate plays a foundational role in the semiconductor device fabrication process. For high electron mobility transistors, various substrate materials have been utilized over time, including silicon (Si), sapphire, silicon carbide, and gallium nitride substrates [9]. Initially, silicon carbide, silicon, and sapphire were the predominant choices as substrates. However, ongoing research is now exploring the feasibility of using diamonds as a substrate, showing promising potential for future applications. In recent years, there has been a notable industry preference for growing GaN devices based on silicon (Si) substrates. This preference stems from the widespread availability of Si substrates, which offer many benefits, like cost-effectiveness and scalability. Despite this trend, it is important to note that silicon carbide (SiC) substrates have historically delivered the most impressive results in connection with device productivity and durability for GaN-based HEMTs.

The decision to use Si substrates for GaN growth is driven by practical considerations, such as cost and availability. However, advancements in fabrication techniques and material science are narrowing the performance gap between SiC and Si substrates. As a result, it is anticipated that the exceptional results achieved for SiC substrates will eventually be replicated on Si substrates. This trend underscores the dynamic nature of semiconductor research and the continuous pursuit of optimizing substrate materials for superior device performance and functionality.

Nucleation and Buffer Layer

These layers are usually provided below the channel layer to prevent the cracking of layers that arise due to lattice mismatch. You-Chen Weng proposed a high-pressure GaN buffer layer between the GaN layer and transition layer, which showed improvement in drain current, transconductance, on-state resistance, and electron mobility [10].

Channel Layer

The channel layer in HEMTs is a crucial component that is essential for the proper functioning of the device. It is usually a thin coating of gallium nitride (GaN) material grown along the [0001] direction, a choice to capitalize on the polarization effects. Within the channel layer, the formation of 2DEG occurs, which is fundamental to the transistor's operation. The selection of material for the channel layer is paramount in HEMT design because it directly impacts device characteristics and performance. Gallium nitride is a highly favored material for the channel layer due to its exceptional properties, particularly its high electron mobility. This high mobility results in superior high-frequency performance, making GaN-based HEMTs highly desirable for applications requiring fast switching speeds and high-frequency operation [11, 12].

Barrier Layer

AlGaN serves as a barrier layer with a larger energy bandgap than the channel layer, which can be doped or remain undoped. The polarization field difference between the AlGaN and the GaN results in the creation of a potential well at their junction. This well captures electrons, forming a channel within it [1]. The barrier layer helps confine charge carriers. This confinement is essential for achieving high electron mobility and enhancing device performance [13].

NORMALLY-ON HEMT

The concept of normally-on HEMT lies in conducting current even when 0V is supplied at the gate. It is desirable for certain applications, like communication systems and RF systems. The normally-on behavior offers advantages in many applications where rapid response times and low power consumption are critical [14]. However, there are many challenges related to normally-on behavior, which include reliability, thermal management, and current leakage [15].

The threshold voltage (V_{th}) plays a pivotal role in characterizing and operating High Electron Mobility Transistors (HEMTs). It is the voltage between the gate and the source at which HEMT begins to conduct significant current between its source-drain terminals. Generally, HEMTs are normally-on because their threshold voltage is negative. A HEMT's threshold voltage is affected by several variables, such as the materials used in its construction, the physical layout of the device, and the doping profile. For instance, the heterojunction design, gate electrode material, and epitaxial layer selection can all impact the threshold voltage in GaN-based HEMTs [16 - 18].

We have simulated the conventional normally-on structure of HEMT on Silvaco software. The device comprises a 25 nm thick Al0.32GaN layer and a 1.475 μm thick GaN layer. The length of the gate (L_g) is 2μm. The gate-to-source distance (Lgs) and gate-to-drain distance ($L_{gd)}$ are 3μm and 3μm. The passivation layer used is Si_3N_4. The transfer characteristics are obtained at V_{DS} =1V, as shown in Fig. (3). The gate voltage is ramped from -8V to 1V. From Fig. (3), we have observed that HEMT has a negative value of V_{th}, which means it is normally off. Similarly, the drain characteristics are visible in Fig. (4), with drain voltage ramped from 0V to 15V with a step size of 0.5V for different values of V_{gs} [1].

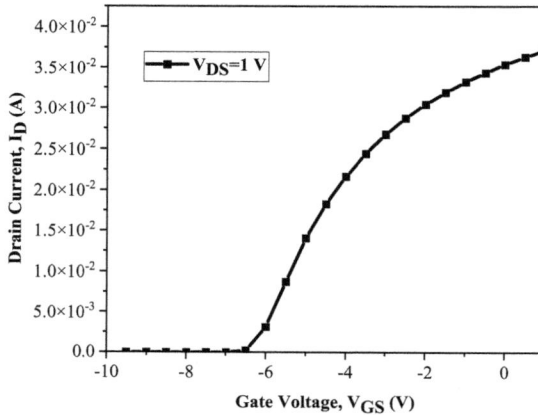

Fig. (3). I_D-V_{GS} curve of HEMT showing negative threshold voltage.

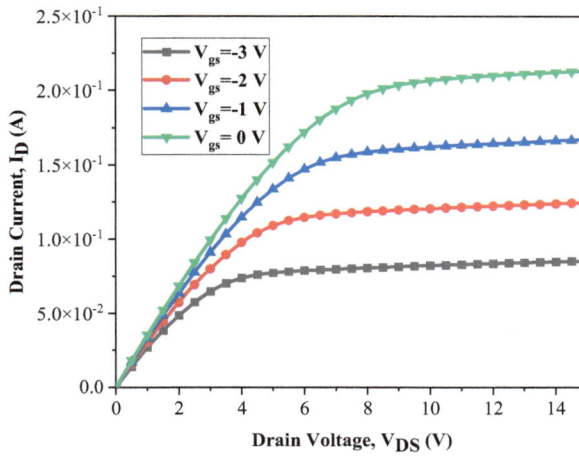

Fig. (4). I_D-V_{DS} curve of normally-on HEMT.

NORMALLY-OFF HEMT

A normally-off HEMT (High Electron Mobility Transistor) remains non-conductive when no biasing is applied to its gate terminal. It stands in contrast to conventional transistors, which are typically "normally-on" [19].

The "normally-off" behavior of HEMTs is achieved through careful engineering of the device structure, often involving specific semiconductor materials and heterojunction interfaces. In these transistors, the conduction channel is normally blocked when 0V is given to the gate, and it only becomes conductive when a suitable voltage is supplied to the gate. This feature is especially useful in many electronic applications as it increases safety, lowers standby power consumption, and increases dependability. Because of their rapid switching speed and high-frequency performance, normally-off HEMTs are widely used in high-speed,

high-capacity applications, such as microwave circuits, amplifiers, and switches. Their normally-off behavior improves overall system efficiency and streamlines circuit design [20, 21].

APPLICATIONS

At present, normally off HEMTs find applications across various fields due to their unique characteristics and performance advantages. Some notable uses include the following:

Power Electronics

Electronic devices involving converters, inverters, and motor drives use normally-off HEMTs. Because of their high-frequency capabilities and fast switching, they are appropriate for high-intensity applications, such as industrial machines, E-vehicles, and renewable energy systems [22, 23].

Wireless Communications

Wireless communication systems use normally-off HEMTs in low-noise and high-frequency amplifiers. They are crucial parts of radar, satellite communication, and cellular base stations because of their excellent linearity and capacity to function at RF frequencies [2 - 5, 22, 23].

Microwave and Radio Frequency

Because of their rapid switching speed and outstanding high-frequency performance, normally-off HEMTs are frequently used in microwave and Radio Frequency (RF) applications. They are used in microwave amplifiers, switches, mixers, and oscillators [23, 24].

Automotive Electronics

HEMTs serve as a crucial part of power management, motor control, and vehicle safety systems in the field of automotive electronics. They can withstand high power levels and perform dependably under challenging conditions, which makes them perfect for automotive applications, such as engine control units, Advanced Driver Assistance Systems (ADAS), and electric power steering [22 - 24].

Aerospace and Defence

Radar, electronic warfare, missile guidance, and communication systems are among the applications in aerospace and defense that use normally-off HEMTs. These applications have very high expectations, which are presently met by HEMT [22 - 24]

NORMALLY OFF TECHNIQUES

Normally-off techniques are methods used in the design and fabrication of HEMT to ensure that it remains in an off state under zero voltage bias. This contrasts normally-on transistors, which are conductive by default and require a negative bias to turn off. There are different techniques to achieve E-mode operation, which are presented as follows:

Recessed Gate Technology

Recessed gate technology is a technique used in fabricating HEMTs to improve their performance and achieve specific device characteristics, such as normally-off behavior. In a recessed gate structure, the gate electrode of the transistor is positioned below the surface of the semiconductor material, rather than on top of it. The gate electrode is recessed into a cavity or trench formed in the semiconductor material in a recessed gate structure. This cavity can be created using etching techniques or other fabrication methods. The main aim of gate recessing is to shift the V_{GS} to the positive side. Fig. (**5**) depicts the structure of the HEMT with the gate being recessed in a barrier layer. The above technique reduces the selective density of electrons under the gate electrode. This transistor type offers advantages, such as reduced gate leakage, improved breakdown voltage, and enhanced thermal stability. Furthermore, the recessed gate HEMT exhibits reduced self-heating effects, which results in higher reliability and longer lifespan. The recessed gate HEMT design configuration is a potential option for a variety of applications since it not only performs better than traditional HEMTs, but also shows remarkable traits. Because of its innovative T-gate and drain-fiel--plate technology, it has attained better breakdown voltage and drain current capabilities, which enable more effective control of the electric field and better device performance [25, 26].

Fig. (5). HEMT with recessed gate structure.

Further, the gate can be fully recessed and partially recessed in the barrier layer AlGaN. The problem with gate recessing is a very low voltage (Vgs) shift towards the positive side, and sometimes, even the threshold voltage is negative. This technique is combined with other normally-off techniques to get a higher value of positive threshold voltage (3-4V) [27, 28].

Thin Barrier Layer

This configuration also confronts the AlGaN/GaN proximity to the gate electrode to achieve E-mode operation. A thin barrier layer of the AlGaN layer is used in this technique so that the gate electrode quickly depletes the 2DEG region and achieves the off-state. Fig. (**6**) describes the E-mode state achieved using a thin barrier layer. Using a narrow $Al_{0.4}5Ga_{0.55}N$ layer indicates precise engineering of material thickness to achieve desired device characteristics, such as improved performance or reduced leakage currents [29, 30]. Using a thin barrier layer affects the 2DEG layer density, which could be compensated by mole fraction variation.

Fig. (6). HEMT with thin AlGaN layer.

Buried p-region

Buried p-region is a technique used for the enhancement-mode operation in HEMT. This concept refers to a specific region within the channel layer (GaN) where the material is doped to be p-type, meaning it contains positively charged "holes" as majority carriers. Fig. (**7**) shows the P-GaN layer buried in the GaN channel layer. The transistor's threshold voltage (V_{th}) can shift towards the positive side if the buried p-region is present. With the decrease in size of the transistor, short channel effects increase considerably. By regulating the dissemination of the E-field inside the transistor structure, the buried P-GaN region helps to lessen these effects [31].

Designers can modify V_{th} to meet particular device requirements and performance targets by varying the doping and thickness of the buried p-area. This control over V_{th} is crucial to ensure appropriate switching behavior and optimize device operation [32, 33].

Fig. (7). P-GaN normally-off HEMT structure.

Introducing the buried p-region in the channel or barrier layer lifts the E_c (conduction band) above the forbidden fermi level, creating an off state of the HEMT. P-GaN is used as a buried region below the gate electrode in the channel layer. The width and distance of P-GaN affect the conduction layer of electrons. The V_{th} increases with the increase in the width of P-GaN, which has been proven with various simulated results [1].

Buried P-GaN regions can also be employed in forming field plates, which are structures that increase the device breakdown voltage and durability of GaN devices, especially in intensive applications. By creating a P-GaN layer below the device's active region, engineers can implement field plate designs that reduce electric field crowding and improve device ruggedness. In certain device configurations, a buried P-GaN region can help reduce leakage currents and improve device efficiency, especially at high voltages. By incorporating a P-GaN layer with appropriate doping profiles, designers can minimize undesirable leakage paths and enhance device reliability [34 - 37].

The buried p-region in a HEMT enables it to operate in a normally-off state, which reduces power consumption and improves safety. However, to further lower the leakage current and increase the V_{th}, an insulating layer is grown beneath the gate. The structure of MIS-HEMT with a p-doped GaN buried layer is displayed in Fig. (**8**). This insulating layer serves to eliminate or minimize current leakage, thereby improving the performance and durability of the HEMT device. The insulating layers used are silicon nitride, silicon oxide, and hafnium oxide. The quality of the insulating layer and its dimensions affect the device parameters, such as threshold voltage.

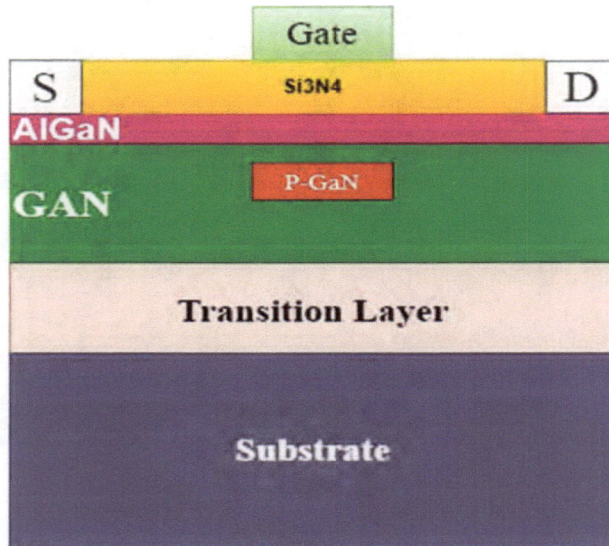

Fig. (8). MIS-HEMT with P-GaN buried layer.

P-GaN Gate

Among all of the techniques discussed above, the P-GaN Gate is the most effective and has been commercialized successfully. P-GaN gate HEMTs show great potential for high-intensity applications, thanks to their inherent normally-off behavior and durable gate reliability. Fig. (**9**) displays the structure of P-GaN cap HEMT, which contains an Mg^+ doped GaN layer. This technique introduces the p-doped GaN layer in between the gate electrode and the AlGaN layer. It lessens the concentration of electrons present in the channel by raising the E_c of the AlGaN layer using p-doped GaN. This method describes a hole injection effect to achieve the E-mode characteristics of the device, leading to its designation as a gate injection transistor. When the V_{GS} is zero, the electron layer beneath the gate becomes fully depleted, preventing drain current flow. As the gate voltage increases up to the forward voltage of the p-n junction, the GIT functions as a FET. Beyond this voltage, additional holes flow into the channel from the p-GaN. These transported holes attract equivalent electrons from the source, maintaining charge balance in the channel. The aggregated electrons with high mobility are propelled by the drain voltage, while the infused holes remain near the gate due to their significantly lower mobility [38 - 45].

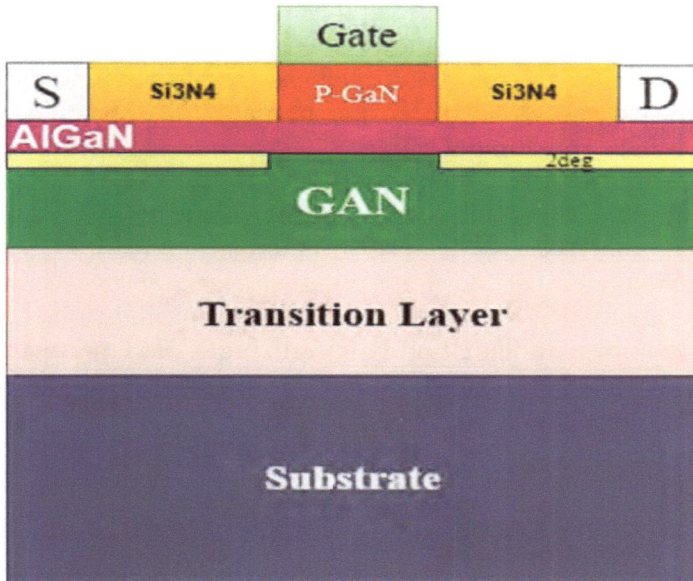

Fig. (9). P-GaN cap HEMT structure.

Researchers have proposed novel device structures to enhance the V_{th} of the above-mentioned HEMT devices, one of which is the gate-source bridge structure. This design improvement helps raise the device's threshold voltage. However, this design also presents a tradeoff: a wider bridge reduces the area of the 2DEG channel, increasing resistance. These factors must be carefully balanced to optimize the device's performance for its intended application. P-GaN gate and recessed gate techniques can be modified and combined to further improve the device characteristics in terms of V_{th} and saturation current. The threshold voltage moves out near 1V using the gate injection or recessed gate technique. Research is ongoing to adopt composite-type structures to increase the V_{th} above 3-4V. Recessing the p-doped GaN layer improves the V_{th}, but it affects the channel electron concentration and increases gate leakage current. Fig. (**10**) shows the P-

GaN recessed structure. Using a dielectric layer above P-GaN can improve the V_{th} and lower leakage current as it can share the voltage [42 - 48].

Fluorine Implantation

Fluorine implantation plays an imperative role in semiconductor fabrication, especially in the context of HEMTs [45]. By introducing fluorine ions into targeted areas of the HEMT structure, engineers can achieve specific modifications tailored to the device's performance and reliability. By implanting fluorine ions, the upliftment of the E_C occurs above the fermi energy level, causing an E-mode condition. Fig. **(11)** describes the structure of HEMT with F⁻ inserted in the channel. This technique is commonly used for threshold voltage adjustment, where fluorine ions are strategically placed to tune the device's turn-off behavior [53]. Additionally, fluorine implantation serves as a method for surface passivation, reducing surface defects and enhancing overall device quality. It also allows for doping profile modifications, influencing carrier mobility and electrical characteristics. Furthermore, engineers utilize fluorine implantation for interface engineering in heterostructure devices, like HEMTs, minimizing interface traps and improving carrier transport properties across semiconductor layers [54]. These applications highlight the versatility and significance of fluorine implantation in optimizing HEMT performance for high-intensity frequency and power applications [45 - 54].

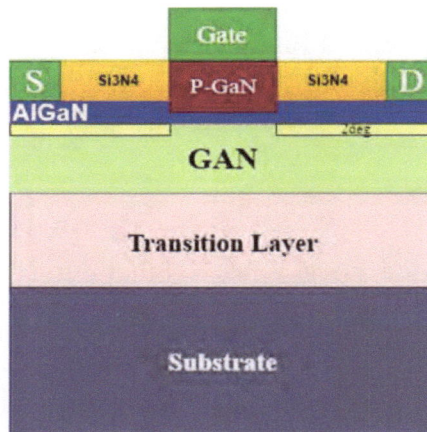

Fig. (10). P-GaN recessed structure.

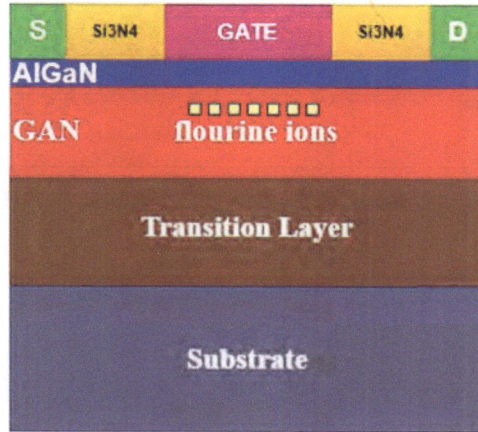

Fig. (11). HEMT structure showing fluorine implantation in the channel layer.

CONCLUSION

The development of GaN-based HEMTs marks a significant milestone in semiconductor technology. GaN-based HEMTs have emerged due to their outstanding features, such as intense E-field, mobility, and thermal stability, surpassing conventional silicon-based devices. Techniques, like recessed gate, P-GaN gate, fluorine implantation, and composite gate structures have been instrumental in enabling normally-off operation in HEMTs, opening doors for their widespread adoption in advanced electronics equipment. Significant advancements have been achieved, but challenges persist, including optimizing device performance, managing process intricacies, and ensuring consistent fabrication outcomes. Future research directions are expected to concentrate on refining device designs, exploring novel materials and fabrication methodologies, and further enhancing the overall efficiency and dependability of normally-off HEMTs. Overall, normally-off HEMTs hold substantial promise for meeting the evolving requirements of modern electronics, spanning from high-speed communication systems to power electronics. They are poised to make substantial contributions to advancing semiconductor technology in the foreseeable future.

KEY POINTS

1. This chapter enables an understanding of the importance of large bandgap semiconductors, like gallium nitride and SiC, in the advancement of next-generation power devices.
2. It analyzes the benefits of gallium nitride-based HEMTs over traditional silicon-based devices in terms of higher current, temperature, and frequency of operation.
3. It further describes the technological principles and working mechanisms of

HEMT and 2DEG formation.

4. The chapter also evaluates different techniques used in E-mode gallium nitride HEMTs, including the recessed gate technique, p-GaN gate, and fluorine implantation, and their impact on device characteristics.

5. It finally studies and applies the various normally-off techniques in combination form to enhance device parameters, like threshold voltage and 2DEG concentration, in GaN HEMT.

ACKNOWLEDGEMENTS

The authors would like to thank the Department of Electronics and Communication Engineering, Dr. B.R Ambedkar National Institute of Technology, Jalandhar, Punjab, for providing valuable support to carry out this study.

REFERENCES

[1]　S. Hamady, *New concepts for Normally-Off Power Gallium Nitride (GaN) High Electron Mobility Transistor (HEMT).* Micro and Nanotechnologies/Microelectronics, Universite Toulouse III Paul Sabatier, 2014.

[2]　A. Udabe, I. Baraia-Etxaburu, and D.G. Diez, "Gallium nitride power devices: a state of the art review", *IEEE Access,* vol. 11, pp. 48628-48650, 2023.
[http://dx.doi.org/10.1109/ACCESS.2023.3277200]

[3]　S.N. Shuji Nakamura, "GaN growth using GaN buffer layer", *Jpn. J. Appl. Phys.,* vol. 30, no. 10A, pp. L1705-L1707, 1991.
[http://dx.doi.org/10.1143/JJAP.30.L1705]

[4]　U.K. Mishra, Shen Likun, T.E. Kazior, and Yi-Feng Wu, "GaN-Based RF Power Devices and Amplifiers", *Proc. IEEE,* vol. 96, no. 2, pp. 287-305, 2008.
[http://dx.doi.org/10.1109/JPROC.2007.911060]

[5]　A. Jarndal, A.Z. Markos, and G. Kompa, "Improved modeling of GaN HEMTs on Si substrate for design of RF power amplifiers", *IEEE Trans. Microw. Theory Tech.,* vol. 59, no. 3, pp. 644-651, 2011.
[http://dx.doi.org/10.1109/TMTT.2010.2095034]

[6]　T. Mimura, "High electron mobility transistors for microwave and millimeter-wave applications", *IEEE Trans. Microw. Theory Tech.,* vol. 50, no. 3, pp. 780-792, 2002.
[http://dx.doi.org/10.1109/22.989961]

[7]　C.P. Wen, "High-electron-mobility transistors (HEMTs)—Device physics and design considerations", *IEEE Trans. Electron Dev.,* vol. 53, no. 5, pp. 1040-1049, 2006.
[http://dx.doi.org/10.1109/TED.2006.873783]

[8]　N. Chugh, S. Haldar, M. Bhattacharya, and R.S. Gupta, *Potential and electric field analysis of field plated AlGaN/GaN HEMT for high voltage applications using 2-D analytical approach.* vol. Vol. 105857. Microelectronics Journal, 2023.

[9]　K.K. Chu, P.C. Chao, M.T. Pizzella, R. Actis, D.E. Meharry, K.B. Nichols, R.P. Vaudo, X. Xu, J.S. Flynn, J. Dion, and G.R. Brandes, "9.4-W/mm Power Density AlGaN–GaN HEMTs on Free-Standing GaN Substrates", *IEEE Electron Device Lett.,* vol. 25, no. 9, pp. 596-598, 2004.
[http://dx.doi.org/10.1109/LED.2004.833847]

[10]　Y.C. Weng, M.Y. Hsiao, C.H. Lin, Y.P. Lan, and E.Y. Chang, "Effect of high-pressure GaN nucleation layer on the performance of AlGaN/GaN HEMTs on Si substrate", *Materials (Basel),* vol.

16, no. 9, p. 3376, 2023.
[http://dx.doi.org/10.3390/ma16093376] [PMID: 37176258]

[11]   J. Kuzmík, "AlGaN/GaN HEMTs with Low-Resistance AlN/GaN Superlattice Buffer Layers", *IEEE Electron Device Lett.*, vol. 38, no. 9, pp. 1217-1220, 2017.

[12]   S.J. Pearton, "High electron mobility transistors (hemts)—a review of device physics and applications", *IEEE Trans. Electron Dev.*, vol. 60, no. 10, pp. 3072-3083, 2013.

[13]   A. Chakraborty, and A.K. Panda, "Recent developments in high electron mobility transistors: A review", *J. Electron. Mater.*, vol. 49, no. 6, pp. 1183-1197, 2020.

[14]   S. Rajan, "Normally-on GaN-based high electron mobility transistors", *IEEE Trans. Electron Dev.*, vol. 60, no. 6, pp. 1853-1863, 2013.
[http://dx.doi.org/10.1109/TED.2013.2255313]

[15]   U. Radhakrishna, "Design and optimization of normally-on AlGaN/GaN HEMTs for power switching applications", *IEEE Trans. Electron Dev.*, vol. 65, no. 5, pp. 1882-1890, 2018.
[http://dx.doi.org/10.1109/TED.2018.2828329]

[16]   H. Yang, "Threshold voltage extraction method for GaN HEMTs based on dynamic current-voltage characteristics", *IEEE Trans. Electron Dev.*, vol. 66, no. 10, pp. 4208-4213, 2019.
[http://dx.doi.org/10.1109/TED.2019.2921558]

[17]   A. Chini, and M.A. Peroni, "Design and optimization of AlGaN/GaN HEMTs for microwave power applications", *Proc. IEEE*, vol. 96, no. 8, pp. 1285-1304, 2008.
[http://dx.doi.org/10.1109/JPROC.2008.926587]

[18]   S. Huang, "Threshold voltage control in AlGaN/GaN high-electron-mobility transistors by CF4 plasma treatment", *IEEE Electron Device Lett.*, vol. 36, no. 3, pp. 240-242, 2015.
[http://dx.doi.org/10.1109/LED.2015.2391675]

[19]   P.D. Ye, and S.M. Sze, *Normal-off Gallium Nitride HEMTs: From Technology to Circuit Applications.* CRC Press, 2020.

[20]   K. Mizuno, "Normally-Off HEMT with Inverted p-Gate AlGaN/GaN Structure", *IEEE Electron Device Lett.*, vol. 30, no. 11, pp. 1156-1158, 2009.
[http://dx.doi.org/10.1109/LED.2009.2031232]

[21]   N. Mandal, "Normally-Off AlGaN/GaN HEMTs: Challenges and Prospects", *IEEE Trans. Electron Dev.*, vol. 62, no. 9, pp. 2843-2850, 2015.
[http://dx.doi.org/10.1109/TED.2015.2468643]

[22]   S.O. Kasap, *Principles of Electronic Materials and Devices.* McGraw-Hill Education, 2006.

[23]   P.D. Ye, and S.M. Sze, *Normal-off Gallium Nitride HEMTs: From Technology to Circuit Applications.* CRC Press, 2020.

[24]   P.D. Ye, "High-performance normally-off AlGaN/GaN high electron mobility transistors with field-plates", *IEEE Electron Device Lett.*, vol. 32, no. 5, pp. 648-650, 2011.
[http://dx.doi.org/10.1109/LED.2011.2109866]

[25]   W. Saito, Y. Takada, M. Kuraguchi, K. Tsuda, and I. Omura, "Recessed-gate structure approach toward normally off high-Voltage AlGaN/GaN HEMT for power electronics applications", *IEEE Trans. Electron Dev.*, vol. 53, no. 2, pp. 356-362, 2006.
[http://dx.doi.org/10.1109/TED.2005.862708]

[26]   A. Varghese, P. Das, and S. Tallur, "A complete analytical model for MOS-HEMT biosensors: capturing the effect of stern layer and charge screening on sensor performance", *IEEE Trans. Electron Dev.*, vol. 66, no. 6, pp. 2683-2690, 2019.
[http://dx.doi.org/10.1109/TED.2019.2910109]

[27]   M.N.A. Aadit, "High electron mobility transistors: performance analysis, research trend and applications", *IEEE Access*, vol. 5, pp. 23536-23552, 2017.

[http://dx.doi.org/10.1109/ACCESS.2017.2765610]

[28] S. Zhu, H. Jia, T. Li, Y. Tong, Y. Liang, X. Wang, T. Zeng, and Y. Yang, "Novel High-Energ-
-Efficiency AlGaN/GaN HEMT with High Gate and Multi-Recessed Buffer", *Micromachines (Basel),*
vol. 10, no. 7, p. 444, 2019.
[http://dx.doi.org/10.3390/mi10070444] [PMID: 31269635]

[29] A. Chakroun, "Normally-off AlGaN/GaN MOS-HEMT using ultra-thin Al0.45Ga0.55N barrier layer",
Proceedings of the IEEE International Conference on Semiconductor Electronics (ICSE), vol. 14, p.
1600836, 2017.
[http://dx.doi.org/10.1002/pssa.201600836]

[30] M.A. Khan, Q. Chen, C.J. Sun, J.W. Yang, M. Blasingame, M.S. Shur, and H. Park, "Enhancement
and depletion mode GaN/AlGaN heterostructure field effect transistors", *Appl. Phys. Lett.,* vol. 68, no.
4, pp. 514-516, 1996.
[http://dx.doi.org/10.1063/1.116384]

[31] Y. Uemoto, *Characteristics of AlGaN/GaN high electron mobility transistors with a buried p-GaN
layer, vol. 24, no. 12, pp. 771-773, 2016.IEEE Electron Device Lett.,* vol. 24, no. 12, pp. 771-773,
2016.

[32] H. Li, "Design and characterization of AlGaN/GaN high electron mobility transistors with a buried p-
type layer", *IEEE Trans. Electron Dev.,* vol. 63, no. 3, pp. 1042-1047, 2016.

[33] K.W. Yung, "Enhanced breakdown voltage and switching speed of AlGaN/GaN HEMTs with a buried
P layer", *IEEE Electron Device Lett.,* vol. 34, no. 7, pp. 919-921, 2013.

[34] Y. Uemoto, K. Tsubaki, H. Matsuo, Y. Muramoto, H. Ueno, and T. Tanaka, "Improved gate-leakage
characteristics of GaN-based high electron mobility transistors with a buried p-type layer", *Appl. Phys.
Lett.,* vol. 81, no. 7, pp. 1269-1271, 2002.

[35] C.K. Sun, J.P. Ibbetson, J.C. Perng, K. Doverspike, and U.K. Mishra, "Buried p-type layers in
GaN/AlGaN/GaN high electron mobility transistors", *Appl. Phys. Lett.,* vol. 71, no. 5, pp. 590-592,
1997.

[36] S.J. Pearton, C.R. Abernathy, F. Ren, J.W. Lee, J. Lin, J.M. Zavada, and L.F. Eastman, "Improved
isolation in AlGaN/GaN HEMTs using buried p-type layers", *Electron. Lett.,* vol. 40, no. 2, pp. 134-
136, 2004.

[37] X. Hu, G. Simin, J. Yang, M. Asif Khan, R. Gaska, and M.S. Shur, "Enhancement mode AlGaN/GaN
HFET with selectively grown pn junction gate", *Electron. Lett.,* vol. 36, no. 8, pp. 753-754, 2000.
[http://dx.doi.org/10.1049/el:20000557]

[38] K. Umesh, P. Parikh, and Y. Wu, "AlGaN/GaN HEMTs-an overview of device operation and
applications", *Proc. IEEE,* vol. 90, pp. 1022-1031, 2000.

[39] P. Singh, and D.S. Yadav, "Impact of work function variation for enhanced electrostatic control with
suppressed ambipolar behavior for dual gate L-TFET", *Curr. Appl. Phys.,* vol. 44, pp. 90-101, 2022.
[http://dx.doi.org/10.1016/j.cap.2022.09.014]

[40] R. Ranjan, P. Kumar, and N. Kumar, *Demonstration and Performance Assessment of Dopant Free
TFET Including Lattice Heating and Temperature Effects.* Silicon, 2024, pp. 1-8.

[41] Y. Uemoto, M. Hikita, H. Ueno, H. Matsuo, H. Ishida, M. Yanagihara, T. Ueda, T. Tanaka, and D.
Ueda, "Gate injection transistor (GIT)-a normally-off AlGaN/GaN power transistor using conductivity
modulation", *IEEE Trans. Electron Dev.,* vol. 54, no. 12, pp. 3393-3399, 2007.
[http://dx.doi.org/10.1109/TED.2007.908601]

[42] Z. Wang, J. Nan, Z. Tian, P. Liu, Y. Wu, and J. Zhang, "Review on main gate characteristics of p-type
GaN gate high-electron-mobility transistors", *Micromachines (Basel),* vol. 15, no. 1, p. 80, 2023.
[http://dx.doi.org/10.3390/mi15010080] [PMID: 38258199]

[43] P. Singh, *Ultra Thin Finger-Like Source Region-Based TFET: Temperature Sensor.IEEE Sensors*

Letters..
[http://dx.doi.org/10.1109/LSENS.2024.3390689]

[44] N. Kumar, C.P. García, A. Dixit, A. Rezaei, and V. Georgiev, "Charge dynamics of amino acids fingerprints and the effect of density on FinFET-based Electrolyte-gated sensor", *Solid-State Electron.,* vol. 210, p. 108789, 2023.
[http://dx.doi.org/10.1016/j.sse.2023.108789]

[45] I. Hwang, J. Oh, H.S. Choi, J. Kim, H. Choi, J. Kim, S. Chong, J. Shin, and U-I. Chung, "Source-connected p-GaN gate HEMTs for increased threshold voltage", *IEEE Electron Device Lett.,* vol. 34, no. 5, pp. 605-607, 2013.
[http://dx.doi.org/10.1109/LED.2013.2249038]

[46] Hwang, "Forward bias gate breakdown mechanism in enhancement-mode p-GaN Gate AlGaN/GaN high-electron mobility transistors", *IEEE Electron Device Lett.,* vol. 36, pp. 1001-1003, 2013.

[47] N. Kumar, *Insights into the Ultra-Steep Subthreshold Slope Gate-all-around Feedback-FET for memory and sensing applications,* pp. 617-620, 2023.*in 2023 IEEE Nanotechnology Materials and Devices Conference (NMDC),* pp. 617-620, 2023.
[http://dx.doi.org/10.1109/NMDC57951.2023.10343913]

[48] P. Singh, and D.S. Yadav, "Assessment of temperature and ITCs on single gate L-shaped tunnel FET for low power high frequency application", *Engineering Research Express,* vol. 6, no. 1, p. 015319, 2024.
[http://dx.doi.org/10.1088/2631-8695/ad32b0]

[49] J. Li, *Normally-off AlGaN/AlN/GaN HEMT with a composite recessed gate,* vol. 161, p. 107064, 2022.*Superlattices and Microstructures,* vol. 161, p. 107064, 2022.
[http://dx.doi.org/10.1016/j.spmi.2021.107064]

[50] P. Singh, and D.S. Yadav, *Design and investigation of f-shaped tunnel fet with enhanced analog/rf parameters.* Silicon, 2021, pp. 1-16.

[51] Yong Cai, Yugang Zhou, K.J. Chen, and K.M. Lau, "High-performance enhancement-mode AlGaN/GaN HEMTs using fluoride-based plasma treatment", *IEEE Electron Device Lett.,* vol. 26, no. 7, pp. 435-437, 2005.
[http://dx.doi.org/10.1109/LED.2005.851122]

[52] K.J. Chen, "Physics of fluorine plasma ion implantation for GaN normally-off HEMT technology", *Electron Devices Meeting (IEDM),* pp. 19-4, 2011.
[http://dx.doi.org/10.1109/IEDM.2011.6131585]

[53] P. Singh, and D.S. Yadav, "Performance analysis of ITCs on analog/RF, linearity and reliability performance metrics of tunnel FET with ultra-thin source region", *Appl. Phys., A Mater. Sci. Process.,* vol. 128, no. 7, p. 612, 2022.
[http://dx.doi.org/10.1007/s00339-022-05741-4]

[54] N. Kumar, R. Dhar, C. Pascual Garcia, and V.P. Georgiev, "A novel computational framework for simulations of bio-field effect transistors", *ECS Trans.,* vol. 111, no. 1, pp. 249-260, 2023.
[http://dx.doi.org/10.1149/11101.0249ecst]

Advanced Semiconductor Sensing Technologies: Materials and Design Challenges at the Nanoscale

Shreya[1,*]**, Peeyush Phogat**[1]**, Ranjana Jha**[1] **and Sukhvir Singh**[1]

Research Lab for Energy Systems, Department of Physics, Netaji Subhas University of Technology, Delhi, India

Abstract: This chapter delves into the intricate realm of semiconductor devices for sensing applications, offering a comprehensive and detailed exploration. It begins with a foundational examination of semiconductor sensing principles, elucidating the fundamental mechanisms that underpin these advanced technologies. The chapter then transitions to the pivotal role of nanoscale materials in enhancing sensing capabilities, emphasizing how these materials revolutionize sensor performance. A meticulous examination of the design considerations for crafting nanoscale semiconductor sensing devices follows, addressing architectural nuances, integration challenges, and concerns related to power consumption and efficiency. The chapter further provides an in-depth discussion on materials synthesis and fabrication techniques, offering an overview of diverse methods for nanomaterial synthesis and the fabrication processes essential for creating these sophisticated devices. Highlighting recent advancements in semiconductor sensing technologies, the chapter unveils state-of-the-art developments and emerging trends. Insightful case studies and real-world applications illustrate these advancements, showcasing how theoretical concepts translate into practical solutions. An in-depth analysis of the challenges and opportunities within the field outlines current obstacles, proposes potential solutions, and envisions future prospects, providing a comprehensive understanding of the landscape. Through engaging case studies, the chapter demonstrates how innovative solutions can be implemented to overcome existing challenges. Further exploration of the critical aspects of testing and characterization of nanoscale semiconductor sensing devices emphasizes the importance of rigorous evaluation. A spectrum of characterization techniques is covered, ensuring a thorough understanding of reliability and durability assessments, ultimately providing readers with a well-rounded and detailed perspective on the future of semiconductor sensing technologies.

Keywords: Advancements, Challenges, Fabrication, Nanomaterials, Sensors, Semiconductors.

* **Corresponding author Shreya:** Research Lab for Energy Systems, Department of Physics, Netaji Subhas University of Technology, Delhi, India; E-mail: shreyasharma.aug15@gmail.com

Ashish Raman, Prabhat Singh, Naveen Kumar & Ravi Ranjan (Eds.)
All rights reserved-© 2025 Bentham Science Publishers

INTRODUCTION

In the intricate tapestry of scientific exploration and technological innovation, sensing technologies emerge as a pivotal thread weaving through diverse realms. Sensors play a crucial role in today's world, serving as the sensory organs that enable the digital transformation of various industries. The significance of sensors lies in their potential to collect real-time data, providing valuable insights that drive efficiency, innovation, and improved decision-making across diverse sectors [1].

In healthcare, sensors play a crucial role in wearable devices, monitoring vital signs, tracking physical activity, and detecting early signs of medical conditions [2]. This promotes proactive healthcare management and personalized treatment plans. Agriculture benefits from sensors optimizing irrigation through soil moisture measurements, monitoring crop health, and employing precision farming for increased yield [3]. The automotive industry relies heavily on sensors, like LiDAR, radar, and cameras for advanced driver assistance and autonomous vehicles, enhancing safety and enabling self-driving capabilities [4, 5].

In manufacturing, sensors contribute to process automation, predictive maintenance, and quality control, boosting productivity and reducing operational costs [6]. In smart cities, sensors monitor and manage urban infrastructure, fostering sustainability and improving the quality of life for residents [7]. The Internet of Things (IoT) heavily depends on sensors, forming interconnected devices for seamless communication, leading to smart homes, grids, and industries [8]. As technology evolves, sensors' significance will grow, influencing advancements in various sectors, like artificial intelligence, machine learning, and data analytics.

In the realm of sensing technologies, semiconductor devices have emerged as integral components, representing a technological frontier that has redefined the landscape of detection and measurements [9, 10]. Semiconductors are fundamental components in the field of sensors and sensing technologies, serving as the building blocks for various sensor types. Semiconductor-based sensors leverage the unique electrical properties of semiconductors to detect and measure physical changes, allowing for precise and reliable sensing across a wide range of applications [11].

One prominent example of semiconductor sensing technology is Metal Oxide Semiconductor (MOS) structures in gas sensing. In these sensors, a semiconductor material is coated with a metal oxide, and the electrical conductivity of the semiconductor changes in the presence of specific gases [12]. Such gas sensors find applications in environmental monitoring, industrial safety, and even in smart homes for detecting gas leaks [13, 14]. Another notable application is the use of semiconductor devices in temperature sensors [15]. Thermistors, which are temperature-sensitive resistors, are made from semiconductor materials. As the temperature changes, the resistance of the semiconductor material also changes proportionally, allowing for accurate temperature measurements [16]. These sensors are extensively used in climate control systems, weather stations, and various electronic devices.

Semiconductors are integral to the development of photodetectors and photodiodes also, which are used in light sensors. The semiconductor material in these sensors generates a current when exposed to light and the current is directly related to the amount of light falling on the sensor. This principle is employed in applications, such as ambient light sensors in smartphones, cameras, and automatic lighting systems [17]. Furthermore, semiconductor-based sensors are crucial in pressure sensing technologies. For instance, piezoresistive sensors use semiconductor materials that change resistance in response to mechanical stress, allowing for accurate pressure measurements. These sensors are commonly found in automotive applications, medical devices, and industrial equipment [18].

The role of semiconductors in sensing technologies is significant because of their versatility, scalability, and ability to integrate with electronic systems. Advances in semiconductor manufacturing processes have led to the development of miniaturized sensors, which are also energy-efficient, making them suitable for diverse applications in today's interconnected and data-driven world [19]. Expanding upon the advanced sensing technologies within the realm of semiconductor applications, the intricacies of innovation that have propelled this field to the forefront of scientific inquiry are important to understand. The integration of nanomaterials with cutting-edge technologies emerges as a transformative force as we delve into the intricate world of advanced semiconductor sensing. The integration of nanomaterials into semiconductor sensing devices unleashes a spectrum of possibilities, fostering innovation in both design and functionality. This chapter serves as a beacon for researchers, engineers, and enthusiasts, exploring the intricate interplay between materials, design, and nanoscale challenges in semiconductor sensing technologies.

FUNDAMENTALS OF SEMICONDUCTOR SENSING

Semiconductor sensing relies on the unique electrical properties of semiconductor materials to detect changes in physical quantities, like light, pressure, temperature, gas concentration, and more. Semiconductor sensors employ various detection mechanisms based on the interaction between charge carriers and the external stimulus. The fundamentals of semiconductor sensing include the knowledge of material properties, doping, charge carriers, detection mechanisms, transduction, calibration, and signal processing. Overall, semiconductor sensing relies on a deep understanding of semiconductor physics and materials science principles to design sensors that are sensitive, selective, and reliable in industries, such as healthcare, automotive, environmental monitoring, and consumer electronics [19]. Here, we briefly discuss various types of semiconductor sensors, the mechanisms involved in various types of sensors, and their response time, along with the effects of environmental influences.

Types of Semiconductor Sensing Technologies

Semiconductor sensing technologies encompass a broad spectrum of methods for detecting and measuring physical quantities using semiconductor materials.

Resistive Sensing

Resistive sensing is a fundamental semiconductor sensing technology that relies on changes in electrical resistance to detect and measure physical quantities. This approach exploits the intrinsic electrical properties of semiconductor materials, particularly their sensitivity to external stimuli, such as temperature, pressure, light, or gas concentration [20]. In resistive sensing, the basic principle involves the modulation of electrical resistance within a semiconductor material in response to an applied stimulus, as depicted in Fig. (**1a**). Semiconductor materials exhibit a characteristic resistance that varies with factors, like temperature, mechanical stress, or chemical environment. This variation arises from changes in the concentration and mobility of charge carriers within the semiconductor lattice structure [21].

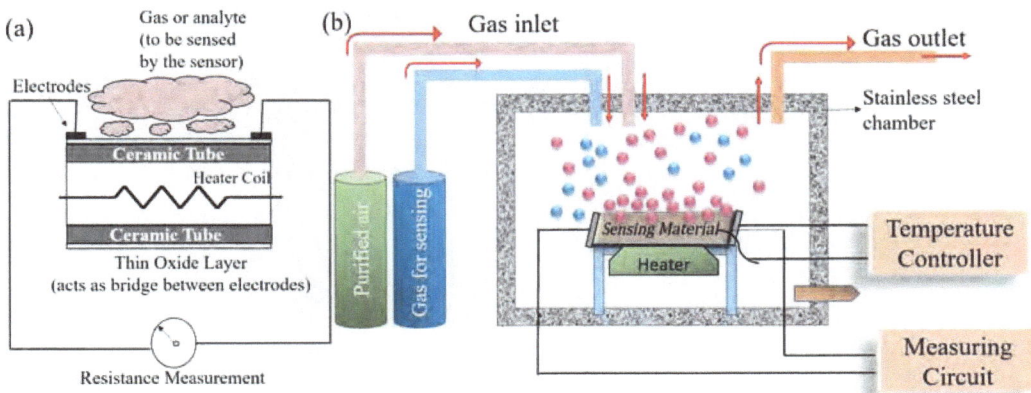

Fig. (1). Schematic diagram showing the (**a**) mechanism of resistive sensing and (**b**) metal oxide semiconductor gas sensor.

In MOS gas sensors, the conductivity of the semiconductor material is altered by the interaction between gas molecules and the semiconductor surface, as shown in Fig. (1b). It leads to a change in resistance, providing a measurable indication of the gas concentration [22, 23]. Thermistors, another type of resistive sensor, demonstrate a strong temperature-dependent resistance behavior. As the temperature increases, the thermal energy promotes more charge carriers into the semiconductor's conduction band, reducing resistance. Conversely, at lower temperatures, fewer charge carriers are available for conduction, resulting in higher resistance [24].

Resistive sensors offer several advantages, including high sensitivity, wide operating temperature ranges, and compatibility with standard electronic circuitry. Moreover, their simplicity allows for easy integration into various sensing applications without the need for complex signal-processing techniques. However, they may exhibit nonlinear behavior and require calibration to ensure accurate measurements over their operating range [25].

Capacitive Sensing

Capacitive sensing is a semiconductor-based technology leveraging changes in capacitance to detect and quantify physical properties. These sensors are appreciated for their sensitivity, minimal power requirements, and versatility across sectors from consumer electronics to industrial automation [26]. At its core, capacitive sensing relies on capacitance modulation, where a sensor's capacitance shifts in response to an external trigger, often due to alterations in dielectric properties or geometry. Jin *et al* . introduced a multi-functional capacitive sensor fabricated using Polydimethylsiloxane (PDMS) and Ag nanowire electrodes. It has been fabricated by selective oxygen plasma treatment method and could sense different stimuli [27].

Typically, semiconductor-based capacitive sensors feature conductive materials separated by a dielectric layer, constituting a capacitor [28, 29]. For example, in capacitive touch sensors employed in consumer electronics, like smartphones, the interaction with a conductive object, like a finger, modifies the capacitance between electrodes, signaling touch presence and location [30]. Similarly, capacitive humidity sensors exploit capacitance changes arising from moisture absorption or desorption by a hygroscopic dielectric layer. These sensors are integral to environmental monitoring, HVAC controls, and industrial process supervision for precise humidity assessment. Furthermore, in automotive contexts, capacitive proximity sensors detect nearby objects, facilitating Advanced Driver Assistance Systems (ADAS) and parking aids [31].

Optical Sensing

Optical sensing, a semiconductor technology, employs light to detect and measure various physical parameters. This method capitalizes on the interaction between semiconductor materials and photons to capture, process, and analyze optical signals. Optical sensors offer rapid response times and high sensitivity and immunity to electromagnetic interference, making them ideal for applications spanning telecommunications, medical diagnostics, environmental monitoring, and industrial automation [32]. At the heart of optical sensing lies the principle of light-matter interaction within semiconductor materials. Utilizing materials, like Si, Ge, GaAs, or InP, which possess unique optical properties, such as absorption, emission, and transmission of light, semiconductor-based optical sensors are structured [33].

A prevalent type of semiconductor optical sensor is the photodiode, converting incident light into electrical current through the photovoltaic effect. Photodiodes find utility in ambient light sensing, optical communication systems, and photovoltaic solar cells [34, 35]. Another vital semiconductor optical sensor is the Charge-coupled Device (CCD), which is employed in imaging and optical signal capture [36]. Beyond photodiodes and CCDs, semiconductor-based optical sensors encompass various other technologies, like phototransistors, Photonic Integrated Circuits (PICs), and optoelectronic sensors, as shown in Fig. (**2**) [37, 38]. These sensors utilize mechanisms, like photoconductivity, photoluminescence, or photonic resonance to detect and measure optical signals. Moreover, semiconductor optical sensors integrate advanced signal processing techniques, like wavelength multiplexing, time-domain modulation, and frequency-domain analysis, to enhance performance and functionality. This makes the optical sensors capable of detecting subtle changes in light intensity, wavelength, polarization, and phase [39].

Piezoresistive Sensing

Piezoresistive sensing exploits resistance modulation in response to mechanical stress or pressure, leveraging semiconductor materials' intrinsic properties. This enables the detection and quantification of physical parameters, like pressure, force, or strain [40]. These sensors offer high sensitivity, wide dynamic range, and integration with circuitry, making them versatile for applications from automotive systems to biomedical devices [41]. At the heart of piezoresistive sensing lies piezoresistance, where a semiconductor's electrical resistance changes under mechanical stress due to alterations in its crystal lattice structure, affecting charge carrier mobility and concentration. Piezoresistive sensing is widely used in pressure sensors for automotive tire pressure monitoring, industrial process control, and medical blood pressure measurement, offering advantages, like high sensitivity and fast response times. Additionally, these sensors are vital in force, acceleration, and strain sensing applications, contributing to automotive safety, aerospace, and structural health monitoring [42]. Fig. (**3**) depicts the materials, structures, performance, processes, and applications of flexible pressure sensors.

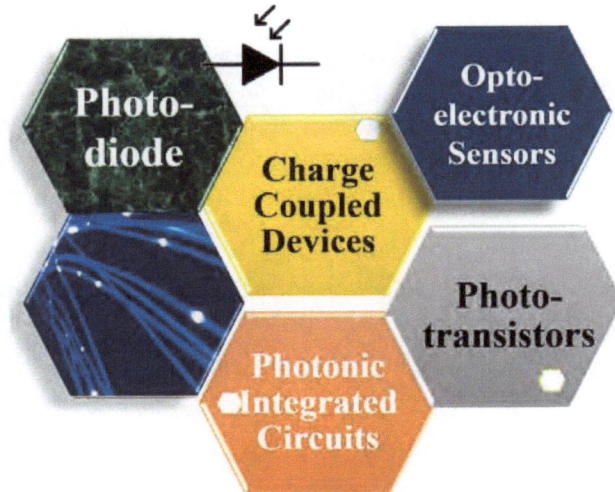

Fig. (2). Types of optical sensors.

Chemical Sensing

Chemical sensing, within the realm of semiconductor technology, involves the detection and measurement of chemical species through interactions between semiconductor materials and target analytes. Chemical sensors based on semiconductor technology offer advantages, such as high selectivity, sensitivity, and rapid response times, making them invaluable tools for applications in environmental monitoring, industrial safety, and healthcare diagnostics. The fundamental principle underlying semiconductor-based chemical sensing is the interaction between the target analyte and the semiconductor material, leading to changes in its electrical, optical, or mechanical properties. These changes are typically manifested as alterations in the conductivity or resistance of the semiconductor, resulting from adsorption, desorption, or chemical reactions at the sensor's surface [44].

One of the most common types of semiconductor chemical sensors is the Metal Oxide Semiconductor (MOS) sensor, which detects changes in electrical conductivity when exposed to specific gases [45]. Another important semiconductor chemical sensing technology is the electrochemical sensor, which utilizes semiconductor-based electrodes immersed in an electrolyte solution to detect changes in chemical concentration through electrochemical reactions [46, 47, 48]. Additionally, semiconductor Quantum Dots (QDs) are emerging as promising platforms for chemical sensing due to their unique optical properties. By functionalizing QD surfaces with specific receptors, they can be tailored to detect various chemical analytes with high sensitivity and selectivity [49].

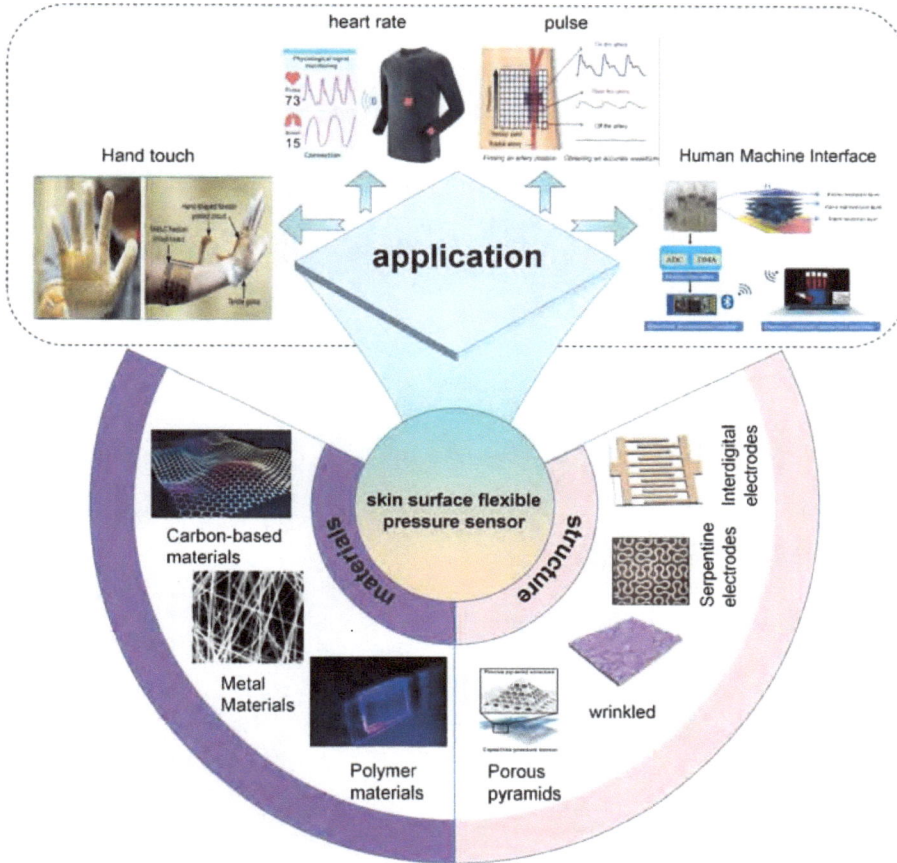

Fig. (3). Materials, structures, performance, processes, and applications of flexible pressure sensors [43].

Biomedical Sensing

Biomedical sensing in semiconductor technology involves detecting and analyzing biological parameters or biomolecules for healthcare, diagnostics, and biotechnology. It utilizes semiconductor materials and fabrication techniques to create sensors that detect physiological signals, biomarkers, and pathogens with high sensitivity and specificity. These sensors offer advantages, like miniaturization, multiplexing, and compatibility with integrated electronics, enabling diverse clinical and research applications. They operate on principles, such as electrochemical, optical, or impedance sensing, to detect and quantify biological signals or analytes [50].

Electrochemical biosensors use semiconductor electrodes to measure changes in electrical current or potential resulting from biochemical reactions with target analytes, as shown in Fig. (**4**). They find applications in glucose monitoring, disease diagnosis, and neuroscience research [51]. Semiconductor technology enables the development of wearable and implantable biomedical sensors for continuous monitoring of parameters, like heart rate and blood pressure. These sensors provide real-time health monitoring and personalized healthcare solutions, crucial for point-of-care testing, remote patient monitoring, and drug discovery. They offer rapid, sensitive, and cost-effective solutions for disease detection, treatment monitoring, and drug efficacy evaluation, advancing precision medicine and personalized healthcare [52].

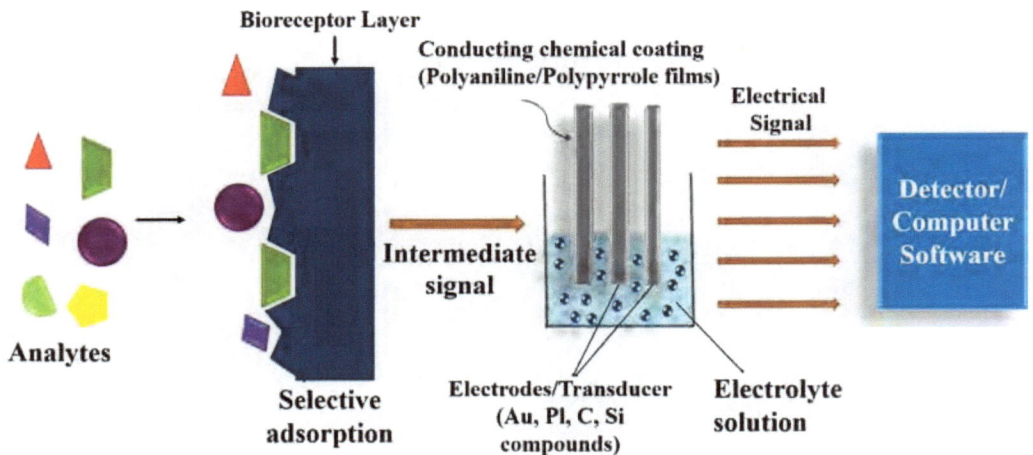

Fig. (4). Schematic diagram of electrochemical biosensing.

Quantum Sensitivity in Semiconductor Sensors

Semiconductor sensors exhibit quantum sensitivity, detecting physical quantities with unparalleled precision due to quantum mechanics. This sensitivity arises from the unique quantum properties of semiconductor materials, allowing manipulation and detection of individual quanta of energy or charge carriers [53].

In semiconductor sensors, quantum sensitivity is evident in discrete electron energy levels within the material's band structure, described by the Schrodinger equation, as follows:

$$-\frac{\hbar^2}{2m^*}\nabla^2\psi + V(r)\psi = E\psi \tag{1}$$

Where, h and m* are the reduced Plank constant and the effective mass of the electron in the semiconductors, respectively; ψ represents the wave function of the electron, $V(r)$ is the potential energy experienced by the electron in the semiconductor material, and E represents the energy eigenvalue associated with the electron's quantized energy level.

When stimulated by external factors, like light or heat, the sensor absorbs or emits photons or phonons in quantized packets of energy, enabling the detection of minute environmental changes. Semiconductor sensors exploit the quantized nature of charge carriers, like electrons and holes. Quantum effects govern their movement, leading to phenomena, such as tunneling and ballistic transport, thus resulting in ultra-high sensitivity in detecting electrical signals. Examples include quantum dot sensors and Single-electron Transistors (SETs) [54]. Quantum dots, nanoscale structures with discrete energy levels, detect single photons or electrons, ideal for applications, like quantum computing. SETs manipulate individual electrons for sensitive charge and current detection, applicable in fields, like quantum information processing [55].

Selective Binding Mechanisms

The selective binding mechanism refers to the capability of the sensor to specifically interact with target molecules or analytes, while excluding interference from other substances in the surrounding environment. This mechanism is crucial for achieving high selectivity, sensitivity, and accuracy in the detection and measurement of specific chemical or biological species. Selective binding mechanisms are often employed in semiconductor-based chemical and biological sensors, where the sensor surface is functionalized with recognition elements that exhibit affinity for the target analyte. These recognition elements can include antibodies, enzymes, aptamers, or molecularly imprinted polymers, depending on the nature of the analyte and the desired level of selectivity [56].

In chemical or biomedical sensing applications, the selective binding mechanism involves the adsorption or absorption of target molecules onto the surface of the semiconductor sensor or the detection of specific biomolecules, such as proteins, DNA, or cells, respectively [57]. The sensor surface may be functionalised with specific chemical receptors or antibodies that interact selectively with the target analyte through chemical interactions, as shown in Fig. (**5**). This selective binding leads to changes in the electrical, optical, or mechanical properties of the sensor, which can be measured and correlated with the concentration of the target analyte.

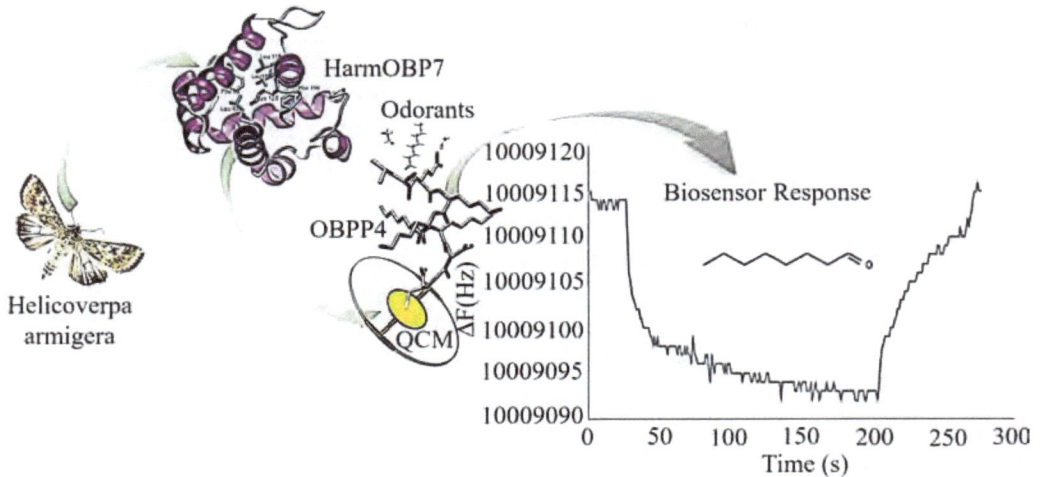

Fig. (5). A highly selective peptide (derived from the HarmOBP7 aldehyde binding site) based biosensor along with the selective binding mechanism.

The selective binding mechanism is essential for minimizing false positives and false negatives in sensor measurements, particularly in complex biological or environmental samples containing multiple analytes. By ensuring that only the target analyte binds to the sensor surface, while other substances are excluded or do not interfere with the detection process, selective binding mechanisms enhance the accuracy and reliability of semiconductor sensors [58].

Temporal Dynamics and Response Time

Temporal dynamics and response time are critical parameters in semiconductor sensors, defining how quickly and accurately a sensor can detect changes in the environment and provide a corresponding output signal. These aspects are especially crucial in applications requiring real-time monitoring, rapid feedback, or precise control. Temporal dynamics refers to how the sensor's output changes over time in response to variations in the input stimulus. This includes aspects, such as the sensor's settling time, transient response, and stability over time. Semiconductor sensors should exhibit fast response times and minimal transient effects to accurately capture dynamic changes in the measured parameter. For example, in environmental monitoring, sensors must be able to rapidly detect fluctuations in pollutant levels or temperature to provide timely alerts or control actions [59].

Response time, on the other hand, specifically quantifies how quickly a sensor achieves a specified level of output change in response to a step change. It is typically measured as the time taken for the sensor's output to reach a certain value (*i.e.*, 90% or 95%) of the final value after a sudden change in the input. A shorter response time indicates a faster sensor, capable of detecting and responding to changes in the environment more quickly [60]. First order response model describes the transient response of a sensor to a step change in the input stimulus by using a first-order differential equation of the form:

$$\frac{dy(t)}{dt} = -\frac{1}{\tau} \cdot (y(t) - y_{ss}) \tag{2}$$

Where, $y(t)$ is the sensor output at time t, τ is the time constant of the sensor, which represents the time taken for the sensor's output to reach 63.2% of its final value, and y_{ss} is the steady-state value of the sensor output. This equation can be used to quantify the sensor's response time based on its time constant.

Rise time (t_r) is another parameter used to quantify the speed of a sensor's response. It is defined as the time taken for the sensor's output to rise from 10% to 90% of its final value in response to a step input, as shown in Fig. (**6a** and **b**) [61]. It can be calculated using the following equation:

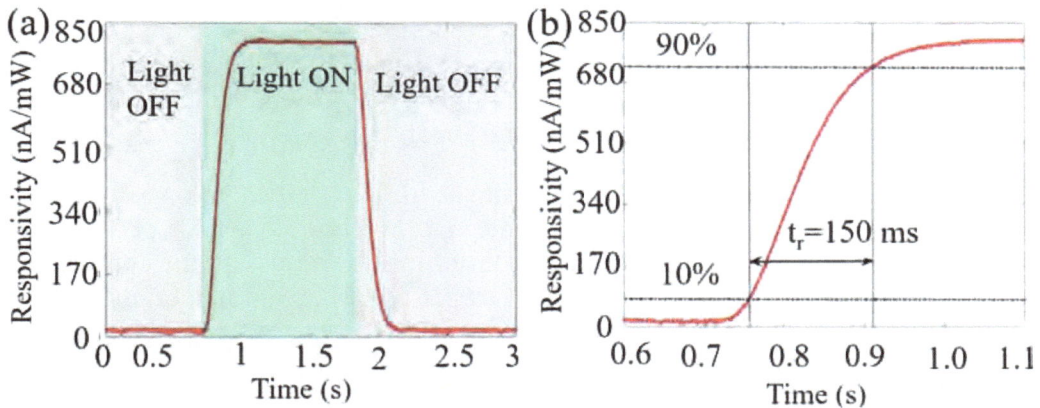

Fig. (**6**). (**a**) Responsivity of a photodetector over one cycle and (**b**) rise time calculation from the response.

$$t_r = t_{90} - t_{10} \tag{3}$$

Where, t_{90} is the time at which the output reaches 90% of its final value and t_{10} is the time at which the output reaches 10% of its final value. Rise time provides a measure of how quickly the sensor responds to changes in the input stimulus.

Achieving fast response times and optimal temporal dynamics in semiconductor sensors requires careful design considerations, including sensor materials, geometry, signal processing techniques, and environmental conditions. For example, reducing the size of sensor elements or optimizing their geometry can minimize thermal mass and increase responsiveness. Additionally, improving the efficiency of signal processing algorithms can enhance the sensor's ability to extract relevant information from noisy or time-varying signals [61].

The choice of semiconductor material also plays a crucial role in determining temporal dynamics and response time. For instance, certain materials may exhibit faster charge carrier mobility or higher sensitivity to changes in the environment, leading to quicker sensor responses. Moreover, advancements in semiconductor fabrication technologies, such as Micro-electromechanical Systems (MEMS) and nanotechnology, enable the development of sensors with reduced dimensions and improved temporal performance [62]. In practical applications, the temporal dynamics and response time of semiconductor sensors must be optimized to meet specific requirements dictated by the application's dynamics, accuracy constraints, and environmental conditions.

NANOMATERIALS IN SEMICONDUCTOR SENSING

Nanomaterials have emerged as key building blocks in semiconductor sensing technology, offering a wide range of functionalities that enhance sensor performance across various applications. These nanoscale materials exhibit unique properties (physical, chemical, and electronic) owing to their small size, high aspect ratio, and quantum confinement [63, 64].

Nanomaterials can be engineered to exhibit specific properties tailored for sensing applications. This tunability allows for the optimization of sensor performance, including sensitivity, selectivity, and response times by adjusting the composition, size, and morphology of the nanomaterials [65]. Nanomaterials can be functionalized with various molecules, receptors, or ligands to enhance sensor selectivity and enable the detection of specific analytes. Functionalization provides molecular recognition capabilities, improving sensor specificity and reducing interference from non-target substances.

There is a diverse range of nanomaterials employed in semiconductor sensing, each with its own set of advantages and applications. From nanoparticles and nanowires to nanotubes and quantum dots, these nanomaterials offer unparalleled opportunities for sensing a multitude of analytes, including gases, chemicals, biomolecules, and environmental pollutants. By harnessing the unique properties of nanomaterials, semiconductor sensors can achieve exceptional sensitivity, rapid response times, and robust performance, making them indispensable tools for a wide range of industries and research fields.

Metal and Noble Metals

In the realm of semiconductor sensing, metal and noble metal nanomaterials represent a versatile class of materials with exceptional sensing capabilities. These nanomaterials, including platinum (Pt), silver (Ag), palladium (Pd), and gold (Au), exhibit unique properties that make them highly desirable for sensing applications across various domains [66].

Au and Ag nanoparticles are among the most extensively studied metal nanomaterials for sensing purposes. Their remarkable optical properties, including Surface Plasmon Resonance (SPR), make them ideal candidates for label-free detection methods, such as surface-enhanced Localized Surface Plasmon Resonance (LSPR) and Raman Scattering (SERS) sensing, as shown in Fig. (**7a**). These nanomaterials can be functionalized with specific receptors to selectively bind target analytes, enabling sensitive and selective detection in applications ranging from medical diagnostics to environmental monitoring, as shown in Fig. (**7b**) [67, 68]. Noble metal nanomaterials, for instance, Pt and Pd, are renowned for their catalytic activity and stability, making them indispensable for gas sensing applications. Pt and Pd nanoparticles are commonly employed as sensing elements in catalytic gas sensors, where they facilitate the detection of gases, for *e.g.*, hydrogen, carbon monoxide, and nitrogen oxides [69].

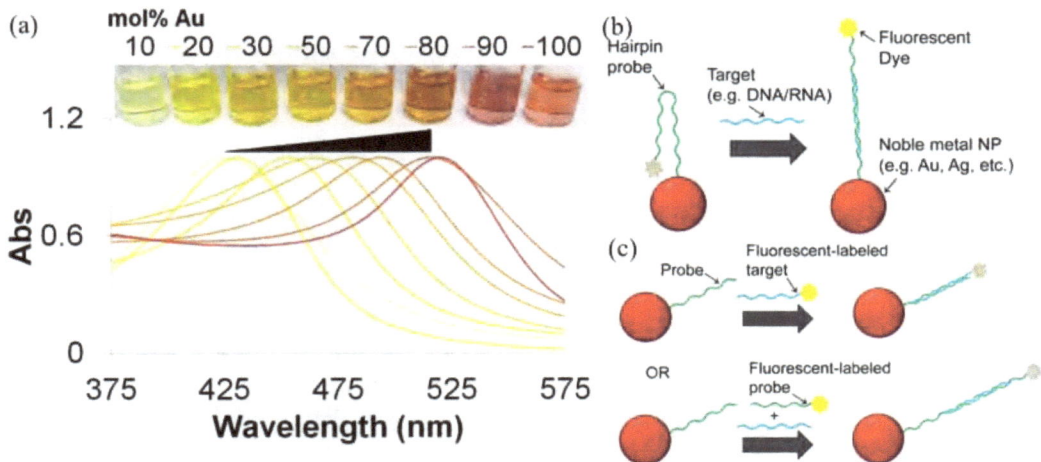

Fig. (7). (**a**) Example of LSPR modulation through different NP compositions; (**b**) molecular nanobeacons; (**c**) nanoprobes [70].

In addition to their intrinsic properties, metal and noble metal nanomaterials can be tailored and functionalized to enhance their sensing performance. Surface modification techniques, such as functionalization with organic ligands or self-assembled monolayers, allow for the selective binding of target analytes and improved sensor specificity. Moreover, the integration of metal nanomaterials into semiconductor-based sensor platforms enables the development of miniaturized, portable sensing devices with enhanced performance characteristics.

Metal Oxide Nanoparticles

Metal oxide nanoparticles play a pivotal role in semiconductor sensing, offering exceptional sensitivity, selectivity, and versatility for detecting a wide range of analytes. Metal oxide nanoparticles, including titanium dioxide (TiO_2), zinc oxide (ZnO), tin dioxide (SnO_2), and iron oxide (Fe_2O_3), exhibit unique electronic, optical, and chemical characteristics that make them well-suited for sensing a wide range of analytes.

Titanium dioxide nanoparticles (TiO_2 NPs) are extensively utilized in gas sensing applications due to their high surface area, excellent stability, and sensitivity to various gases. TiO_2-based gas sensors operate *via* the adsorption and desorption of gas molecules on the nanoparticle surface, leading to changes in electrical conductivity or other measurable properties. These sensors are particularly effective for detecting reducing gases, such as H_2 and CO, making them valuable for industrial safety and environmental monitoring applications [71].

Zinc oxide nanoparticles (ZnO NPs) are renowned for their semiconducting properties, large surface area, and high electron mobility, making them ideal candidates for chemical sensing, gas sensing, and biosensing applications. ZnO-based sensors exhibit high sensitivity and selectivity towards a wide range of analytes, like volatile organic compounds, gases, and biomolecules. Additionally, ZnO NPs can be functionalized with specific receptors or surface modifiers to enhance sensor specificity and selectivity for target analytes [72].

Tin dioxide nanoparticles (SnO_2 NPs) are widely used in gas sensing applications due to their high sensitivity to gases, like CO, H_2, and CH_4. SnO_2-based gas sensors operate *via* the adsorption of gas molecules onto the nanoparticle surface, leading to changes in electrical conductivity or resistance. These sensors exhibit rapid response times, high sensitivity, and good stability, making them suitable for applications in automotive exhaust monitoring, industrial safety, and indoor air quality monitoring [73].

Iron oxide nanoparticles (Fe_2O_3 NPs), including hematite (α-Fe_2O_3) and magnetite (Fe_3O_4), are employed in gas sensing, chemical sensing, and biomedical sensing applications. Fe_2O_3-based sensors exhibit high sensitivity toward reducing gases, VOCs, and biomolecules, owing to their unique magnetic and catalytic properties. These sensors operate *via* surface reactions between gas molecules and nanoparticle surfaces, leading to changes in electrical conductivity or other measurable parameters. Fe_2O_3 NPs can also be functionalized with specific ligands or receptors for selective detection of target analytes in biomedical and environmental sensing applications [74].

Carbon-based Nanomaterials

In semiconductor sensing, carbon-based materials represent a diverse and promising class of nanomaterials. Their unique properties and capabilities make them invaluable tools for addressing complex sensing challenges in areas, such as environmental monitoring, healthcare diagnostics, industrial process control, and beyond. These materials, including Carbon Nanotubes (CNTs), graphene, Carbon Nanofibers (CNFs), and Carbon Quantum Dots (CQDs), as shown in Fig. (**8**), offer unique electronic, mechanical, and chemical characteristics that make them highly attractive for sensing applications.

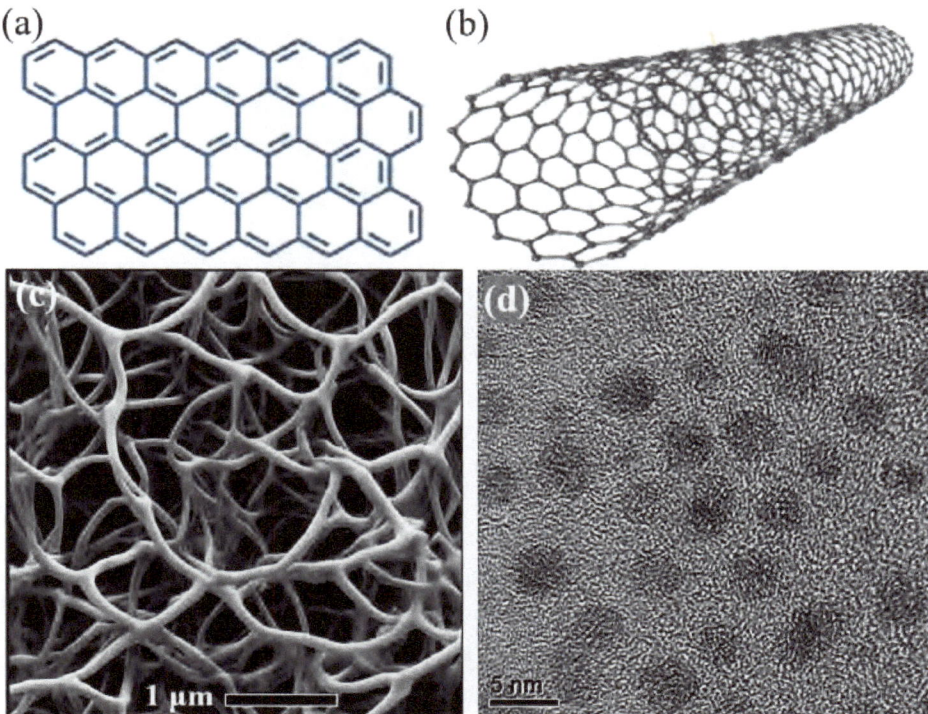

Fig. (**8**). (**a**) Schematic diagram of graphene and (**b**) carbon nanotubes [77]; (**c**) AI-generated FESEM image of carbon nanofibres; (**d**) TEM images of carbon quantum dots [78].

CNTs are one of the widely studied carbon-based materials for sensing as their unique one-dimensional structure enables efficient charge transport, leading to rapid response times and high sensitivity. Functionalization of CNT surfaces with specific receptors or functional groups further enhances sensor selectivity and enables the detection of target analytes with high precision [75]. Graphene, a two-dimensional carbon material consisting of a single layer of sp2-bonded carbon atoms arranged in a honeycomb lattice, has garnered significant attention for sensing applications. Graphene exhibits exceptional electrical conductivity, high surface area, and mechanical strength, making it an ideal platform for gas sensing, biosensing, and environmental monitoring [76].

Carbon Nanofibers (CNFs) are another class of carbon-based materials widely explored for sensing applications. CNFs possess a fibrous morphology with high aspect ratios, large surface area, and excellent electrical conductivity. Their porous structure allows efficient adsorption of analyte molecules, leading to changes in electrical conductivity or other measurable properties [79]. Carbon Quantum Dots (CQDs) are emerging carbon-based nanomaterials that exhibit size-dependent fluorescence properties, making them ideal candidates for fluorescent sensing of ions, biomolecules, and environmental pollutants. Their small size, biocompatibility, and tunable surface chemistry enable sensitive and selective detection of analytes in biological, environmental, and chemical sensing applications. Additionally, CQDs can be integrated into various sensing platforms, including solid-state devices, optical sensors, and biosensors, offering versatile solutions for sensing applications [80].

Transition Metal Dichalcogenides (TMDs)

TMDs are layered materials composed of transition metal atoms sandwiched between chalcogen atoms, such as sulfur (S), selenium (Se), or tellurium (Te). Some of the TMDs are molybdenum disulfide (MoS_2), tungsten disulfide (WS_2), and molybdenum diselenide ($MoSe_2$) [81]. TMDs have emerged as promising nanomaterials for semiconductor sensing, offering a high surface-to-volume ratio, large surface area, and tunable electronic properties that enable sensitive and selective detection of various analytes along with rapid response times. TMDs also exhibit unique electronic band structures, including direct bandgaps in the monolayer regime, which enable efficient charge transport and high sensitivity to changes in the local environment [82]. Fig. (**9**) shows brief details of 2D TMDs, including the preparation methods, structure, and properties.

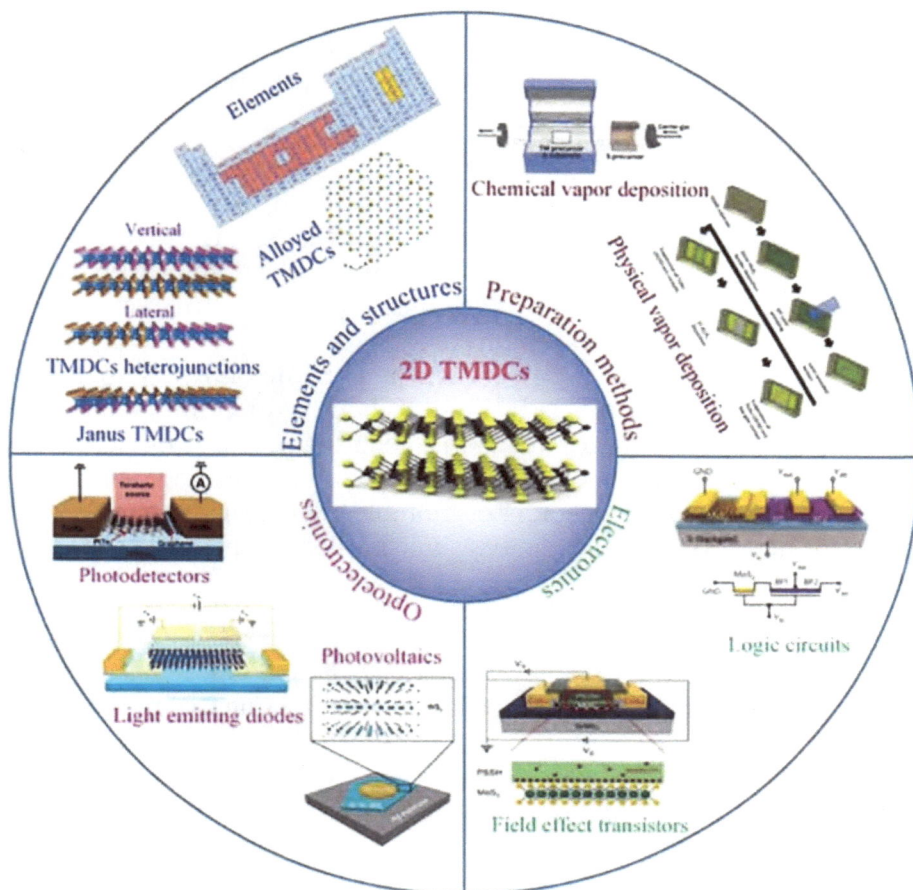

Fig. (9). Brief details of 2D TMDs, including preparation methods, structure and properties [86].

TMD-based sensors operate *via* surface interactions between the target analyte molecules and the TMD surface, leading to changes in electrical conductivity, optical properties, or other measurable parameters. For example, TMD field-effect transistors can detect gas molecules through adsorption-induced changes in the conductance of the TMD channel [83]. TMDs have also shown promise for biosensing as they can detect biomolecular interactions, such as antibody-antigen binding or DNA hybridization. Functionalization of TMD surfaces with specific receptors or biomolecules further enhances sensor selectivity and enables sensitive detection of target analytes in biological samples [84].

TMDs exhibit strong light-matter interactions, making them suitable for optoelectronic sensing applications. TMD-based photodetectors and photonic devices can detect light with high sensitivity and efficiency, enabling applications, such as optical sensing, spectroscopy, and imaging. The unique optical properties of TMDs, including photoluminescence and Raman scattering, provide additional avenues for sensing and detection of analytes in various environments [85].

Polymer and Bio-nanomaterials

Polymer and bio-nanomaterials that encompass a diverse array of polymers, biopolymers, and biomolecules, have gained significant attention in semiconductor sensing due to their biocompatibility and excellent sensor performance and selectivity [87]. Polymer nanomaterials, including conductive polymers, polymer nanoparticles, and nanocomposites, offer excellent mechanical flexibility, tunable electrical properties, and facile functionalization [88]. Conductive polymers, like PANI, PPy, and PEDOT, exhibit intrinsic conductivity and can undergo reversible redox reactions, enabling sensitive detection of gases, biomolecules, and chemical species. Polymer nanoparticles and nanocomposites, functionalized with specific receptors or biomolecules, enhance sensor selectivity and enable the detection of target analytes in complex sample matrices [89].

Biopolymer nanomaterials, including proteins, peptides, DNA, and polysaccharides, offer biocompatibility, specificity, and self-assembly properties, making them ideal for biosensing applications. Biopolymer-based sensors can detect biomolecular interactions, such as protein-protein binding or DNA hybridization [90]. Furthermore, bio-nanomaterials derived from living organisms, such as viruses, bacteria, and cells, offer unique functionalities for sensing applications. Polymer and bio-nanomaterials can be integrated into semiconductor-based sensor platforms, such as Field-effect Transistors (FETs), microfluidic devices, and lab-on-a-chip systems, to enable miniaturized, portable sensing devices with enhanced performance characteristics.

FABRICATION AND TESTING OF NANOSCALE SEMICONDUCTOR SENSORS

Fabrication and testing are crucial stages in the development of nanoscale semiconductor sensors, where careful design and engineering principles are applied to achieve high sensitivity, selectivity, and reliability. The design and architecture of these sensors play a critical role in determining their performance characteristics, including sensitivity, response time, and signal-to-noise ratio. In this section, we explore the key considerations and methodologies involved in fabricating and testing nanoscale semiconductor sensors, focusing on the design principles, fabrication techniques, and characterization methods used to optimize the sensor's performance.

Architecture and Layout

The architecture and layout of nanoscale semiconductor sensors are critical factors that directly influence their performance, sensitivity, and reliability. The design of sensor architecture involves careful consideration of various parameters, including sensor geometry, electrode configuration, and interconnectivity. Additionally, the layout of sensor components plays a crucial role in optimizing signal transduction, minimizing noise, and enhancing detection capabilities. A key aspect of sensor architecture is the selection of appropriate materials and fabrication techniques to achieve the desired sensor characteristics. A sensor fabricated using semiconductor thin film typically consists of several essential components, including the substrate, semiconductor thin film, electrodes, contacts, encapsulation layer, transducer, and electronic circuitry. Some of the sensors fabricated by different researchers are shown in Fig. (**10**), in which some of the above-mentioned components are used.

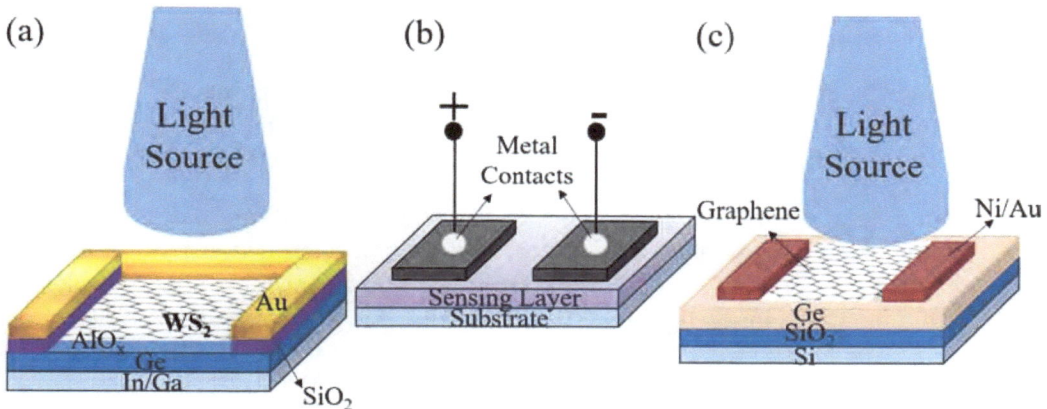

Fig. (10). (**a**) Schematic diagram of the $WS_2/AlO_x/Ge$ heterojunction device [91]; (**b**) schematic structure of the planar/resistive gas sensor [92]; (**c**) 3D schematic illustration of photodetector [93].

The substrate provides a solid foundation for the sensor and can be made of various materials, like silicon, glass, or flexible polymers. Semiconductor thin film is the active sensing element of the sensor. It is usually made of semiconductor nanomaterials, such as nanowires, nanotubes, or thin films of materials, like zinc oxide (ZnO), tin dioxide (SnO_2), or silicon (Si). These nanomaterials provide a high aspect ratio, enhancing sensitivity to target analytes. Electrodes are used to apply an electric field to a semiconductor nanomaterial thin film and measure the resulting electrical response. They can be made of materials, like gold, platinum, or Indium Tin Oxide (ITO). Contacts are used to connect the electrodes to the external circuitry for signal processing and readout.

An encapsulation layer protects the semiconductor nanomaterial thin film and electrodes from environmental factors, such as dust, moisture, and chemical contaminants. It helps ensure the stability and longevity of the sensor. In some cases, a transducer may be incorporated into the sensor design to convert the physical or chemical changes detected by the semiconductor nanomaterial thin film into an electrical signal. This signal can then be processed and analyzed by the electronic circuitry. Electronic circuitry includes components, such as amplifiers, filters, and analog-to-digital converters, which are used to process and analyze the electrical signals generated by the sensor. The circuitry may also include microcontrollers or digital signal processors for data processing and communication with external devices.

By integrating these components, sensors based on semiconductor nanomaterial thin films can detect a wide range of physical, chemical, and biological analytes with high sensitivity and selectivity. The layout of sensor components is carefully designed to maximize sensor performance, while minimizing parasitic effects, such as capacitive coupling and leakage currents.

Fabrication Techniques for Nanoscale Semiconductor Sensing Devices

Nanoscale Thin-film Deposition

Deposition is a critical process in the fabrication of semiconductor sensing devices, enabling the creation of uniform and precisely controlled sensing layers with tailored properties. Various deposition techniques are employed to deposit semiconductor materials onto substrates, allowing for the fabrication of thin films with nanoscale dimensions. One of the primary methods used for nanoscale thin film deposition is Physical Vapor Deposition (PVD). In PVD, material is evaporated from a solid source in a vacuum chamber and then condensed onto a substrate to form a thin film. Techniques, such as thermal evaporation, e-beam evaporation, and sputtering, are employed in PVD [94].

Chemical Vapor Deposition (CVD) is another widely used technique for nanoscale thin film deposition. In CVD, precursor gases are introduced into a reaction chamber, where they react to form a thin film on the substrate surface, as depicted in Fig. (**11**). CVD processes can be classified into various types, including thermal CVD, plasma-enhanced CVD, and Atomic Layer Deposition (ALD) [95]. In addition to PVD and CVD, other techniques, such as Molecular Beam Epitaxy (MBE) and Pulsed Laser Deposition (PLD), are also utilized for nanoscale thin film deposition. Each of these techniques offers unique advantages in terms of film quality, deposition rate, and control over film properties, allowing for precise fabrication of nanoscale semiconductor films for sensing applications.

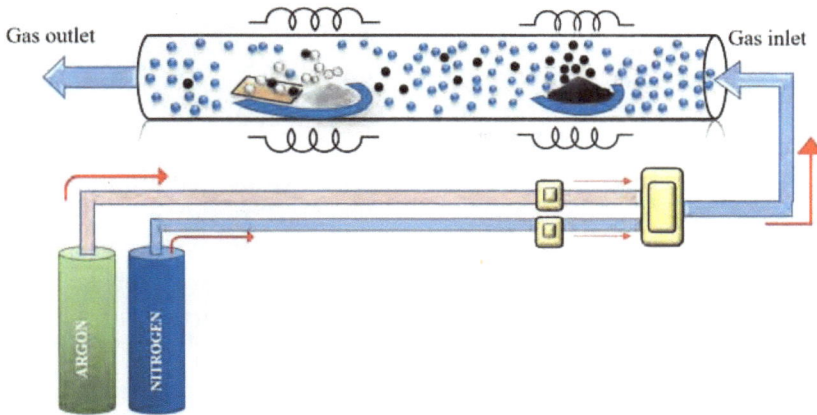

Fig. (11). Schematic diagram of nanoscale film deposition by double zone thermal CVD [95].

Some other techniques, like spin coating, dip coting, and spray pyrolysis, can also be utilised for the deposition of thin films of nanomaterials. These techniques require the synthesis of nanomaterials as a first step and then using the as-synthesized materials for the deposition of thin films. Spin coating involves depositing a thin film of liquid precursor onto a substrate and then spinning the substrate at high speeds to spread the liquid evenly over the surface, as shown in Fig. (**12**). While it is typically used for depositing polymer or organic thin films, it can also be adapted for depositing the thin films of semiconductor materials. Spin coating is particularly useful for creating thin films with thicknesses in the nanometer range, making it suitable for nanoscale sensor fabrication [96].

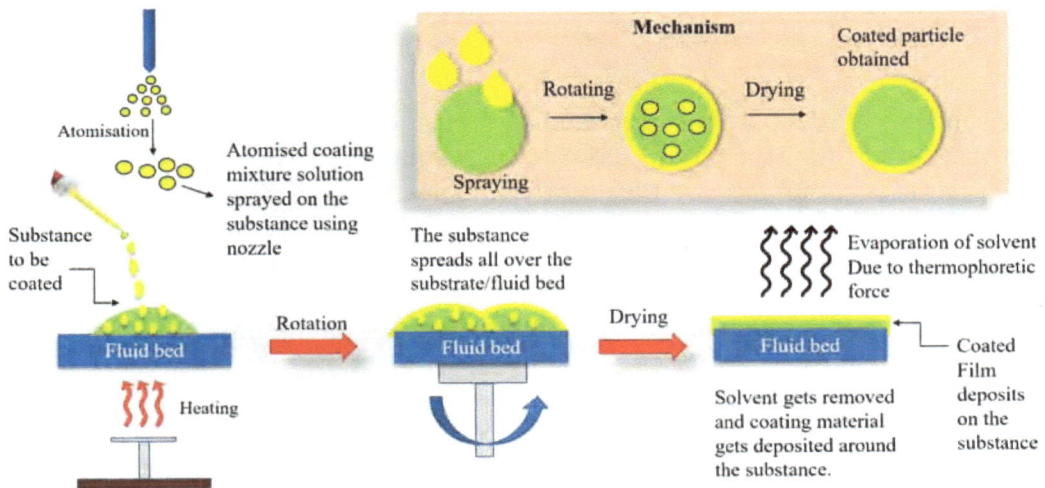

Fig. (12). Schematic diagram of nanoscale film deposition by spin coating.

Dip coating entails the immersion of a substrate into the precursor solution and then withdrawing it at a controlled rate, as depicted in Fig. (**13**). As the substrate is withdrawn, a thin film of the precursor material is deposited onto its surface [97]. Spray pyrolysis involves spraying a precursor solution or suspension onto a heated substrate, where it decomposes or reacts to form a thin film. This technique is particularly useful for depositing thin films of oxide or semiconductor materials at relatively low temperatures. Spray pyrolysis offers advantages, such as high throughput, scalability, and uniform film deposition. While it may not be as commonly used for nanoscale semiconductor sensor fabrication as other techniques, spray pyrolysis can still be employed for certain applications where thin film deposition is required [98].

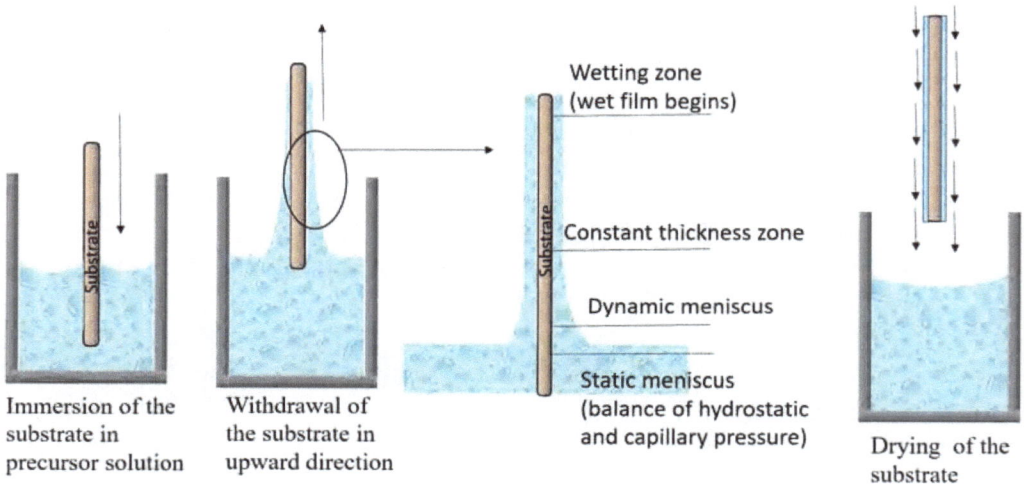

Fig. (13). Schematic diagram of the nanoscale film deposition by dip coating.

The choice of deposition technique depends on factors, like desired film properties, substrate material, and device requirements. For example, thermal evaporation is commonly used for depositing metal films due to its simplicity and high deposition rates, while Plasma-enhanced CVD (PECVD) is preferred for depositing dielectric films with precise control over film stoichiometry and thickness. ALD, on the other hand, offers atomic-level control over film deposition and is well-suited for depositing ultrathin films with excellent conformality and uniformity.

Photolithography for Nanoscale Patterning

Photolithography stands as a cornerstone in the fabrication of nanoscale semiconductor devices, providing the means to precisely define intricate patterns and geometries on semiconductor substrates. This technique, rooted in the principles of photochemistry and optics, enables the creation of nanoscale features essential for semiconductor sensor functionality. At its core, photolithography involves a sequence of steps beginning with the deposition of a photoresist material onto the substrate surface. This photoresist is then selectively exposed to Ultraviolet (UV) light through a photomask, which bears the desired pattern to be transferred onto the substrate. The exposure to UV light induces a chemical change in the photoresist, rendering it either soluble or insoluble in a subsequent development step, as shown in Fig. (**14a**). By carefully designing a photomask, complex patterns and features can be precisely delineated at the nanoscale [99].

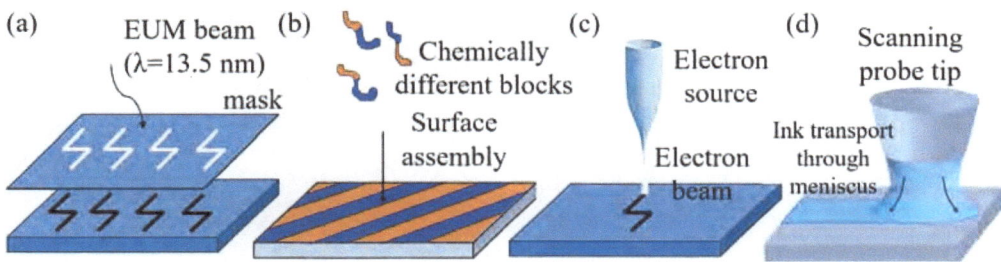

Fig. (14). Overview of high-resolution interface patterning methods: (**a**) UV photolithography, (**b**) block-copolymer lithography, (**c**) electron beam lithography, and (**d**) scanning probe lithography [101].

The success of photolithography in achieving nanoscale patterning hinges on several critical factors. One such factor is the resolution of the photomask, which determines the minimum feature size achievable in the pattern transfer process. Advances in photomask fabrication techniques, such as block copolymer lithography, electron beam lithography, Scanning Probe Lithography (SPL), and laser writing, as shown in Fig. (**14b, c**, and **d**), have pushed the boundaries of resolution to the sub-10 nanometer regime, enabling the creation of nanoscale sensor structures with unparalleled precision. Furthermore, advancements in photolithography techniques, such as Electron Beam Lithography (EBL) and nanoimprint lithography, have extended the capabilities of traditional photolithography to achieve even finer feature sizes and higher resolution [100].

Self-assembly Methods for Nanostructured Materials

Self-assembly methods represent a cutting-edge approach to the fabrication of nanostructured materials, offering a bottom-up approach to creating complex structures with precise control over morphology and functionality. These techniques harness the inherent properties of molecules to spontaneously organize into ordered structures, providing a route to fabricate nanostructured materials for semiconductor sensor applications [102]. Molecular self-assembly involves noncovalent interactions-driven spontaneous organization of molecules into well-defined structures. By carefully selecting precursor molecules and optimizing processing conditions, researchers can manipulate these interactions to guide the self-assembly process and create nanostructured materials with tailored properties [103].

Colloidal self-assembly, which includes hydrothermal and solvothermal routes, sol-gel method, *etc.*, as shown in Fig. (**15a** and **b**), relies on the assembly of colloidal particles into ordered arrangements through processes, such as sedimentation, diffusion, or evaporation-induced assembly. Colloidal particles, like nanoparticles, can self-assemble into close-packed structures, lattices, or superlattices with tunable properties [104]. By controlling parameters, such as particle size, shape, and surface chemistry, researchers can engineer the self-assembly process to create nanostructured materials with tailored properties for semiconductor sensor applications. For example, gold nanoparticles can self-assemble into ordered arrays with plasmonic properties suitable for sensing applications based on localized surface plasmon resonance [105].

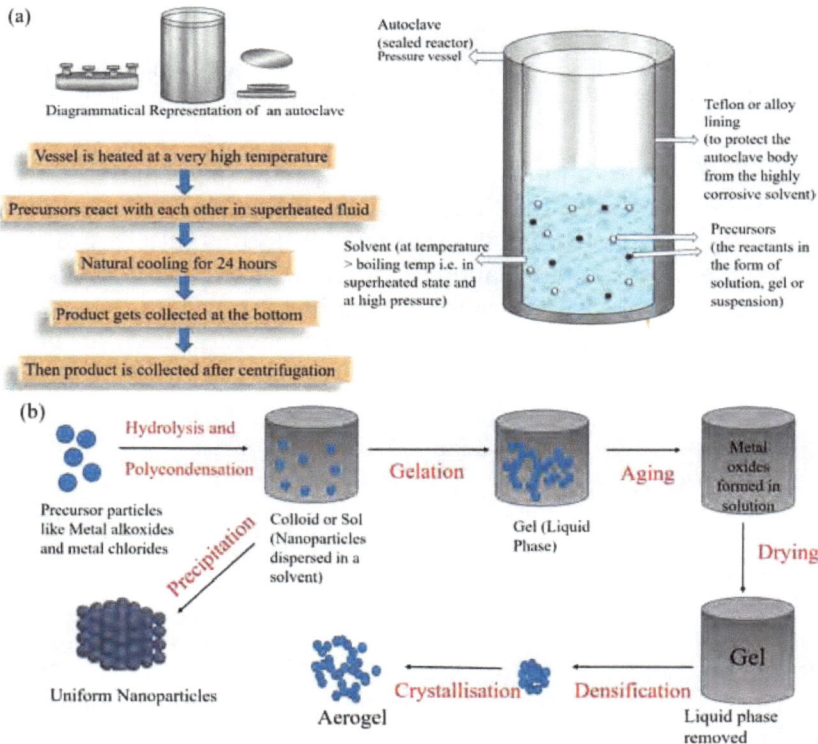

Fig. (15). Schematic diagram of the colloidal self-assembly methods. (**a**) Hydrothermal route, (**b**) sol-gel method.

These self-assembly methods include various methods that offer several advantages for fabricating nanostructured materials for semiconductor sensors. Firstly, they provide a bottom-up approach to material synthesis, enabling precise control over structure, composition, and functionality at the nanoscale. Secondly, self-assembly techniques are highly scalable and cost-effective, making them suitable for large-scale production of nanostructured materials. Thirdly, self-assembled nanostructures often exhibit unique properties, such as high surface area, porosity, and surface chemistry, which are advantageous for sensing applications requiring enhanced sensitivity and selectivity [106].

Characterization Techniques

Characterization techniques are indispensable for comprehensively assessing the structure, properties, and performance of nanoscale semiconductor sensors. These methodologies offer invaluable insights into the intricate details of sensor morphology, composition, and various properties, including electrical, optical, and chemical characteristics. Such thorough analysis empowers researchers to fine-tune device performance and functionality, facilitating the optimization of sensor design and fabrication processes. Some characterization techniques that are used to study the properties of fabricated sensors are discussed further.

Scanning Electron Microscopy (SEM)

It stands as a cornerstone in the arsenal of characterization techniques for nanoscale semiconductor sensing devices. This powerful imaging technique provides important information about the surface morphology, topography, and microstructure of fabricated sensors with unprecedented resolution and detail, as shown in Fig. (**16**). At its core, SEM scans a focused electron beam across the sample surface and detects various signals that are emitted from the electron beam and sample's interaction. These signals include secondary electrons, backscattered electrons, and characteristic X-rays, which carry information about the sample's surface morphology, composition, and crystal structure [107].

SEM offers several distinct advantages for characterizing nanoscale semiconductor sensors. Firstly, SEM provides high-resolution imaging capabilities, allowing researchers to visualize nanoscale features and structures with sub-nanometer resolution. This enables the precise examination of sensor morphology, including features, such as nanoparticle aggregates, thin film morphology, and surface roughness, which are critical for understanding sensor performance. Secondly, SEM allows for the examination of samples under a wide range of conditions, including variable pressure, temperature, and sample orientation. This versatility enables researchers to characterize nanoscale semiconductor sensors under conditions relevant to their intended application environment, providing insights into sensor behavior and performance under realistic operating conditions [108].

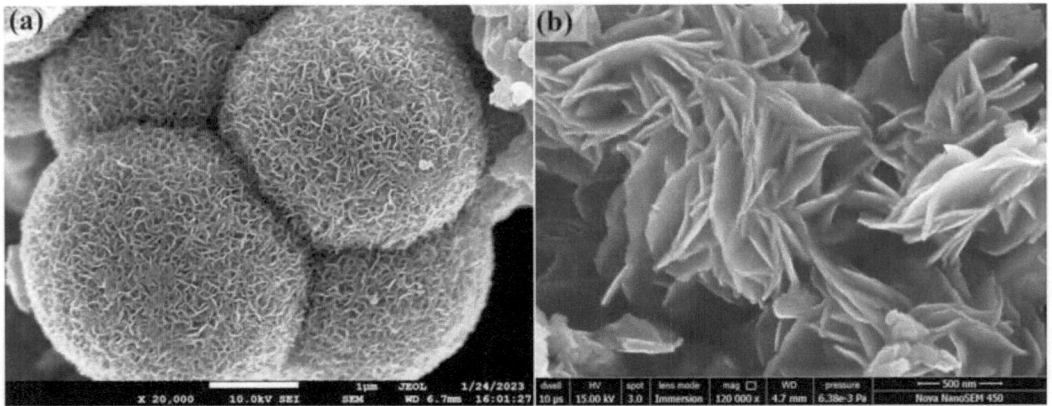

Fig. (16). SEM images of (**a**) carbon-MoS$_2$ core-shell structures and (**b**) WO$_3$ nanosheets.

Moreover, SEM can be combined with other analytical techniques, such as Energy-dispersive X-ray Spectroscopy (EDS) to provide elemental composition analysis of nanoscale sensor materials. By detecting characteristic X-rays emitted from the sample during SEM imaging, EDS enables researchers to identify and quantify the elemental composition of sensor materials, aiding in material characterization and quality control. Additionally, SEM offers the capability for 3D imaging and reconstruction of nanoscale sensor structures using techniques, such as focused ion beam milling, electron tomography, *etc*. These advanced imaging techniques allow researchers to visualize and analyze the internal structure and morphology of nanoscale sensors in three dimensions, providing insights into material properties and device architecture.

Transmission Electron Microscopy (TEM)

It serves as a vital tool in the characterization of nanoscale semiconductor sensing devices, offering unparalleled resolution and insight into their internal structure and composition. TEM enables researchers to probe materials at the atomic scale, providing detailed information about crystal structure, defects, and nanostructure morphology. TEM operates on the principle of transmitting a focused electron beam through a thin specimen, interacting with the material to generate high-resolution images and diffraction patterns, as shown in Fig. (**17a and b**) [109].

This technique offers several distinct advantages for characterizing nanoscale semiconductor sensors. First and foremost, TEM provides atomic-scale imaging capabilities, allowing researchers to visualize individual atoms and crystal lattice planes within nanoscale sensor materials. This level of resolution enables the precise examination of material defects, interfaces, and nanostructures, providing critical insights into sensor performance and reliability. Additionally, TEM offers the ability to perform selected area electron diffraction analysis, which provides information about the crystallographic structure and orientation of nanoscale sensor materials [109].

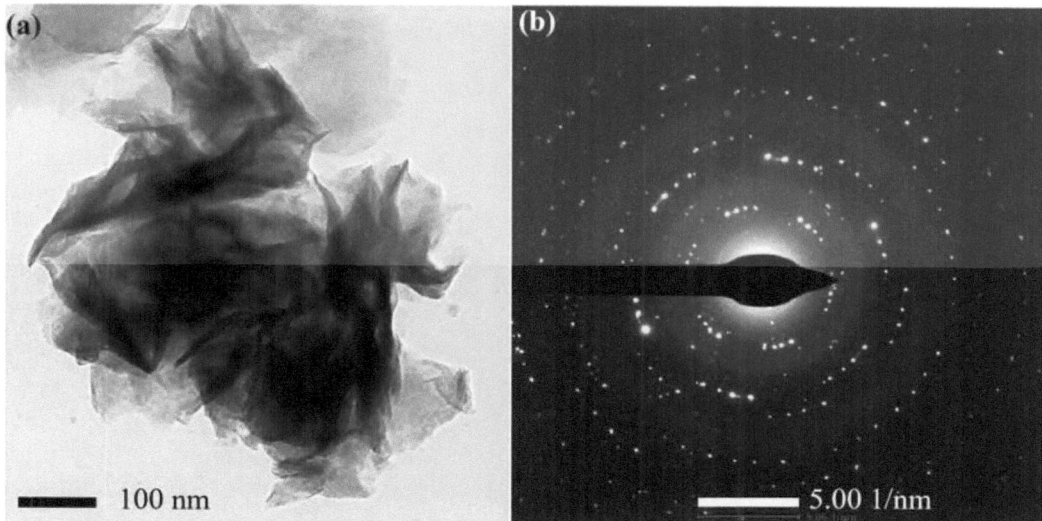

Fig. (17). (**a**) TEM image showing MoS$_2$ nanolayers and (**b**) SAED pattern of WO$_3$ nanosheets.

Furthermore, TEM can be used to characterize the morphology and size distribution of nanoparticles, nanowires, and other nanostructures incorporated into semiconductor sensors. By imaging these nanostructures at high magnification, researchers can assess their size, shape, and spatial distribution, which are critical factors influencing sensor performance. Moreover, TEM can be combined with advanced techniques, such as EDS and Electron Energy Loss Spectroscopy (EELS), to provide elemental composition and chemical bonding information. EDS enables researchers to identify and quantify the elemental composition of nanoscale sensor materials, while EELS provides insights into chemical bonding and electronic structure [110].

Atomic Force Microscopy

This technique has emerged as a critical characterization tool for nanoscale semiconductor sensing devices, providing detailed insights into surface morphology, mechanical properties, and topography with exceptional resolution. This technique operates by scanning a sharp tip attached to a flexible cantilever across the sample surface, measuring the interaction forces between the tip and the surface to generate high-resolution images and maps.

One of the primary advantages of Atomic Force Microscopy (AFM) lies in its ability to provide topographic imaging of nanoscale sensor surfaces with sub-nanometer resolution. By measuring the cantilever's deflection as it interacts with the sample surface, AFM creates high-resolution topographic images that reveal features, such as surface roughness, grain boundaries, and nanoparticle aggregates. This capability is invaluable for assessing the quality of nanoscale sensor fabrication and identifying defects or irregularities that may impact device performance.

Furthermore, AFM offers the capability to characterize the mechanical properties of nanoscale sensor materials, including stiffness, elasticity, and adhesion. By applying controlled forces to the sample surface using the AFM tip, researchers can measure mechanical properties at the nanoscale, providing insights into material behavior under different conditions. This is particularly important for understanding the durability and reliability of nanoscale sensor materials in real-world applications.

Additionally, AFM can be used to perform nanomanipulation and nanolithography, enabling researchers to precisely manipulate and pattern nanoscale sensor materials. By applying localized forces to the sample surface, AFM can be used to move, reposition, or remove individual nanoparticles or nanostructures, offering unprecedented control over material assembly and device fabrication. This capability opens up new possibilities for engineering custom-designed nanoscale sensor arrays with tailored properties and functionalities.

Moreover, AFM can be operated in various modes, which include tapping mode, contact mode, and non-contact mode, allowing researchers to optimize imaging conditions for different sample types and surface properties. This versatility makes AFM suitable for characterizing a wide range of nanoscale sensor materials, including thin films, nanoparticles, nanowires, and biomolecules.

X-ray Photoelectron Spectroscopy (XPS)

This is a crucial analytical technique for characterizing nanoscale semiconductor sensing devices. It provides detailed insights into surface composition, chemical bonding, and elemental states with exceptional sensitivity and resolution. XPS operates by irradiating the sample surface with X-rays, causing photoemission of electrons, which, when analyzed, provide information about the elemental composition, chemical states, and electronic structure of nanoscale sensor materials [111].

A primary advantage of XPS is its ability to offer quantitative analysis of surface composition and elemental distribution. By measuring the energy of emitted photoelectrons, XPS enables researchers to identify and quantify the elemental composition of the sample surface, assess material purity, detect surface contaminants, and monitor compositional changes during device fabrication processes. XPS also probes chemical bonding and electronic states within nanoscale sensor materials, providing insights into surface chemistry and reactivity. By analyzing the binding energy of emitted photoelectrons, XPS identifies specific chemical bonds and determines the oxidation states of elements, critical for understanding surface functionalization, interface properties, and chemical reactions at the sensor surface, affecting device performance [111].

XPS also performs in-depth profiling analysis, investigating the elemental composition and chemical states. XPS offers exceptional sensitivity to surface chemistry and elemental states, suitable for characterizing various nanoscale sensor materials, like thin films, nanoparticles, and nanowires. Its versatility, quantitative analysis capabilities, and high spatial resolution make XPS a powerful tool for investigating surface properties and understanding structure-property relationships in nanoscale semiconductor sensing devices [112].

Electrical Characterization

Electrical characterization techniques are fundamental for evaluating the performance and functionality of nanoscale semiconductor sensors. These techniques provide insights into the electrical properties, carrier transport mechanisms, and device behavior under various operating conditions, enabling researchers to optimize device design and performance for specific sensing applications.

One of the primary electrical characterization techniques used for nanoscale semiconductor sensors is current-voltage (I-V) measurements (Fig. **18a**). By varying the applied voltage and measuring the corresponding current in this technique, researchers can characterize device conductivity, determine carrier mobility, and assess the presence of defects or traps. The relationship among current, voltage, and resistance (R) in a semiconductor device can be described by Ohm's law. By analyzing the I-V characteristics of nanoscale semiconductor sensors, researchers can extract valuable information about device performance, including threshold voltage, breakdown voltage, and operating regimes, such as ohmic or non-ohmic behavior [113, 114].

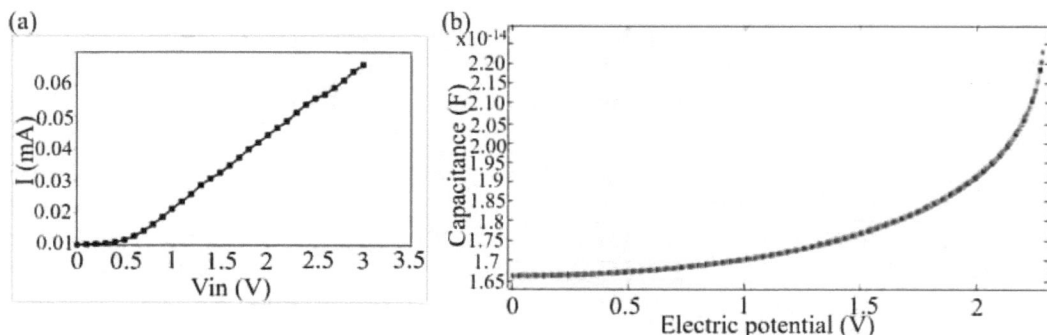

Fig. (18). (a) Sensor I-V characteristic curve [117]; (b) point graph of capacitance *vs.* voltage [118].

Another important electrical characterization technique is capacitance-voltage (C-V) measurements (Fig. **18b**), which provide insights into the semiconductor device's capacitance as a function of applied voltage. C-V measurements are particularly useful for characterizing semiconductor-insulator interfaces, determining carrier concentration, and evaluating the effectiveness of doping profiles in nanoscale sensor devices [115]. The relationship among capacitance (C), voltage (V), and charge (Q) in a semiconductor device can be described by the equation:

$$C=dQ/dV \tag{4}$$

Where, *C, V,* and *Q* are capacitance, applied voltage, and charge stored in the device, respectively. By analyzing the C-V characteristics of nanoscale semiconductor sensors, researchers can gain insights into interface trap density, dielectric properties, and carrier depletion effects, which are critical for optimizing device performance and sensitivity.

Furthermore, impedance spectroscopy is an essential electrical characterization technique that is used to analyze the frequency-dependent response of nanoscale semiconductor sensors. Impedance spectroscopy involves applying an AC voltage signal across the sensor device and measuring the resulting AC current response. By varying the frequency of the applied signal and analyzing the impedance spectrum, researchers can extract information about charge carrier dynamics, interface properties, and device capacitance. The complex impedance (Z) of a semiconductor device can be described by the equation:

$$Z=R+jX \tag{5}$$

Where, R and X are resistance and reactance, respectively, which include both capacitive and inductive components. By analyzing the impedance spectrum, researchers can characterize phenomena, like charge trapping, interface recombination, and frequency-dependent behavior, providing insights into device performance and reliability [116].

In summary, electrical characterization techniques, like I-V and C-V measurements, and impedance spectroscopy are essential for evaluating the electrical properties and performance of nanoscale semiconductor sensing devices. By analyzing the I-V characteristics, C-V profiles, and impedance spectra of semiconductor sensors, researchers can optimize device design, assess material quality, and improve sensor performance for several sensing applications.

Power Consumption and Efficiency

Power consumption and efficiency are critical considerations in the design and operation of nanoscale semiconductor sensing devices, as they directly impact device performance, battery life, and overall system cost. Achieving low power consumption while maintaining high efficiency is essential for prolonging device operation, minimizing energy consumption, and ensuring reliable sensing performance in various applications. The calculation of power consumption is crucial in understanding how efficiently the sensor utilizes energy during its operation [119]. Power consumption can be calculated using the basic principle of electrical power, which is the product of voltage and current. The formula used to calculate power consumption (P) is as follows:

$$P = I \times V \tag{6}$$

Where, P is the power consumption, I is the current flowing through the sensor, and V is the voltage applied across the sensor (in volts, V). All the values should be in SI units. To calculate power consumption, the measurement of the current and voltage values is necessary during the operation of the sensor, which can be obtained by the I-V measurement, as discussed in a previous section.

Reducing operating voltage and optimizing current levels minimizes power consumption in nanoscale semiconductor sensors. Techniques, like low-power CMOS and sub-threshold operation, enable the design of ultra-low-power devices with improved energy efficiency. Implementing signal processing algorithms for data compression and event detection optimizes energy consumption by reducing computational workload and active components in the sensor system [120]. It is required to define specific efficiency metrics that capture the sensor's performance in terms of energy consumption, sensing capability, and overall effectiveness. The energy consumed by the sensor for each sensing event, *i.e.*, E_{event}, or measurement cycle, can be calculated using equation 7.

$$E_{event} = P \times t \tag{7}$$

where, P is the power consumption of the sensor (obtained from electrical characterization data) and t is the duration of the sensing event.

The energy efficiency of the sensor in terms of energy consumed per unit area of sensor coverage can be determined using equation 8.

$$E_{area} = \frac{E_{total}}{A} \tag{8}$$

Where, E_{total} is the total energy consumed by the sensor and A is the working area of the sensor. The sensing performance achieved by the sensor relative to the power consumed during operation can be evaluated and, thereafter, the parameters, such as sensitivity, dynamic range, or accuracy per unit of power consumed, can be calculated. This provides insight into how effectively the sensor converts energy into useful sensing information.

Moreover, optimizing sensor architecture and circuit design can significantly impact power consumption and efficiency. By minimizing parasitic capacitance, resistance, and leakage currents in sensor circuits, researchers can enhance energy efficiency and reduce power dissipation during sensor operation. Techniques, such as voltage scaling, clock gating, and power gating, enable dynamic power management, allowing sensors to operate in low-power modes during idle or standby periods, while conserving energy for active sensing tasks.

Reliability and Durability Assessment

Reliability and durability are critical aspects of nanoscale semiconductor sensors to ensure consistent and long-term performance in various operating conditions. One method for assessing reliability is through reliability engineering, which involves analyzing failure rates and estimating the probability of failure over time. This can be achieved using techniques, such as Weibull analysis, which models the probability density function of failure times [121]. The Weibull distribution is expressed as follows:

$$f(t) = \frac{k}{\lambda} (\frac{t}{\lambda})^{k-1} e^{-(t/\lambda)^k} \tag{9}$$

Where, $f(t)$ is the probability density function, t is time, k is the shape parameter, and λ is the scale parameter. By fitting experimental failure data to the Weibull distribution, researchers can estimate the sensor's reliability and predict failure rates over its lifetime. Accelerated life testing is conducted to simulate the aging process and assess the reliability of the sensor under accelerated conditions [121]. ALT involves subjecting the sensor to elevated stress levels (*e.g.*, temperature, humidity, voltage) to accelerate the occurrence of failure mechanisms.

Durability assessment involves subjecting the sensor to accelerated aging tests to simulate long-term environmental exposure and identify potential failure mechanisms. Accelerated aging tests apply stress factors, such as temperature, humidity, and mechanical strain, to accelerate degradation processes and assess the sensor's resilience. For example, high-temperature storage tests can reveal vulnerability to thermal stress, while humidity cycling tests can assess moisture resistance.

Overall, the reliability and durability assessment of nanoscale semiconductor sensors requires a multidisciplinary approach, combining experimental testing, statistical analysis, and computational modeling. By rigorously evaluating reliability and durability, researchers can ensure the reliability and long-term performance of nanoscale semiconductor sensors in real-world applications.

RECENT DEVELOPMENTS IN SEMICONDUCTOR SENSING TECHNOLOGIES

Semiconductor sensing technologies have witnessed rapid advancements in recent decades, driven by innovations in materials science, fabrication techniques, and signal processing algorithms. These developments have led to enhanced sensing capabilities, improved performance, and expanded applications across various industries. Semiconductor sensing technologies have undergone remarkable advancements, leading to the development of state-of-the-art sensors with unprecedented performance. Some of the notable developments include the following:

1. Miniaturization and Integration: Advances in semiconductor fabrication techniques have enabled the miniaturization and integration of sensing elements, resulting in compact and multifunctional sensor devices. Microelectromechanical Systems (MEMS) technology plays a pivotal role in realizing miniaturized sensors with high sensitivity and selectivity [122].
2. Multi-modal Sensing: State-of-the-art semiconductor sensors often incorporate multiple sensing modalities, allowing for comprehensive analysis of complex environments or biological samples [123]. Integration of optical, electrical, and mechanical sensing mechanisms enables versatile sensor platforms capable of detecting a wide range of stimuli [124].
3. Smart Sensing Systems: Recent developments have focused on the integration of smart features into semiconductor sensors, such as on-chip signal processing, wireless communication, and autonomous operation. These smart sensing

systems offer real-time data, remote monitoring, and adaptive sensing functionalities, enabling advanced applications in healthcare, environmental monitoring, and Internet of Things devices [125].

4. Flexible and Wearable Sensors: Flexible and wearable semiconductor sensors have emerged as a promising area of research, offering conformal integration with human skin and soft tissues for continuous health monitoring and personalized healthcare applications. Flexible electronics and novel materials enable the development of bendable, stretchable, and biocompatible sensor platforms with enhanced comfort and usability [126].

5. High-resolution Imaging: Semiconductor-based imaging sensors have achieved remarkable advancements in terms of resolution, sensitivity, and speed. High-resolution imaging techniques, such as CMOS imagers and CCDs, enable detailed visualization of biological samples, materials, and surfaces for applications in microscopy, spectroscopy, and diagnostics [127].

6. 3D Integration and Packaging: Three-dimensional (3D) integration techniques, such as Through-silicon Vias (TSVs), wafer stacking, and chip-on-chip integration, enable the vertical integration of multiple sensor layers, packaging components, and signal processing circuits within a single device. 3D integration enhances sensor performance, reduces footprint, and enables heterogeneous integration of diverse sensor technologies, leading to compact, multi-functional sensor systems with enhanced capabilities and versatility [128].

7. Integration of Nanophotonics and Plasmonics: Integration of nanophotonic and plasmonic components into semiconductor sensor platforms enables enhanced light-matter interactions, leading to improved sensitivity, resolution, and multiplexing capabilities. Nanophotonic and plasmonic structures, like photonic crystals, metamaterials, and plasmonic nanoparticles, enable efficient light trapping, waveguiding, and surface-enhanced spectroscopy for applications in optical sensing, bioimaging, and environmental monitoring [129].

Applications of Semiconductor Sensors

The application of semiconductor sensing technologies spans across a wide range of industries and disciplines, with numerous case studies demonstrating their effectiveness in solving real-world challenges and driving innovation. Here are listed some applications showcasing the diverse utility of semiconductor sensors.

1. Healthcare and Biomedical Monitoring: Semiconductor sensors play an essential role in healthcare and biomedical fields, enabling non-invasive monitoring of physiological parameters, early disease detection, and personalized medical interventions. Case studies include wearable biosensors

for continuous monitoring of vital signs, lab-on-a-chip devices for point-of-care diagnostics, and implantable sensors for real-time monitoring of chronic conditions, for *e.g.*, diabetes and cardiovascular diseases [130].

2. Environmental Monitoring and Pollution Control: Semiconductor sensors are widely used for environmental monitoring and pollution control, facilitating the detection and quantification of various pollutants, toxins, and hazardous gases in air, water, and soil. Case studies include air quality sensors for urban pollution monitoring, water quality sensors for detecting contaminants in natural water sources, and soil sensors for assessing soil health and fertility in agricultural settings [131].

3. Industrial Process Control and Automation: Semiconductor sensors are employed in industrial process control and automation systems to ensure product quality, optimize manufacturing processes, and enhance operational efficiency. Case studies include gas sensors for detecting leaks and monitoring air quality in industrial facilities, temperature sensors for thermal management and control in manufacturing processes, and pressure sensors for monitoring fluid flow and system integrity in oil and gas pipelines [132].

4. Smart Agriculture and Precision Farming: Semiconductor sensors are revolutionizing agriculture and farming practices by the implementation of different farming techniques and smart agriculture. Case studies include soil moisture sensors for irrigation management, crop health sensors for early detection of pests and diseases, and environmental sensors for monitoring microclimatic conditions and optimizing crop yield and quality [133].

5. Smart Cities and Infrastructure Management: Semiconductor sensors are integral components of smart city initiatives and infrastructure management systems, enabling real-time monitoring and optimization of urban infrastructure and services. Case studies include traffic sensors for congestion management and traffic flow optimization, structural health monitoring sensors for assessing the integrity of bridges and buildings, and smart energy meters for efficient energy distribution and consumption management [134].

6. Consumer Electronics and Internet of Things (IoT) Devices: Semiconductor sensors are ubiquitous in consumer electronics and IoT devices, providing essential functionalities, such as motion detection, environmental sensing, and biometric recognition. Case studies include smartphone sensors for location tracking and activity monitoring, smart home sensors for home automation and energy efficiency, and wearable fitness trackers for personal health monitoring and fitness tracking [135].

Challenges and Opportunities

As sensor devices become smaller and more integrated, the challenges associated with fabrication complexity, material compatibility, and interface engineering increase. Achieving high performance and reliability in miniaturized sensors requires overcoming technical hurdles related to material selection, device integration, and manufacturing processes. Enhancement in the sensitivity and selectivity of semiconductor sensors, particularly in complex and dynamic environments, remains a significant challenge. Improving sensor performance while minimizing cross-reactivity, interference, and noise requires innovative materials, signal-processing algorithms, and sensing mechanisms [136]. Semiconductor sensors must maintain stability and reliability in harsh and unpredictable environmental conditions, including temperature variations, humidity, mechanical stress, and chemical exposure. Ensuring long-term performance and durability in real-world applications necessitates robust sensor designs, materials, and encapsulation techniques.

Environmental influences can significantly impact the performance and reliability of semiconductor sensors, affecting their accuracy, stability, and longevity. These influences can arise from various factors, such as temperature fluctuations, humidity levels, electromagnetic interference, and exposure to harsh chemicals or contaminants. To ensure optimal sensor operation and mitigate the effects of environmental influences, several strategies are employed in semiconductor sensor design and implementation, including temperature compensation, encapsulation and packaging, shielding against electromagnetic interference, humidity control, calibration and drift compensation, and environmental testing and qualification.

Temperature variations can affect the electrical properties of semiconductor materials and lead to drift in sensor readings. To mitigate this influence, temperature compensation techniques are employed, such as incorporating temperature sensors (thermistors or integrated temperature sensors) to monitor and compensate for changes in ambient temperature. Additionally, sensor calibration procedures may be implemented to account for temperature-dependent effects on sensor performance. Encapsulating semiconductor sensors within protective housings or packages helps shield them from environmental factors, such as moisture, dust, and chemical exposure. Hermetic sealing techniques, conformal coatings, and barrier materials are utilized to create a protective barrier around the sensor, ensuring long-term reliability and stability, particularly in harsh operating environments. Semiconductor sensors are susceptible to electromagnetic interference from external sources, such as radio frequency signals, power lines, or electronic devices. To mitigate EMI effects, shielding techniques, such as electromagnetic shielding materials and Faraday cages, may be employed to minimize the impact of external electromagnetic fields on sensor performance.

High humidity levels can affect the electrical properties of semiconductor materials and lead to corrosion or degradation of sensor components. Humidity-resistant materials and moisture-resistant coatings are utilized to protect sensitive sensor elements from moisture ingress. Additionally, desiccants or humidity control systems may be integrated into sensor enclosures to maintain low humidity levels and prevent condensation. Semiconductor sensors may exhibit drift in their output signals over time due to aging, material degradation, or environmental influences. Calibration procedures are implemented to periodically recalibrate sensors and correct for drift, ensuring accurate and reliable measurements. Additionally, drift compensation algorithms may be employed to continuously monitor sensor performance and adjust output signals in real time to maintain accuracy.

Semiconductor sensors undergo rigorous environmental testing and qualification procedures to assess their performance under various operating conditions and environmental stressors. Accelerated aging tests, temperature cycling tests, and exposure to simulated environmental conditions are conducted to evaluate sensor reliability, durability, and robustness before deployment in real-world applications. By implementing these mitigation strategies, semiconductor sensors can withstand a wide range of environmental influences and maintain optimal performance and reliability over extended periods of operation. Continuous advancements in sensor design, materials science, and packaging technologies contribute to the development of sensors capable of withstanding increasingly challenging environmental conditions, while delivering accurate and dependable measurements across diverse applications.

Despite advancements in low-power electronics and energy-efficient sensor designs, power consumption remains a critical challenge, particularly for portable and battery-operated devices. Balancing sensor performance with energy efficiency requires optimizing power consumption, signal processing algorithms, and sensor activation modes [137]. Cost-effective fabrication and scalability of semiconductor sensors for mass production present challenges related to materials cost, manufacturing complexity, and yield optimization. Addressing these challenges requires innovative fabrication techniques, economical practices, and collaborative partnerships across industry and academia [138 - 144].

The emergence of novel materials and nanotechnology offers unprecedented opportunities for enhancing sensor performance, sensitivity, and functionality. Leveraging advances in nanomaterial synthesis, nanostructuring techniques, and nanoelectronics enables the development of next-generation sensors with tailored properties and improved performance. The proliferation of IoT devices and interconnected sensor networks presents opportunities for seamless integration of semiconductor sensors into smart systems and applications [134, 135]. Integration of AI and machine learning algorithms with semiconductor sensors enables pattern recognition, intelligent data processing, and predictive analytics. AI-driven sensor systems can adapt to changing environmental conditions, optimize sensor performance, and uncover insights from complex datasets, unlocking new opportunities for innovation and discovery [125]. Collaborative efforts across disciplines, such as materials science, electronics, biology, and data science, can foster innovation and drive breakthroughs in semiconductor sensing technologies. Interdisciplinary research and collaboration can facilitate the development of holistic solutions to complex challenges, leading to novel sensor designs, applications, and insights. The growing emphasis on sustainability and environmental stewardship presents opportunities for developing eco-friendly and sustainable sensor solutions. From biodegradable sensors to energy-efficient designs, semiconductor sensing technologies can contribute to environmental conservation and resource management, while addressing societal needs and challenges.

CONCLUSION

This book chapter has provided a comprehensive overview of semiconductor sensing technologies, covering fundamental principles, advanced sensing mechanisms, fabrication techniques, recent developments, and real-world applications. Semiconductor sensors play a pivotal role in diverse fields, like healthcare, environmental monitoring, industrial automation, and consumer electronics. By leveraging semiconductor materials, nanotechnology, and advanced fabrication processes, researchers and engineers have made significant strides in enhancing sensor performance, sensitivity, and functionality. Throughout this chapter, we have explored the underlying principles of semiconductor sensing, including resistive, capacitive, optical, and piezoresistive mechanisms, as well as emerging sensing technologies. We have also discussed the importance of sensors in today's world, highlighting their critical role in addressing societal challenges, improving quality of life and driving technological innovation. Moreover, this chapter has delved into the fabrication techniques for nanoscale semiconductor sensors, including thin film deposition, photolithography, and self-assembly methods, along with characterization techniques for assessing sensor performance and reliability. Case studies and applications have illustrated the diverse utility of semiconductor sensors in healthcare, environmental monitoring, industrial process control, agriculture, smart cities, and consumer electronics. Despite the remarkable progress in semiconductor sensing, several challenges remain, including miniaturization complexity, sensitivity optimization, environmental stability, power consumption, and cost scalability. However, these challenges also present opportunities for innovation, driven by advances in materials science, nanotechnology, artificial intelligence, and sustainable sensor solutions. Looking ahead, interdisciplinary collaboration, technological innovation, and sustainable practices are essential for unveiling the full potential of semiconductor sensing technologies and addressing global challenges. By harnessing the collective expertise and ingenuity of researchers, industry stakeholders, policymakers, and the broader scientific community, we can continue to push the boundaries of sensing technology, driving progress and improving lives in the years to come.

KEY POINTS

1. The chapter delves into fundamental semiconductor sensing principles, elucidating resistive, capacitive, optical, and piezoresistive mechanisms.
2. It explores the pivotal role of nanoscale materials in amplifying sensor performance, sensitivity, and functionality.
3. It examines in meticulous detail the design considerations essential for crafting nanoscale semiconductor sensing devices, addressing architectural nuances, integration challenges, and power consumption concerns.
4. It discusses diverse methods for nanomaterial synthesis and fabrication techniques tailored for nanoscale semiconductor sensors.
5. It also unveils state-of-the-art developments and emerging trends in semiconductor sensing technologies, supported by insightful case studies and real-world applications.
6. It further provides an in-depth analysis of challenges and opportunities within the field, proposing potential solutions and envisioning future prospects.
7. It emphasizes the critical aspects of testing and characterization for assessing the reliability and durability of nanoscale semiconductor sensing devices.

REFERENCES

[1]   M. Javaid, A. Haleem, S. Rab, R. Pratap Singh, and R. Suman, "Sensors for daily life: A review", *Sensors International,* vol. 2, p. 100121, 2021.
[http://dx.doi.org/10.1016/j.sintl.2021.100121]

[2]   S. Yao, P. Swetha, and Y. Zhu, "Nanomaterial-enabled wearable sensors for healthcare", *Adv. Healthc. Mater.,* vol. 7, no. 1, p. 1700889, 2018.
[http://dx.doi.org/10.1002/adhm.201700889] [PMID: 29193793]

[3]   H. Yin, Y. Cao, B. Marelli, X. Zeng, A.J. Mason, and C. Cao, "Soil sensors and plant wearables for smart and precision agriculture", *Adv. Mater.,* vol. 33, no. 20, p. 2007764, 2021.
[http://dx.doi.org/10.1002/adma.202007764] [PMID: 33829545]

[4]   I. Bilik, "Comparative analysis of radar and lidar technologies for automotive applications", *IEEE Intell. Transp. Syst. Mag.,* vol. 15, no. 1, pp. 244-269, 2023.
[http://dx.doi.org/10.1109/MITS.2022.3162886]

[5]   J. Steinbaeck, C. Steger, G. Holweg, and N. Druml, "Next generation radar sensors in automotive sensor fusion systems", *2017 Symposium on Sensor Data Fusion: Trends, Solutions, Applications, SDF 2017,* pp. 1-6, 2017.

[http://dx.doi.org/10.1109/SDF.2017.8126389]

[6]   M. Javaid, A. Haleem, R.P. Singh, S. Rab, and R. Suman, "Significance of sensors for industry 4.0: Roles, capabilities, and applications", *Sensors International,* vol. 2, p. 100110, 2021.
[http://dx.doi.org/10.1016/j.sintl.2021.100110]

[7]   M. A. Ramírez-Moreno, "Sensors for sustainable smart cities: A review", *Applied Sciences,* vol. 11, no. 17, p. 8198, 2021.
[http://dx.doi.org/10.3390/app11178198]

[8]   K. Gulati, R.S. Kumar Boddu, D. Kapila, S.L. Bangare, N. Chandnani, and G. Saravanan, "A review paper on wireless sensor network techniques in internet of things (IoT)", *Mater. Today Proc.,* vol. 51, pp. 161-165, 2022.
[http://dx.doi.org/10.1016/j.matpr.2021.05.067]

[9]   P. PHOGAT, Shreya, R. JHA, and S. Singh, ""Electrochemical analysis of thermally treated two dimensional zinc sulphide hexagonal nano-sheets with reduced band gap", *Phys Scr,* vol. 98, no. 12, p. 125962, 2023.
[http://dx.doi.org/10.1088/1402-4896/ad0d93]

[10]   T. Kumar, P. Shreya, P. Phogat, V. Sahgal, and R. Jha, "Surfactant-mediated modulation of morphology and charge transfer dynamics in tungsten oxide nanoparticles", *Phys. Scr.,* vol. 98, no. 8, p. 085936, 2023.
[http://dx.doi.org/10.1088/1402-4896/ace566]

[11]   P. Raju, Q. Li, and J. Baumgartner, "Review—semiconductor materials and devices for gas sensors", *J. Electrochem. Soc.,* vol. 169, no. 5, p. 057518, 2022.
[http://dx.doi.org/10.1149/1945-7111/ac6e0a]

[12]   K.G. Krishna, S. Parne, N. Pothukanuri, V. Kathirvelu, S. Gandi, and D. Joshi, "Nanostructured metal oxide semiconductor-based gas sensors: A comprehensive review", *Sens. Actuators A Phys.,* vol. 341, p. 113578, 2022.
[http://dx.doi.org/10.1016/j.sna.2022.113578]

[13]   J.B.A. Gomes, J.J.P.C. Rodrigues, R.A.L. Rabêlo, S. Tanwar, J. Al-Muhtadi, and S. Kozlov, "A novel internet of things -based plug-and-play multigas sensor for environmental monitoring", *Trans. Emerg. Telecommun. Technol.,* vol. 32, no. 6, p. e3967, 2021.
[http://dx.doi.org/10.1002/ett.3967]

[14]   H. M. Tayyab, Y. Javed, I. Ullah, A. A. Dogar, and B. Ahmed, *Sensor-based gas leakage detector system,* vol. 2, no. 1, p. 28, 2020.*Engineering Proceedings,* vol. 2, no. 1, p. 28, 2020.

[15]   O. Cicek, S. Altindal, and Y. Azizian-Kalandaragh, "A highly sensitive temperature sensor based on Au/Graphene-PVP/ n -Si type schottky diodes and the possible conduction mechanisms in the wide range temperatures", *IEEE Sens. J.,* vol. 20, no. 23, pp. 14081-14089, 2020.
[http://dx.doi.org/10.1109/JSEN.2020.3009108]

[16]   M.Z. Bodic, S.O. Aleksic, V.M. Rajs, M.S. Damnjanovic, and M.G. Kisic, "Thermally coupled thick-film thermistors: main properties and applications", *IEEE Sens. J.,* vol. 23, no. 18, pp. 21010-21017, 2023.
[http://dx.doi.org/10.1109/JSEN.2023.3298224]

[17]   Y. Xu, and Q. Lin, "Photodetectors based on solution-processable semiconductors: Recent advances and perspectives", *Appl. Phys. Rev.,* vol. 7, no. 1, p. 011315, 2020.
[http://dx.doi.org/10.1063/1.5144840]

[18]   A. Waseem, M.A. Johar, M.A. Hassan, I.V. Bagal, A. Abdullah, J-S. Ha, J.K. Lee, and S-W. Ryu, "Flexible self-powered piezoelectric pressure sensor based on GaN/p-GaN coaxial nanowires", *J. Alloys Compd.,* vol. 872, p. 159661, 2021.
[http://dx.doi.org/10.1016/j.jallcom.2021.159661]

[19]   J. Wang, J. Jiang, C. Zhang, M. Sun, S. Han, R. Zhang, N. Liang, D. Sun, and H. Liu, "Energy-

efficient, fully flexible, high-performance tactile sensor based on piezotronic effect: Piezoelectric signal amplified with organic field-effect transistors", *Nano Energy,* vol. 76, p. 105050, 2020.
[http://dx.doi.org/10.1016/j.nanoen.2020.105050]

[20]   S. Mulmi, and V. Thangadurai, "Editors' Choice—review—solid-state electrochemical carbon dioxide sensors: fundamentals, materials and applications", *J. Electrochem. Soc.,* vol. 167, no. 3, p. 037567, 2020.
[http://dx.doi.org/10.1149/1945-7111/ab67a9]

[21]   S. Poncé, W. Li, S. Reichardt, and F. Giustino, "First-principles calculations of charge carrier mobility and conductivity in bulk semiconductors and two-dimensional materials", *Rep. Prog. Phys.,* vol. 83, no. 3, p. 036501, 2020.
[http://dx.doi.org/10.1088/1361-6633/ab6a43] [PMID: 31923906]

[22]   L. X. Ou, M. Y. Liu, L. Y. Zhu, D. W. Zhang, and H. L. Lu, "Recent progress on flexible room-temperature gas sensors based on metal oxide semiconductor", *Nano-Micro Letters,* vol. 14, no. 1, pp. 1-42, 2022.
[http://dx.doi.org/10.1007/s40820-022-00956-9]

[23]   P. PHOGAT, S. Shreya, R. JHA, and S. Singh, "Diffusion controlled features of microwave assisted zns/zno nanocomposite with reduced band gap", *ECS Journal of Solid State Science and Technology,* 2023.

[24]   R.T. Kusuma, S. Aisyah, P.N. Pertami, S.U. Sukmawati, S. Humairah, M. Dahrul, N.P. Har, P.L. Bintari, V. Rahmawaty, M.R.A. Pahlefi, I. Abdurrahman, Sejahtera, Irmansyah, and Irzaman, "Characteristics of thermistors as temperature sensors on the sensor unit (su-6803) and op amp unit (ou-6801)", *AIP Conf. Proc.,* vol. 2320, no. 1, p. 050011, 2021.
[http://dx.doi.org/10.1063/5.0037677]

[25]   R.A.B. John, and A. Ruban Kumar, "A review on resistive-based gas sensors for the detection of volatile organic compounds using metal-oxide nanostructures", *Inorg. Chem. Commun.,* vol. 133, p. 108893, 2021.
[http://dx.doi.org/10.1016/j.inoche.2021.108893]

[26]   J. Qin, L.J. Yin, Y.N. Hao, S.L. Zhong, D.L. Zhang, K. Bi, Y.X. Zhang, Y. Zhao, and Z.M. Dang, "Flexible and stretchable capacitive sensors with different microstructures", *Adv. Mater.,* vol. 33, no. 34, p. 2008267, 2021.
[http://dx.doi.org/10.1002/adma.202008267] [PMID: 34240474]

[27]   H. Jin, "Stretchable dual-capacitor multi-sensor for touch-curvature-pressure-strain sensing", *Scientific Reports,* vol. 7, no. 1, pp. 1-8, 2017.
[http://dx.doi.org/10.1038/s41598-017-11217-w]

[28]   S. Ratan, C. Kumar, A. Kumar, D.K. Jarwal, A.K. Mishra, R.K. Upadhyay, A.P. Singh, and S. Jit, "Room temperature high hydrogen gas response in Pd/TiO 2 /Si/Al capacitive sensor", *Micro & Nano Lett.,* vol. 15, no. 9, pp. 632-635, 2020.
[http://dx.doi.org/10.1049/mnl.2020.0154]

[29]   P.P. Dipti, P. Phogat, Shreya, D. Kumari, and S. Singh, "Fabrication of tunable band gap carbon based zinc nanocomposites for enhanced capacitive behaviour", *Phys. Scr.,* vol. 98, no. 9, p. 095030, 2023.
[http://dx.doi.org/10.1088/1402-4896/acf07b]

[30]   H. Nam, K. H. Seol, J. Lee, H. Cho, and S. W. Jung, "Review of capacitive touchscreen technologies: overview, research trends, and machine learning approaches", *Sensors,* vol. 21, no. 14, p. 4776, 2021.
[http://dx.doi.org/10.3390/s21144776]

[31]   Z. Zhang, M. Chen, S. Alem, Y. Tao, T-Y. Chu, G. Xiao, C. Ramful, and R. Griffin, "Printed flexible capacitive humidity sensors for field application", *Sens. Actuators B Chem.,* vol. 359, p. 131620, 2022.
[http://dx.doi.org/10.1016/j.snb.2022.131620]

[32]   L. Nanver, and T. Knežević, "Optical detectors", *Advances in Semiconductor Technologies,* no. Oct, pp. 211-229, 2022.

[http://dx.doi.org/10.1002/9781119869610.ch11]

[33]   Q. Chen, H. Xu, and C. S. Tan, "Introduction of optical imaging and sensing: materials, devices, and applications", *Optical Imaging and Sensing: Materials, Devices, and Applications,* pp. 1-9, 2023. [http://dx.doi.org/10.1002/9783527835201.ch1]

[34]   H.D. Jabbar, M.A. Fakhri, and M. Jalal AbdulRazzaq, "Gallium nitride –based photodiode: A review", *Mater. Today Proc.,* vol. 42, pp. 2829-2834, 2021. [http://dx.doi.org/10.1016/j.matpr.2020.12.729]

[35]   S. Sharma, P. Phogat, R. Jha, and S. Singh, "Electrochemical and optical properties of microwave assisted mos2 nanospheres for solar cell application", *International Journal of Smart Grid and Clean Energy,* pp. 66-72, 2023. [http://dx.doi.org/10.12720/sgce.12.3.66-72]

[36]   R. Tejas, P. Macherla, and N. Shylashree, "Image sensor—CCD and CMOS", *Lecture Notes in Electrical Engineering,* vol. 887, pp. 455-484, 2023. [http://dx.doi.org/10.1007/978-981-19-1906-0_40]

[37]   C. Wang, X. Zhang, and W. Hu, "Organic photodiodes and phototransistors toward infrared detection: materials, devices, and applications", *Chem. Soc. Rev.,* vol. 49, no. 3, pp. 653-670, 2020. [http://dx.doi.org/10.1039/C9CS00431A] [PMID: 31829375]

[38]   J. Dahiya, P. Phogat, A. Hooda, and S. Khasa, "Investigations of Praseodymium doped LiF-Zn--Bi2O3-B2O3 glass matrix for photonic applications", *AIP Conf. Proc.,* vol. 2995, no. 1, p. 020065, 2024. [http://dx.doi.org/10.1063/5.0178197]

[39]   J.L. Santos, "Optical sensors for industry 4.0", *IEEE J. Sel. Top. Quantum Electron.,* vol. 27, no. 6, pp. 1-11, 2021. [http://dx.doi.org/10.1109/JSTQE.2021.3078126]

[40]   M. Hajj-Hassan, "Direct-dispense polymeric waveguides platform for optical chemical sensors", *Sensors,* vol. 8, no. 12, pp. 7636-7648, 2008. [http://dx.doi.org/10.3390/s8127636]

[41]   A.S. Fiorillo, C.D. Critello, and S.A. Pullano, "Theory, technology and applications of piezoresistive sensors: A review", *Sens. Actuators A Phys.,* vol. 281, pp. 156-175, 2018. [http://dx.doi.org/10.1016/j.sna.2018.07.006]

[42]   J. Li, L. Fang, B. Sun, X. Li, and S. H. Kang, "Review—recent progress in flexible and stretchable piezoresistive sensors and their applications", *J Electrochem Soc,* vol. 167, no. 3, p. 037561, 2020. [http://dx.doi.org/10.1149/1945-7111/ab6828]

[43]   X. Nan, "A review of epidermal flexible pressure sensing arrays", *Biosensors,* vol. 13, no. 6, p. 656, 2023. [http://dx.doi.org/10.3390/bios13060656]

[44]   M. A. Al Mamun, and M. R. Yuce, "Recent progress in nanomaterial enabled chemical sensors for wearable environmental monitoring applications", *Adv Funct Mater,* vol. 30, no. 51, p. 2005703, 2020. [http://dx.doi.org/10.1002/adfm.202005703]

[45]   A. Moumen, G.C.W. Kumarage, and E. Comini, "P-type metal oxide semiconductor thin films: synthesis and chemical sensor applications", In: *Sensors* vol. 22. , 2022, p. 1359.

[46]   J. Baranwal, B. Barse, G. Gatto, G. Broncova, and A. Kumar, "Electrochemical sensors and their applications: A review", *Chemosensors,* vol. 10, no. 9, p. 363, 2022. [http://dx.doi.org/10.3390/chemosensors10090363]

[47]   P. Shreya, Phogat, R. Jha, and S. Singh, "Microwave-synthesized γ-WO3 nanorods exhibiting high current density and diffusion characteristics", *Transition Metal Chemistry,* vol. 1, pp. 1-17, 2023.

[48]   D. Kumari, P. Shreya, P. Phogat, Dipti, S. Singh, and R. Jha, "Enhanced electrochemical behavior of

C@CdS Core-Shell heterostructures", *Mater. Sci. Eng. B,* vol. 301, p. 117212, 2024.
[http://dx.doi.org/10.1016/j.mseb.2024.117212]

[49] P. Phogat, S. Shreya, R. Jha, and S. Singh, "Impedance study of zinc sulphide quantum dots *via* one step green synthesis", *Mater. Sci. Forum,* vol. 1099, pp. 119-125, 2023.
[http://dx.doi.org/10.4028/p-G1CCxq]

[50] R. Khatri, and N.K. Puri, "Electrochemical biosensor utilizing dual-mode output for detection of lung cancer biomarker based on reduced graphene oxide-modified reduced-molybdenum disulfide multi-layered nanosheets", *J. Mater. Res.,* vol. 37, no. 8, pp. 1451-1463, 2022.
[http://dx.doi.org/10.1557/s43578-022-00546-w]

[51] K. Sinha, Z. Uddin, H.I. Kawsar, S. Islam, M.J. Deen, and M.M.R. Howlader, "Analyzing chronic disease biomarkers using electrochemical sensors and artificial neural networks", *Trends Analyt. Chem.,* vol. 158, p. 116861, 2023.
[http://dx.doi.org/10.1016/j.trac.2022.116861]

[52] J.F. Hernández-Rodríguez, D. Rojas, and A. Escarpa, "Electrochemical sensing directions for next-generation healthcare: trends, challenges, and frontiers", *Anal. Chem.,* vol. 93, no. 1, pp. 167-183, 2021.
[http://dx.doi.org/10.1021/acs.analchem.0c04378] [PMID: 33174738]

[53] J.P. Dowling, and K.P. Seshadreesan, "Quantum optical technologies for metrology, sensing, and imaging", *J. Lightwave Technol.,* vol. 33, no. 12, pp. 2359-2370, 2015.
[http://dx.doi.org/10.1109/JLT.2014.2386795]

[54] W.P. Schleich, K.S. Ranade, C. Anton, M. Arndt, M. Aspelmeyer, M. Bayer, G. Berg, T. Calarco, H. Fuchs, E. Giacobino, M. Grassl, P. Hänggi, W.M. Heckl, I-V. Hertel, S. Huelga, F. Jelezko, B. Keimer, J.P. Kotthaus, G. Leuchs, N. Lütkenhaus, U. Maurer, T. Pfau, M.B. Plenio, E.M. Rasel, O. Renn, C. Silberhorn, J. Schiedmayer, D. Schmitt-Landsiedel, K. Schönhammer, A. Ustinov, P. Walther, H. Weinfurter, E. Welzl, R. Wiesendanger, S. Wolf, A. Zeilinger, and P. Zoller, "Quantum technology: from research to application", *Appl. Phys. B,* vol. 122, no. 5, p. 130, 2016.
[http://dx.doi.org/10.1007/s00340-016-6353-8]

[55] P. Huang, "Dephasing of exchange-coupled spins in quantum dots for quantum computing", *Adv. Quantum Technol.,* vol. 4, no. 11, p. 2100018, 2021.
[http://dx.doi.org/10.1002/qute.202100018]

[56] F. Abebe, P. Perkins, R. Shaw, and S. Tadesse, "A rhodamine-based fluorescent sensor for selective detection of Cu2+ in aqueous media: Synthesis and spectroscopic properties", *J. Mol. Struct.,* vol. 1205, p. 127594, 2020.
[http://dx.doi.org/10.1016/j.molstruc.2019.127594] [PMID: 32601506]

[57] J.F. Wang, R.N. Bian, T. Feng, K.F. Xie, L. Wang, and Y.J. Ding, "A highly sensitive dual-channel chemical sensor for selective identification of B4O72", *Microchem. J.,* vol. 160, p. 105676, 2021.
[http://dx.doi.org/10.1016/j.microc.2020.105676]

[58] A. Marikutsa, M. Rumyantseva, E. A. Konstantinova, and A. Gaskov, "The key role of active sites in the development of selective metal oxide sensor materials",
[http://dx.doi.org/10.3390/s21072554]

[59] G.W. Hunter, S. Akbar, S. Bhansali, M. Daniele, P.D. Erb, K. Johnson, C-C. Liu, D. Miller, O. Oralkan, P.J. Hesketh, P. Manickam, and R.L. Vander Wal, "Editors' Choice—critical review—a critical review of solid state gas sensors", *J. Electrochem. Soc.,* vol. 167, no. 3, p. 037570, 2020.
[http://dx.doi.org/10.1149/1945-7111/ab729c]

[60] M.J. Tierney, and H.O.L. Kim, "Electrochemical gas sensor with extremely fast response times", *Anal. Chem.,* vol. 65, no. 23, pp. 3435-3440, 1993.
[http://dx.doi.org/10.1021/ac00071a017]

[61] S.Y. Kim, K.H. Kim, and Y. Kim, "Comparative study on pressure sensors for sloshing experiment", *Ocean Eng.,* vol. 94, pp. 199-212, 2015.

[http://dx.doi.org/10.1016/j.oceaneng.2014.11.014]

[62] A.S. Algamili, M.H.M. Khir, J.O. Dennis, A.Y. Ahmed, S.S. Alabsi, S.S. Ba Hashwan, and M.M. Junaid, "A review of actuation and sensing mechanisms in mems-based sensor devices", *Nanoscale Res. Lett.,* vol. 16, no. 1, p. 16, 2021.
[http://dx.doi.org/10.1186/s11671-021-03481-7] [PMID: 33496852]

[63] P. Phogat, Shreya, R Jha, and S. Singh, *Optical and Microstructural Study of Wide Band Gap ZnO@ZnS Core–Shell Nanorods to be Used as Solar Cell Applications,* 2023, pp. 419-429.

[64] Shreya, P. Phogat, R Jha, and S. Singh, *Elevated Refractive Index of MoS2 Amorphous Nanoparticles with a Reduced Band Gap Applicable for Optoelectronic,* 2023, pp. 431-439.

[65] C. Dong, R. Zhao, L. Yao, Y. Ran, X. Zhang, and Y. Wang, "A review on WO3 based gas sensors: Morphology control and enhanced sensing properties", *J. Alloys Compd.,* vol. 820, p. 153194, 2020.
[http://dx.doi.org/10.1016/j.jallcom.2019.153194]

[66] Sreeprasad, T. S., & Pradeep, T. "Noble metal nanoparticles", *Springer handbook of nanomaterials,* pp. 303-388, 2013.

[67] J. Yi, and Y. Xianyu, "Gold nanomaterials-implemented wearable sensors for healthcare applications", *Adv. Funct. Mater.,* vol. 32, no. 19, p. 2113012, 2022.
[http://dx.doi.org/10.1002/adfm.202113012]

[68] S.J. Young, Y-H. Liu, Z-D. Lin, K. Ahmed, M.D.N.I. Shiblee, S. Romanuik, P.K. Sekhar, T. Thundat, L. Nagahara, S. Arya, R. Ahmed, H. Furukawa, and A. Khosla, "Multi-walled carbon nanotubes decorated with silver nanoparticles for acetone gas sensing at room temperature", *J. Electrochem. Soc.,* vol. 167, no. 16, p. 167519, 2020.
[http://dx.doi.org/10.1149/1945-7111/abd1be]

[69] J. Chen, X. Liu, G. Zheng, W. Feng, P. Wang, J. Gao, J. Liu, M. Wang, and Q. Wang, "Detection of glucose based on noble metal nanozymes: mechanism, activity regulation, and enantioselective recognition", *Small,* vol. 19, no. 8, p. 2205924, 2023.
[http://dx.doi.org/10.1002/smll.202205924] [PMID: 36509680]

[70] G. Doria, "Noble Metal Nanoparticles for Biosensing Applications", *Sensors,* vol. 12, no. 2, pp. 1657-1687, 2012.
[http://dx.doi.org/10.3390/s120201657]

[71] X. Tian, X. Cui, T. Lai, J. Ren, Z. Yang, M. Xiao, B. Wang, X. Xiao, and Y. Wang, "Gas sensors based on TiO2 nanostructured materials for the detection of hazardous gases: A review", *Nano Materials Science,* vol. 3, no. 4, pp. 390-403, 2021.
[http://dx.doi.org/10.1016/j.nanoms.2021.05.011]

[72] V.S. Bhati, M. Hojamberdiev, and M. Kumar, "Enhanced sensing performance of ZnO nanostructures-based gas sensors: A review", *Energy Rep.,* vol. 6, pp. 46-62, 2020.
[http://dx.doi.org/10.1016/j.egyr.2019.08.070]

[73] Q. Ren, X. Zhang, Y. Wang, M. Xu, J. Wang, Q. Tian, K. Jia, X. Liu, Y. Sui, C. Liu, J. Yun, J. Yan, W. Zhao, and Z. Zhang, "Shape-controlled and stable hollow frame structures of SnO and their highly sensitive NO2 gas sensing", *Sens. Actuators B Chem.,* vol. 340, p. 129940, 2021.
[http://dx.doi.org/10.1016/j.snb.2021.129940]

[74] G. Li, Z. Ma, Q. Hu, D. Zhang, Y. Fan, X. Wang, X. Chu, and J. Xu, "PdPt nanoparticle-functionalized α-Fe 2 O 3 hollow nanorods for triethylamine sensing", *ACS Appl. Nano Mater.,* vol. 4, no. 10, pp. 10921-10930, 2021.
[http://dx.doi.org/10.1021/acsanm.1c02377]

[75] M.N. Norizan, M.H. Moklis, S.Z. Ngah Demon, N.A. Halim, A. Samsuri, I.S. Mohamad, V.F. Knight, and N. Abdullah, "Carbon nanotubes: functionalisation and their application in chemical sensors", *RSC Advances,* vol. 10, no. 71, pp. 43704-43732, 2020.
[http://dx.doi.org/10.1039/D0RA09438B] [PMID: 35519676]

[76]   G. Yildiz, M. Bolton-Warberg, and F. Awaja, "Graphene and graphene oxide for bio-sensing: General properties and the effects of graphene ripples", *Acta Biomater.,* vol. 131, pp. 62-79, 2021. [http://dx.doi.org/10.1016/j.actbio.2021.06.047] [PMID: 34237423]

[77]   A. Hashmi, V. Nayak, K.R.B. Singh, B. Jain, M. Baid, F. Alexis, and A.K. Singh, "Potentialities of graphene and its allied derivatives to combat against SARS-CoV-2 infection", *Materials Today Advances,* vol. 13, p. 100208, 2022. [http://dx.doi.org/10.1016/j.mtadv.2022.100208] [PMID: 35039802]

[78]   A. Dager, takashi Uchida, toru Maekawa, and M. tachibana, "Synthesis and characterization of Mono-disperse carbon Quantum Dots from fennel Seeds: photoluminescence analysis using", *Mach. Learn..* [http://dx.doi.org/10.1038/s41598-019-50397-5] [PMID: 31570739]

[79]   J. Meng, T. Liu, C. Meng, Z. Lu, and J. Li, "Porous carbon nanofibres with humidity sensing potential", *Microporous Mesoporous Mater.,* vol. 359, p. 112663, 2023. [http://dx.doi.org/10.1016/j.micromeso.2023.112663]

[80]   M.J. Molaei, "Principles, mechanisms, and application of carbon quantum dots in sensors: a review", *Anal. Methods,* vol. 12, no. 10, pp. 1266-1287, 2020. [http://dx.doi.org/10.1039/C9AY02696G]

[81]   A. Yadav, Shreya, and N.K. Puri, *Preliminary Observations of Synthesized WS2 and Various Synthesis Techniques for Preparation of Nanomaterials*, 2023, pp. 546-556.

[82]   S. Aftab, M.Z. Iqbal, and Y.S. Rim, "Recent advances in rolling 2D TMDs nanosheets into 1d tmds nanotubes/nanoscrolls", *Small,* vol. 19, no. 1, p. 2205418, 2023. [http://dx.doi.org/10.1002/smll.202205418] [PMID: 36373722]

[83]   R. Canton-Vitoria, K. Sato, Y. Motooka, S. Toyokuni, Z. Liu, and R. Kitaura, "Field-effect transistor antigen/antibody-TMDs sensors for the detection of COVID-19 samples", *Nanoscale,* vol. 15, no. 9, pp. 4570-4580, 2023. [http://dx.doi.org/10.1039/D2NR06630K] [PMID: 36762571]

[84]   R. Khatri, and N.K. Puri, "Electrochemical studies of biofunctionalized MoS 2 matrix for highly stable immobilization of antibodies and detection of lung cancer protein biomarker", *New J. Chem.,* vol. 46, no. 16, pp. 7477-7489, 2022. [http://dx.doi.org/10.1039/D2NJ00540A]

[85]   K. F. Mak, and J. Shan, "Photonics and optoelectronics of 2D semiconductor transition metal dichalcogenides", *Nature Photonics,* vol. 10, no. 4, pp. 216-226, 2016. [http://dx.doi.org/10.1038/nphoton.2015.282]

[86]   M. Wu, Y. Xiao, Y. Zeng, Y. Zhou, X. Zeng, L. Zhang, and W. Liao, "Synthesis of two-dimensional transition metal dichalcogenides for electronics and optoelectronics", *InfoMat,* vol. 3, no. 4, pp. 362-396, 2021. [http://dx.doi.org/10.1002/inf2.12161]

[87]   K.M. Aguilar-Pérez, M.S. Heya, R. Parra-Saldívar, and H.M.N. Iqbal, "Nano-biomaterials in-focus as sensing/detection cues for environmental pollutants", *Case Studies in Chemical and Environmental Engineering,* vol. 2, p. 100055, 2020. [http://dx.doi.org/10.1016/j.cscee.2020.100055]

[88]   K. Spychalska, D. Zajac, S. Baluta, K. Halicka, and J. Cabaj, "Functional polymers structures for (Bio)sensing application—a review", *Polymers,* vol. 12, p. 1154, 2020. [http://dx.doi.org/10.3390/polym12051154]

[89]   Y. Wang, A. Liu, Y. Han, and T. Li, "Sensors based on conductive polymers and their composites: A review", *Polym. Int.,* vol. 69, no. 1, pp. 7-17, 2020. [http://dx.doi.org/10.1002/pi.5907]

[90]   S. Er, "Amino acids, peptides, and proteins: implications for nanotechnological applications in biosensing and drug/gene delivery", In: *Nanomaterials* vol. 11. , 2021, p. 3002.

[91]   D. Wu, J. Guo, C. Wang, X. Ren, Y. Chen, P. Lin, L. Zeng, Z. Shi, X.J. Li, C.X. Shan, and J. Jie, "Ultrabroadband and high-detectivity photodetector based on WS 2 /Ge heterojunction through defect engineering and interface passivation", *ACS Nano*, vol. 15, no. 6, pp. 10119-10129, 2021. [http://dx.doi.org/10.1021/acsnano.1c02007] [PMID: 34024094]

[92]   P. Bhattacharyya, D. Acharyya, and K. Dutta, *Resistive and capacitive measurement of nanostructured gas sensors.*, 2019, pp. 25-62. [http://dx.doi.org/10.1007/978-3-319-98708-8_2]

[93]   F. Yang, H. Cong, K. Yu, L. Zhou, N. Wang, Z. Liu, C. Li, Q. Wang, and B. Cheng, "Ultrathin broadband germanium–graphene hybrid photodetector with high performance", *ACS Appl. Mater. Interfaces,* vol. 9, no. 15, pp. 13422-13429, 2017. [http://dx.doi.org/10.1021/acsami.6b16511] [PMID: 28361534]

[94]   K. Bobzin, T. Brögelmann, C. Kalscheuer, B. Yildirim, and T. Liang, "Deposition of a nanocomposite (Ti, Al, Si)N coating with high thickness by high-speed physical vapor deposition", *Materialwiss. Werkstofftech.,* vol. 51, no. 3, pp. 297-312, 2020. [http://dx.doi.org/10.1002/mawe.201900103]

[95]   A. Shreya, *Yadav, R. Khatri, N. Jain, A. Bhandari, and N. K. Puri.* Double Zone Thermal CVD and Plasma Enhanced CVD Systems for Deposition of Films/Coatings with Eminent Conformal Coverage, 2023, pp. 273-283. [http://dx.doi.org/10.1007/978-981-16-9523-0_31]

[96]   H. Soon Min, "Thin films deposited by spin coating technique: review", *Article in Pakistan Journal of Chemistry,* 2021.

[97]   Q.Y. Li, Z-F. Yao, Y. Lu, S. Zhang, Z. Ahmad, J-Y. Wang, X. Gu, and J. Pei, "Achieving high alignment of conjugated polymers by controlled dip-coating", *Adv. Electron. Mater.,* vol. 6, no. 6, p. 2000080, 2020. [http://dx.doi.org/10.1002/aelm.202000080]

[98]   S.R. Sriram, S.R. Parne, N. Pothukanuri, and D.R. Edla, "Prospects of spray pyrolysis technique for gas sensor applications – A comprehensive review", *J. Anal. Appl. Pyrolysis,* vol. 164, p. 105527, 2022. [http://dx.doi.org/10.1016/j.jaap.2022.105527]

[99]   J. Fan, and L. Qian, "Quantum dot patterning by direct photolithography", *Nature Nanotechnology,* vol. 17, no. 9, pp. 906-907, 2022. [http://dx.doi.org/10.1038/s41565-022-01187-0]

[100]   Y. Hong, D. Zhao, J. Wang, J. Lu, G. Yao, D. Liu, H. Luo, Q. Li, and M. Qiu, "Solvent-free nanofabrication based on ice-assisted electron-beam lithography", *Nano Lett.,* vol. 20, no. 12, pp. 8841-8846, 2020. [http://dx.doi.org/10.1021/acs.nanolett.0c03809] [PMID: 33185450]

[101]   A. Singh, A. Shi, and S.A. Claridge, "Nanometer-scale patterning of hard and soft interfaces: from photolithography to molecular-scale design", *Chem. Commun. (Camb.),* vol. 58, no. 94, pp. 13059-13070, 2022. [http://dx.doi.org/10.1039/D2CC05221K] [PMID: 36373584]

[102]   J. Shi, and M. Wang, "Self-assembly methods for recently reported discrete supramolecular structures based on terpyridine", *Chem. Asian J.,* vol. 16, no. 24, pp. 4037-4048, 2021. [http://dx.doi.org/10.1002/asia.202101136] [PMID: 34672098]

[103]   L. R. MacFarlane, H. Shaikh, J. D. Garcia-Hernandez, M. Vespa, T. Fukui, and I. Manners, "Functional nanoparticles through π-conjugated polymer self-assembly", *Nature Reviews Materials,* vol. 6, no. 1, pp. 7-26, 2020. [http://dx.doi.org/10.1038/s41578-020-00233-4]

[104]   Z. Li, Q. Fan, and Y. Yin, "Colloidal self-assembly approaches to smart nanostructured materials",

Chem. Rev., vol. 122, no. 5, pp. 4976-5067, 2022.
[http://dx.doi.org/10.1021/acs.chemrev.1c00482] [PMID: 34747588]

[105] T. Kang, J. Zhu, X. Luo, W. Jia, P. Wu, and C. Cai, "Controlled self-assembly of a close-packed gold octahedra array for SERS sensing exosomal MicroRNAs", *Anal. Chem.,* vol. 93, no. 4, pp. 2519-2526, 2021.
[http://dx.doi.org/10.1021/acs.analchem.0c04561] [PMID: 33404216]

[106] C. Cummins, R. Lundy, J.J. Walsh, V. Ponsinet, G. Fleury, and M.A. Morris, "Enabling future nanomanufacturing through block copolymer self-assembly: A review", *Nano Today,* vol. 35, p. 100936, 2020.
[http://dx.doi.org/10.1016/j.nantod.2020.100936]

[107] T.E. Davies, H. Li, S. Bessette, R. Gauvin, G.S. Patience, and N.F. Dummer, "Experimental methods in chemical engineering: Scanning electron microscopy and X -ray ultra-microscopy— SEM and XuM", *Can. J. Chem. Eng.,* vol. 100, no. 11, pp. 3145-3159, 2022.
[http://dx.doi.org/10.1002/cjce.24405]

[108] K. Akhtar, S.A. Khan, S.B. Khan, and A.M. Asiri, "Scanning electron microscopy (SEM) and transmission electron microscopy (TEM) for materials characterization", In: *Materials Characterization Using Nondestructive Evaluation (NDE) Methods,* 2018, pp. 17-43.

[109] B.J. Inkson, "Scanning electron microscopy (SEM) and transmission electron microscopy (TEM) for materials characterization", *Methods,* no. Jan, pp. 17-43, 2016.
[http://dx.doi.org/10.1016/B978-0-08-100040-3.00002-X]

[110] C. Turquat, H.J. Kleebe, G. Gregori, S. Walter, and G.D. Sorarù, "Transmission electron microscopy and electron energy-loss spectroscopy study of nonstoichiometric silicon-carbon-oxygen glasses", *J. Am. Ceram. Soc.,* vol. 84, no. 10, pp. 2189-2196, 2001.
[http://dx.doi.org/10.1111/j.1151-2916.2001.tb00986.x]

[111] J.C. Dupin, D. Gonbeau, P. Vinatier, and A. Levasseur, "Systematic XPS studies of metal oxides, hydroxides and peroxides", *Phys. Chem. Chem. Phys.,* vol. 2, no. 6, pp. 1319-1324, 2000.
[http://dx.doi.org/10.1039/a908800h]

[112] P.S. Bagus, E.S. Ilton, and C.J. Nelin, "The interpretation of XPS spectra: Insights into materials properties", *Surf. Sci. Rep.,* vol. 68, no. 2, pp. 273-304, 2013.
[http://dx.doi.org/10.1016/j.surfrep.2013.03.001]

[113] R. W. Fransiska, E. M. P. Septia, W. K. Vessabhu, W. Frans, W. Abednego, and Hendro, "Electrical power measurement using arduino uno microcontroller and LabVIEW", *Proc. of 2013 3rd Int. Conf. on Instrumentation, Communications, Information Technol., and Biomedical Engineering: Science and Technol. for Improvement of Health, Safety, and Environ., ICICI-BME,* pp. 226-229, 2013.
[http://dx.doi.org/10.1109/ICICI-BME.2013.6698497]

[114] Y.I. Abdulkarim, "Design and study of a metamaterial based sensor for the application of liquid chemicals detection", *J. Mater. Res. Technol.,* vol. 9, no. 5, pp. 10291-10304, 2020.
[http://dx.doi.org/10.1016/j.jmrt.2020.07.034]

[115] J.O. Bodunrin, D.A. Oeba, and S.J. Moloi, "Current-voltage and capacitance-voltage characteristics of cadmium-doped p-silicon Schottky diodes", *Sens. Actuators A Phys.,* vol. 331, p. 112957, 2021.
[http://dx.doi.org/10.1016/j.sna.2021.112957]

[116] A. Labidi, C. Jacolin, M. Bendahan, A. Abdelghani, J. Guerin, K. Aguir, and M. Maaref, "Impedance spectroscopy on WO gas sensor", *Sens. Actuators B Chem.,* vol. 106, no. 2, pp. 713-718, 2005.
[http://dx.doi.org/10.1016/j.snb.2004.09.022]

[117] A. A. S. Mohammed, W. A. Moussa, and E. Lou, "High sensitivity mems strain sensor: design and simulation", *Sensors,* vol. 8, no. 4, pp. 2642-2661, 2008.
[http://dx.doi.org/10.3390/s8042642]

[118] A.D. Singh, and R.M. Patrikar, "Development of nonlinear electromechanical coupled macro model

for electrostatic MEMS cantilever beam", *IEEE Access,* vol. 7, pp. 140596-140605, 2019.
[http://dx.doi.org/10.1109/ACCESS.2019.2943422]

[119] M.A. Elgailani, A.H.H. Al-Masoodi, N.B. Sariff, and N. Abdulrahman, "Light dependent resistor sensor used for optimal power consumption for indoor lighting system", *2021 2nd International Conference on Smart Computing and Electronic Enterprise: Ubiquitous, Adaptive, and Sustainable Computing Solutions for New Normal, ICSCEE,* pp. 237-242, 2021.
[http://dx.doi.org/10.1109/ICSCEE50312.2021.9498097]

[120] M. Amirinasab, S. Shamshirband, A. T. Chronopoulos, A. Mosavi, and N. Nabipour, "Energy-efficient method for wireless sensor networks low-power radio operation in internet of things", *Electronics,* vol. 9, pp. 320-, 2020.
[http://dx.doi.org/10.3390/electronics9020320]

[121] C.W. Zhang, "Weibull parameter estimation and reliability analysis with zero-failure data from high-quality products", *Reliab. Eng. Syst. Saf.,* vol. 207, p. 107321, 2021.
[http://dx.doi.org/10.1016/j.ress.2020.107321]

[122] A.A.M. Faudzi, Y. Sabzehmeidani, K. Suzumori, J. Sultan Yahya Petra, and K. Datuk Keramat, "Application of micro-electro-mechanical systems (MEMS) as sensors: A review", *Journal of Robotics and Mechatronics,* vol. 32, no. 2, pp. 281-288, 2020.
[http://dx.doi.org/10.20965/jrm.2020.p0281]

[123] Z. Huang, C. Lv, Y. Xing, and J. Wu, "Multi-modal sensor fusion-based deep neural network for end-to-end autonomous driving with scene understanding", *IEEE Sens. J.,* vol. 21, no. 10, pp. 11781-11790, 2021.
[http://dx.doi.org/10.1109/JSEN.2020.3003121]

[124] Y. Lu, K. Xu, L. Zhang, M. Deguchi, H. Shishido, T. Arie, R. Pan, A. Hayashi, L. Shen, S. Akita, and K. Takei, "Multimodal plant healthcare flexible sensor system", *ACS Nano,* vol. 14, no. 9, pp. 10966-10975, 2020.
[http://dx.doi.org/10.1021/acsnano.0c03757] [PMID: 32806070]

[125] N. Ha, K. Xu, G. Ren, A. Mitchell, and J.Z. Ou, "Machine learning-enabled smart sensor systems", *Adv. Intell. Syst.,* vol. 2, no. 9, p. 2000063, 2020.
[http://dx.doi.org/10.1002/aisy.202000063]

[126] M. Mathew, S. Radhakrishnan, A. Vaidyanathan, B. Chakraborty, and C. S. Rout, "Flexible and wearable electrochemical biosensors based on two-dimensional materials: Recent development", *Analytical and Bioanalytical Chemistry,* vol. 413, no. 3, pp. 727-762, 2020.

[127] R. Turchetta, "Complementary metal-oxide-semiconductor (CMOS) sensors for high-performance scientific imaging", *High Performance Silicon Imaging: Fundamentals and Applications of CMOS and CCD Sensors,* no. Jan, pp. 289-317, 2020.
[http://dx.doi.org/10.1016/B978-0-08-102434-8.00010-6]

[128] J.H. Lau, "3D IC integration and 3D IC packaging, semiconductor", *Adv. Packag.,* pp. 343-378, 2021.

[129] J. Xavier, D. Yu, C. Jones, E. Zossimova, and F. Vollmer, "Quantum nanophotonic and nanoplasmonic sensing: towards quantum optical bioscience laboratories on chip", *Nanophotonics,* vol. 10, no. 5, pp. 1387-1435, 2021.
[http://dx.doi.org/10.1515/nanoph-2020-0593]

[130] P. Mohankumar, J. Ajayan, T. Mohanraj, and R. Yasodharan, "Recent developments in biosensors for healthcare and biomedical applications: A review", *Measurement,* vol. 167, p. 108293, 2021.
[http://dx.doi.org/10.1016/j.measurement.2020.108293]

[131] S. L. Ullo, and G. R. Sinha, "Advances in smart environment monitoring systems using IoT and sensors", *Sensors,* vol. 20, no. 11, p. 3113, 2020.
[http://dx.doi.org/10.3390/s20113113]

[132] Y. Jiang, S. Yin, J. Dong, and O. Kaynak, "A review on soft sensors for monitoring, control, and

optimization of industrial processes", *IEEE Sens. J.,* vol. 21, no. 11, pp. 12868-12881, 2021.
[http://dx.doi.org/10.1109/JSEN.2020.3033153]

[133] P. Sanjeevi, S. Prasanna, B. Siva Kumar, G. Gunasekaran, I. Alagiri, and R. Vijay Anand, "Precision agriculture and farming using Internet of Things based on wireless sensor network", *Trans. Emerg. Telecommun. Technol.,* vol. 31, no. 12, p. e3978, 2020.
[http://dx.doi.org/10.1002/ett.3978]

[134] R. Fedele, and M. Merenda, "An IoT System for Social Distancing and Emergency Management in Smart Cities Using Multi-Sensor Data", In: *Algorithms* vol. 13. , 2020, p. 254.
[http://dx.doi.org/10.3390/a13100254]

[135] J. Passos, "Wearables and internet of things (IoT) technologies for fitness assessment: A systematic review", *Sensors,* vol. 21, no. 16, p. 5418, 2021.
[http://dx.doi.org/10.37766/inplasy2021.6.0041]

[136] S. Nasiri, and M.R. Khosravani, "Progress and challenges in fabrication of wearable sensors for health monitoring", *Sens. Actuators A Phys.,* vol. 312, p. 112105, 2020.
[http://dx.doi.org/10.1016/j.sna.2020.112105]

[137] M.R. Khosravani, and T. Reinicke, "3D-printed sensors: Current progress and future challenges", *Sens. Actuators A Phys.,* vol. 305, p. 111916, 2020.
[http://dx.doi.org/10.1016/j.sna.2020.111916]

[138] P. Shreya, P. Phogat, R. Jha, and S. Singh, "Enhanced electrochemical performance and charge-transfer dynamics of 2D MoS 2 /WO 3 nanocomposites for futuristic energy applications", *ACS Appl. Nano Mater.,* vol. 7, no. 8, pp. 8593-8611, 2024.
[http://dx.doi.org/10.1021/acsanm.3c06017]

[139] P. Shreya, P. Phogat, R. Jha, and S. Singh, "Carbon nanospheres-induced enhanced capacitive dynamics in C/WS2/WO3 nanocomposites for high-performance electrochemical capacitors", *Mater. Sci. Eng. B,* vol. 304, p. 117390, 2024.
[http://dx.doi.org/10.1016/j.mseb.2024.117390]

[140] S. Rai, Shreya, P. Phogat, R. Jha, S. Singh, "Hydrothermal synthesis and characterization of selenium-doped MoS2 for enhanced optoelectronic properties", *MATEC Web of Conferences,* vol. 393, p. 01008, 2024.
[http://dx.doi.org/10.1051/matecconf/202439301016]

[141] A. Rai, P. Phogat, S. Shreya, R. Jha, and S. Singh, "Microwave assisted zinc sulphide quantum dots for energy device applications", *MATEC Web ofConferences,* vol. 393, p. 01011, 2024.
[http://dx.doi.org/10.1051/matecconf/202439301011]

[142] A. Sharma, S. Shreya, P. Phogat, R. Jha, and S. Singh, "Hydrothermally synthesized NiS2 and NiSO4(H2O)6 nanocomposites and its characterizations", *MATEC Web of Conferences,* vol. 393, p. 01016, 2024.
[http://dx.doi.org/10.1051/matecconf/202439301016]

[143] S. Shreya, P. Phogat, S. Singh, and R. Jha, "Reduction mechanism of hydrothermally synthesized wide band gap ZnWO4 nanorods for HER application", *MATEC Web of Conferences,* vol. 393, p. 01004, 2024.
[http://dx.doi.org/10.1051/matecconf/202439301004]

[144] P. Phogat, S. Shreya, R. Jha, and S. Singh, "Phase transition of thermally treated polyhedral nano nickel oxide with reduced band gap", *MATEC Web of Conferences,* vol. 393, p. 01001, 2024.
[http://dx.doi.org/10.1051/matecconf/202439301001]

Semiconductor Nanoscale Devices, 2025, 263-285

Engineering TFET Biosensors: Design Optimization, Analytical Modeling, and Radiation Considerations

Priyanka Goma[1,*] and **Ashwani K. Rana**[1]

[1] Department of Electronics and Communication Engineering, National Institute of Technology, Hamirpur, Himachal Pradesh, India

Abstract: This chapter provides a thorough examination of the key factors influencing the development and functionality of Tunnel Field-effect Transistor (TFET) biosensors. It focuses on three main areas: design techniques, analytical modeling for DNA detection, and the impact of radiation-induced effects, particularly X-rays, on TFET sensitivity. Commencing with an overview of TFET biosensors and their importance in biomedical and environmental sensing, the chapter delves into the complexities of design strategies aimed at enhancing sensor performance. It scrutinizes various design methodologies, such as material selection, device architecture, and surface functionalization, highlighting their effects on sensitivity, selectivity, and stability. Following this, the chapter investigates tailored analytical modeling approaches for TFET biosensors in DNA detection applications. It elucidates the theoretical foundations and numerical methods governing DNA sensing mechanisms, encompassing electrostatics modeling, charge transport simulations, and device-level simulations. Practical insights into amalgamating analytical models with empirical data enable the refinement of TFET biosensors for DNA detection, enhancing their precision and dependability. Moreover, the chapter delves into the repercussions of ionizing radiation, specifically X-rays, on TFET biosensor performance. It explores radiation-induced phenomena, such as shifts in threshold voltage, damage to gate oxide, and alterations in sensitivity, elucidating their implications for sensor functionality in radiation-rich settings. Strategies for mitigating these effects and bolstering sensor resilience are discussed to ensure consistent operation across diverse application scenarios.

Keywords: Analytical modeling, Biomedical sensing, DNA detection, Material selection, Radiation-induced effects, Tensitivity, TFET biosensors.

[*] **Corresponding author Priyanka Goma:** Department of Electronics and Communication Engineering, National Institute of Technology, Hamirpur, Himachal Pradesh, India; E-mail: priyanka@nith.ac.in

Ashish Raman, Prabhat Singh, Naveen Kumar & Ravi Ranjan (Eds.)

INTRODUCTION

In the rapidly evolving landscape of sensor technology, tunnel field effect-based biosensors have emerged as pivotal instruments, poised to revolutionize the field with their distinctive blend of capabilities [1 - 10]. These biosensors, characterized by their low power consumption, exceptional sensitivity, and seamless integration with Complementary Metal Oxide Semiconductor (CMOS) technology, have garnered significant attention as versatile platforms for modern sensing applications. In an era marked by the escalating demand for miniature, high-performance biosensing solutions capable of detecting a diverse array of analytes with precision and reliability, TFET biosensors stand out as promising candidates poised to meet these evolving needs. As we embark on a journey through the intricate realm of TFET biosensor engineering, our exploration focuses on three core pillars: design optimization, analytical modeling, and the nuanced consideration of radiation effects. These pillars serve as the cornerstones of TFET biosensor development, guiding efforts to enhance performance, understand underlying physics, and fortify sensors against environmental challenges. Through the investigation of these aspects, we aim to contribute to the betterment of biosensing technology, paving the way for innovative solutions that transcend existing boundaries.

Design optimization consists of a myriad of strategies aimed at tailoring TFET biosensors to achieve optimal performance characteristics. From refining sensitivity and selectivity to minimizing power consumption and fabrication complexity, design optimization strategies play a pivotal role in bolstering the capabilities of TFET biosensors, empowering them to meet the diverse requirements of modern sensing applications. In parallel, our exploration delves into the realm of analytical modeling, offering insights into the underlying physical mechanisms that govern TFET operation. Analytical models serve as invaluable tools for predicting device behavior, unraveling intricate electrical phenomena, and informing design decisions with precision. By leveraging these models, the readers are expected to gain a deeper understanding of TFET biosensor performance under various operating conditions, empowering researchers to refine designs and optimize sensor performance.

Furthermore, the presented investigation extends to the realm of radiation considerations, recognizing the importance of fortifying TFET biosensors against environmental challenges. In an era where radiation exposure is increasingly prevalent in diverse applications, from medical imaging to space exploration, understanding the effects of radiation on biosensor performance is paramount. By exploring strategies to enhance radiation tolerance and mitigate radiation-induced effects, we strive to ensure the reliability and robustness of TFET biosensors in demanding operational environments.

Through this multidimensional exploration, we aim not only to advance the state-of-the-art in TFET biosensor engineering, but also to catalyze transformative advancements in biosensing technology as a whole. By unraveling the intricacies of design optimization, analytical modeling, and radiation considerations, we endeavor to push the boundaries of sensor performance, enabling the development of innovative solutions that address the evolving needs of society.

DESIGN TECHNIQUES FOR TFET BIOSENSORS

Designing Tunnel Field Effect Transistor (TFET) biosensors requires a nuanced understanding of the intricate interplay between device engineering and biosensing principles. TFET biosensors offer remarkable advantages, such as low-power operation, high sensitivity, and compatibility with ongoing CMOS technology, which make them highly promising in a variety of applications, such as healthcare, environmental monitoring, and beyond. In this exploration of design techniques for TFET biosensors, the key strategies aimed at optimizing device performance and enhancing sensitivity, selectivity, and reliability are mentioned. From material selection and device geometry optimization to surface functionalization and signal amplification, each technique plays a vital role in shaping the capabilities and effectiveness of TFET biosensors [7 - 15]. Through this examination, the authors aim to provide insights and practical guidance to researchers and engineers seeking to leverage TFET biosensors for innovative biosensing solutions. Designing TFET biosensor devices involves several key techniques to optimize their performance for specific applications. Here are listed some essential design techniques.

Material Selection

The choice of semiconductor material is critical for TFET biosensors. III-V compound semiconductors, like InAs and GaSb, are often preferred due to their high carrier mobility, which allows for efficient charge transport. Additionally, 2D materials, such as MoS_2 and WSe_2, offer advantages, like tunable bandgaps and large surface-to-volume ratios, making them suitable for biosensing applications. In one of the studies [11, 12], highly sensitive ovarian cancer biomarker detection was made possible with the help of InGaAs/Si heterojunction-based TFET biosensor.

Device Geometry Optimization

The geometry of the TFET, including the physical gate length, gate metal width, and thickness, significantly impacts its performance. For example, shorter channel lengths can enhance sensitivity by reducing the distance charges needed to tunnel, while optimizing the width and thickness can improve signal-to-noise ratio and current levels. Researchers have investigated different device geometries to improve the sensitivity of biosensor devices based on tunneling mechanisms. Among these, a notable study [13] investigated the utilization of an I-shaped TFET structure, specifically addressing steric and trap-related issues.

Surface Functionalization

Surface functionalization involves modifying the TFET surface to enable selective binding of target biomolecules [14]. This is achieved by attaching specific molecules, such as antibodies or DNA probes, onto the surface using linker molecules or self-assembled monolayers. Functionalization enhances the sensor's specificity by ensuring that only the desired analytes bind to the surface.

Gate Dielectric Engineering

The function of the gate dielectric layer is critical in TFET operation as it regulates tunneling behavior. A study [15] evaluated the Ferro-TFET biosensor against state-of-the-art biosensors, showcasing superior performance. By utilizing ferroelectric material as the oxide, it enhanced subthreshold swing, gate control, and suitability for low-power biosensor design. High-k dielectric materials are frequently used to improve the electric field effect, leading to enhanced sensitivity to charge variations induced by biomolecular binding.

Bioreceptor Integration

Bioreceptors are molecules immobilized on the TFET surface that selectively capture target analytes. Optimizing their density and orientation is crucial for maximizing binding efficiency and improving sensor performance. Moreover, the selection of a bioreceptor depends on the particular analyte being detected and the desired levels of sensitivity and specificity required for the biosensor.

Signal Amplification

Signal amplification techniques are employed to enhance the detection sensitivity of TFET biosensors. This may involve incorporating amplifying circuits on-chip or utilizing external amplification methods, such as enzymatic amplification or the use of nanoparticle-based signal enhancement strategies.

Noise Reduction

Minimizing noise sources is essential for improving the signal-to-noise ratio of TFET biosensors. Techniques, such as optimizing the device structure to reduce thermal noise, operating at lower temperatures, and using low-noise readout circuits, help mitigate noise effects and improve sensor performance.

Microfluidic Integration

Integrating microfluidic channels with TFET biosensors allows for precise control of sample delivery and manipulation. Microfluidic systems enhance sensor performance by reducing sample volume, improving response time, and enabling multiplexed detection of multiple analytes in a single device.

On-chip Reference Electrode

Including an on-chip reference electrode provides a stable reference potential for accurate sensing. This helps compensate for environmental variations and ensures reliable and reproducible measurements.

Packaging and Encapsulation

Robust packaging and encapsulation techniques are essential to protect TFET biosensors from environmental factors, such as moisture, temperature fluctuations, and mechanical stress. Proper packaging ensures the long-term stability and reliability of the biosensor device.

ANALYTICAL MODELING OF THE SURFACE POTENTIAL OF TFET BIOSENSORS FOR DNA DETECTION

Biosensors serve an important role in a variety of disciplines, including healthcare, environmental monitoring, and biotechnology, since they enable the rapid and accurate detection of biomolecules. TFET biosensors have received a lot of attention as an emerging biosensing technology because of their ability to detect ultra-sensitively and with low power [16 - 19]. TFETs, which use quantum tunneling rather than thermal emission, provide notable advantages over traditional MOSFET-based biosensors, including lower power consumption and increased sensitivity. However, the intricate interplay of quantum tunneling, semiconductor physics, and biomolecular interactions in TFET biosensors needs the development of a complete analytical model to better understand device behavior and guide optimization efforts.

DNA detection and analysis are critical in a wide range of sectors, including medical research, forensics, and personalized medicine. Traditional DNA identification methods, such as Polymerase Chain Reaction (PCR) and fluorescence-based tests, are often time-consuming, labor-intensive, and necessitate sophisticated instrumentation. Furthermore, these approaches may lack the necessary sensitivity to detect low-abundance DNA targets, inhibiting their usefulness in certain applications. These issues can be addressed with the help of TFET biosensing technology. So far, a lot of simulation-based research work has been discussed in the literature that considers DNA detection by utilizing tunneling transistors. However, only a few studies are there that show analytical modeling of surface and drain potential for DNA biomolecules. A work proposed by R. Priyanka*et al..* [20] presented a comparison of tunneling and inversion phenomena for DNA detection. It showed that TFET-based detection provides 63% better sensitivity to DNA than inversion mode FET devices. Another work [21] showed the importance of negative capacitance or ferroelectric layer in the TFET devices for improved detection of negatively charged biomolecules, such as DNA molecules. By using this layer, the overall sensitivity of the biosensor can be enhanced significantly. Furthermore, a study by R. Narang*et al..* [22] provided evidence of the drain-current parameter as a sensing metric and also an analytical modeling approach to its calculation by considering process-induced variations. The negatively charged biomolecular detection and its verification by using suitable models were also presented. Despite the potential of DNA TFET biosensors, there are considerable research gaps in understanding their underlying principles and optimizing their performance for practical use. The existing literature focuses mostly on device manufacturing and experimental validation, with little effort paid to establish comprehensive analytical models specific to DNA sensing. As a result, the primary goal of our research was to propose an analytical model that explains the fundamental principles regulating DNA detection in TFET biosensors.

The proposed TFET biosensor employs a novel device structure tailored for DNA detection. The schematic diagram of the proposed DGTFET biosensor is illustrated in Fig. (**1a**). The device comprises a strained silicon-germanium (strained Si-Ge with 35% Ge fraction) optimized pocket and gate configurations to facilitate efficient tunneling and biomolecular interaction. The device features a dual-material (tunneling, TG with work function 3.9 eV and auxiliary, AG with work function 4.1 eV) double-gate architecture to enhance control over the channel and improve sensing performance. Additionally, plasma-based doping is strategically employed in the source (p^+) and drain (n) regions to mitigate ambipolarity commonly associated with TFET devices.

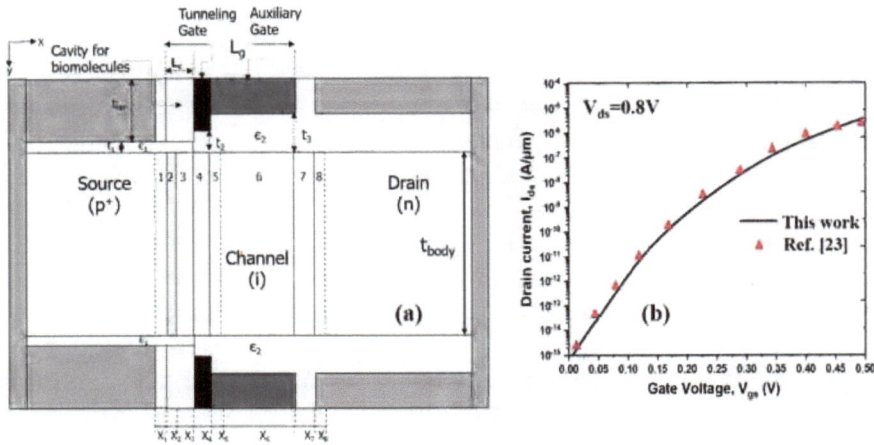

Fig. (1). (a) Structure of proposed DGTFET biosensor with multiple regions for analytical modeling (L_g=50 nm, t_{cav}=5 nm, L_c=7 nm, t_3=3 nm, t_2=1.8 nm, t_{body}=10 nm, ϵ_2= ϵ_{SiO2}, $\epsilon1$= ϵ_{TiO2}); (b) calibration of the proposed device with experimental data [23].

Non-local tunneling models, incorporating field and concentration-dependent mobility, non-local band-to-band tunneling, Fermi statistics, Shockley-Read-Hall recombination rates, and bandgap narrowing effects, are employed to accurately simulate charge transport. Furthermore, the simulation framework accounts for the presence of DNA biomolecules through appropriate charge assignments and biomolecular binding kinetics. The simulations are done in a TCAD environment. Parameters, such as biomolecule charge and size, are sourced from experimental data to ensure realistic simulation results. A 'K' (dielectric constant) value of 1 with no charge represents an empty cavity, whereas negative charge densities ranging from -10^{11} cm^{-2} to -10^{12} cm^{-2} indicate the presence of bound DNA biomolecules within the cavity. The device is validated with the experimental data [23] shown in Fig. (**1b**).

To characterize the drain current in the described biosensor, Landauer's formula [24] is applied using a non-local method. This formula relies on tunneling probability and considers the band structure of the material. Determining tunneling probability is inherently quantum mechanical and involves solving the Schrödinger equation, a computationally demanding task. To simplify calculations for our semiconductor device, approximations are employed. The Wentzel-Kramer-Brillouin (WKB) method is utilized to calculate tunneling probability, as it is independent of material and band structure. The formula is represented as equation (1):

$$I_{drain} = \frac{2q}{\hbar} \int\limits_{E_{C,Q_r}(Channel)}^{E_{V,P_r}(Source)} \left(f_{P_r}(E_r) - f_{Q_r}(E_r) \right) T_{BTB} dE_r \tag{1}$$

Here, I_{drain} is the drain current calculated over the source to channel region by using the energy band concept. T_{BTB} is the source-to-channel tunneling probability at points P_r to Q_r equal to $e^{-2\gamma}$ (γ depends upon the potential barrier between the source and channel interface). The energies of source and channel regions contribute to the fermi statistics, where the corresponding Fermi energies are given by:

$$f_{P_r}(E_r) = \frac{1}{1+e^{(E_i-E_{F,Source})/kT}}; f_{Q_r}(E_r) = \frac{1}{1+e^{(E_i-E_{F,Channel})/kT}} \tag{2}$$

and

$$\gamma = \int\limits_{x_{P_r}}^{x_{Q_r}} \left| \sqrt{\frac{2m(P(x)-E_i)}{\hbar^2}} \right| dx \tag{3}$$

The points x_{Q_r} and x_{P_r} are taken as E_g/qE_{avg} and 0, respectively, for simpler calculations of the above integral. E_{avg} is determined by taking the magnitude of electric fields in the lateral and vertical directions of the biosensor. To calculate T_{BTB}, the term $(P(x)-E_i)$ should be known; therefore, $P(x)$ is written in terms of the surface potential in equation (3):

$$P(x) = E_g + q\psi_{sp}(x) \tag{4}$$

The integral in equation (3) when solved gives $\gamma = \dfrac{2\sqrt{2mE_g^{\,3}}}{3qE_{avg}\hbar}$ and the tunneling probability thus becomes

$$T_{BBT} = e^{\left(-4\frac{\sqrt{2mE_g^{\,3}}}{3qE_{avg}\hbar} \right)} \tag{5}$$

All these values are substituted in Landauer's equation, giving the drain current as:

$$I_{drain} = \frac{2qkT}{\hbar} T_{BBT} \ln\left| \frac{(1+I_a)(1+I_b)}{(1+I_c)(1+I_d)} \right| \tag{6}$$

where $I_a = e^{\left(\frac{E_{F,Source} - E_{C,Channel}}{kT}\right)}$; $I_b = e^{\left(\frac{E_{F,Channel} - E_{V,Source}}{kT}\right)}$; $I_c = e^{\left(\frac{E_{C,Source} - E_{V,Source}}{kT}\right)}$;

$I_d = e^{\left(\frac{E_{F,Channel} - E_{C,Channel}}{kT}\right)}$

The surface potential term taken for the calculation of I_d is also calculated by using 2D Poisson distribution at eight regions of the biosensor device. Various parameters calculated during the model formulation are presented in the Appendix section for reference. The areas consist of depletion regions for the source and drain, a strained Si-Ge pocket region, the space beneath the cavity, thickness below the dual gates, the depletion zone at the interface of these gates, and an underlap region near the drain. The underlap region serves to restrict the diffusion of high carrier density beyond the channel region towards the drain terminal, thereby mitigating the effects of ambipolarity [25]. Additionally, optimizing the work function of the drain metal induces asymmetric doping, further addressing potential ambipolarity concerns. Provided below is 2D Poisson's equation utilized to find the solution of ψ_{sp}:

$$\partial^2 \varphi / \partial x^2 + \partial^2 \varphi / \partial y^2 = q N_k / \varepsilon_{si}, k = 1 \ to \ 8 \tag{7}$$

The term N_s equals $-N_d$ for region 2 as the pocket is doped, whereas for all the other regions, the term N_s equals N_A. The 2D potential solution is obtained under the assumption that the potential profile varies parabolically through the thickness of the channel given by:

$$\varphi_k(x, y) = A_{1k}(x) + A_{2k}(x)y + A_{3k}(x)y^2 \tag{8}$$

The coefficients in equation (2) $(A_{1k}(x), A_{2k}(x) \ and \ A_{3k}(x))$ are determined by applying the boundary conditions, which are provided as:

$$\varphi_k(x, y)\big|_{y=0} = \varphi_k(x, 0) = \varphi_k(x, t_{body})$$

$$\frac{d\varphi_k(x, y)}{dy}\bigg|_{y=0} = \left(\frac{\beta_k}{t_{body}}\right)\left(\varphi_{sk}(x) - V_{gseff,k}\right)$$

$$\frac{d\varphi_k(x, y)}{dy}\bigg|_{y=t_{body}} = \left(\frac{\beta_k}{t_{body}}\right)\left(V_{gseff,k} - \varphi_{sk}(x)\right) \tag{9}$$

$\beta_k = C_k / C_{Channel}, C_2 = C_3 = \left(\dfrac{\varepsilon_1 \varepsilon_c}{\varepsilon_c t_1 + \varepsilon_1 t_c}\right), C_1 = (2/\pi)C_2, C_8 = 2C_7 / \pi, C_4 = (\varepsilon_2/t_3), C_5 = C_6 = C_7 = (\varepsilon_2/t_2) =$

$C_{Channel} = (\varepsilon_{si}/t_{body}), \varepsilon_1 = 80\varepsilon_0, \varepsilon_2 = 3.9\varepsilon_0$

The surface potential is given by . The model of surface potential incorporates the fringing field effect [26] occurring due to the TG and AG. The capacitances reflecting this effect are C_1 and C_8. On the other hand, the capacitances generated due to the impact of biomolecules are given by C_2 and C_3. The effect of DNA can be included in the flat-band voltage by the term qN_{DNA}/C, where N_{DNA} is the charge on DNA biomolecules and C is the cavity capacitance. The effective gate to source voltage becomes:

$$V_{gseff,2} = V_{gs} - \left(V_{flatb,2} - \left(qN_{DNA}/C_2\right)\right) \text{ and } V_{gseff,3} = V_{gs} - \left(V_{flatb,3} - \left(qN_{DNA}/C_2\right)\right)$$

All these values when used can give the values of the coefficients $A_{1k}(x)$, $A_{2k}(x)$, and $A_{3k}(x)$. On replacing these coefficients in equation (7), the 1D surface potential solution becomes:

$$\partial^2 \varphi_{sk}(x)/\partial x^2 - \lambda^2 \varphi_{sk}(x) - \alpha_k = 0 \text{ for } k=1 \text{ to } 8 \tag{10}$$

The continuity of potential and electric flux is utilized at each region's boundary to ascertain the solution of equation (10) to equation (11) as:

$$\varphi_{sk}(x) = Ue^{\lambda_k x} + Ve^{-\lambda_k x} + \sigma_k \text{ ;for } k=1 \text{ to } 8 \tag{11}$$

The boundary conditions are:

$$\varphi_{s1}(x=0) = V_{bi,source} = -\left(kT/q\right)\ln\left(N_a/n_i\right); \varphi_{s8}\left(x = \sum_{k=1}^{8} X_k\right) = V_{bi,drain} + V_{ds}, \text{where} V_{bi,drain} = \left(kT/q\right)\ln\left(N_A/n_i\right)$$

$$\varphi_{sk}(x)\Big|_{x=\sum_{l=1}^{k} X_l} = \varphi_{s(k+1)}(x)\Big|_{x=\sum_{l=1}^{k} X_l} ; \quad k=1 \text{ to } 7$$

$$\frac{d\varphi_{sk}(x)}{dx}\Big|_{x=\sum_{l=1}^{k} X_l} = \frac{d\varphi_{s(k+1)}(x)}{dx}\Big|_{x=\sum_{l=1}^{k} X_l} ; \quad k=1 \text{ to } 7$$

X_1(source), X_5(channel), and X_8(drain) represent the depletion regions.

The dielectric constant of a DNA molecule is 8 with variable charge densities [20]. Threshold voltage and ON-current levels differ for charged biomolecules due to their impact on the flat-band voltage. Fig. (2a) illustrates their effect on the energy band along with the linear potential profile, while Fig. (2b) presents the transfer characteristics of DNA biomolecules within the nanocavity. It could be observed from Fig. (2b) that I_d is reduced with the increase in the intensity of negative charge on DNA biomolecules and is least for the absence of biomolecule in the cavity. The significant shift measured in the proposed device serves as a sensitivity parameter. Threshold voltage and drain current are chosen as sensitivity metrics for charged biomolecules [27]. The expressions for these quantities are given in equation (12).

When DNA becomes immobilized in the nanocavity, there is a modulation of dielectric properties. This change in the drain current occurs as the number of carriers transported from the source to the drain decreases, primarily influenced by the negative charge of DNA. As the charge density on DNA rises, the coupling between the gate and tunneling junction weakens, leading to a reduction in drain current, especially with higher levels of negative charge on the biomolecules. This phenomenon is shown in Fig. (3a-b). The modeled and simulated drain current data closely match each other.

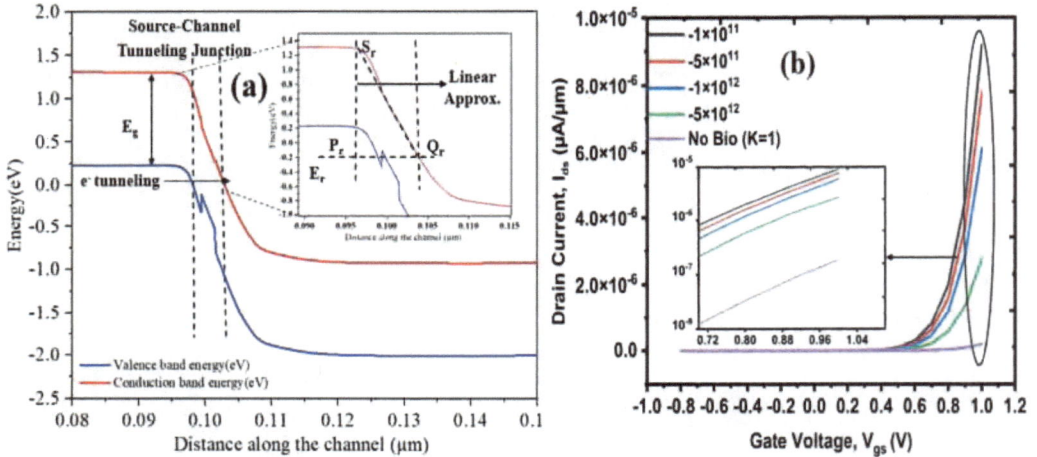

Fig. (2). (a) Energy band diagram for the proposed device along with potential profile showing linear approximation for the surface potential model; (b) transfer characteristics showing the effect of DNA on the drain current.

$$\Delta V_{th,DNA} = V_{th}\big|_{K=1} - V_{th}\big|_{DNA}\;;\; S_{I_{ON},DNA} = \left(I_{ON}\big|_{K=DNA} - I_{ON}\big|_{K=1}\right)\Big/I_{ON}\big|_{K=1} \qquad (12)$$

Fig. (4a) shows the variation in ON-current sensitivity with varying negative charges on DNA biomolecules. The modeled value of maximum ON-current sensitivity is approximately 48.38, whereas the simulated value is 34, indicating a relatively lower sensitivity due to the assistance provided by the pocket region in supplying the majority of carriers for tunneling at the source-channel interface. However, the V_{th} sensitivity is 308.9 mV, which is considered significant. Fig. (4b) shows that increasing negative charge on DNA biomolecules leads to a corresponding increase in threshold voltage. Therefore, significant V_{th} sensitivity for negatively charged DNA biomolecules is expected for detection in the proposed biosensor.

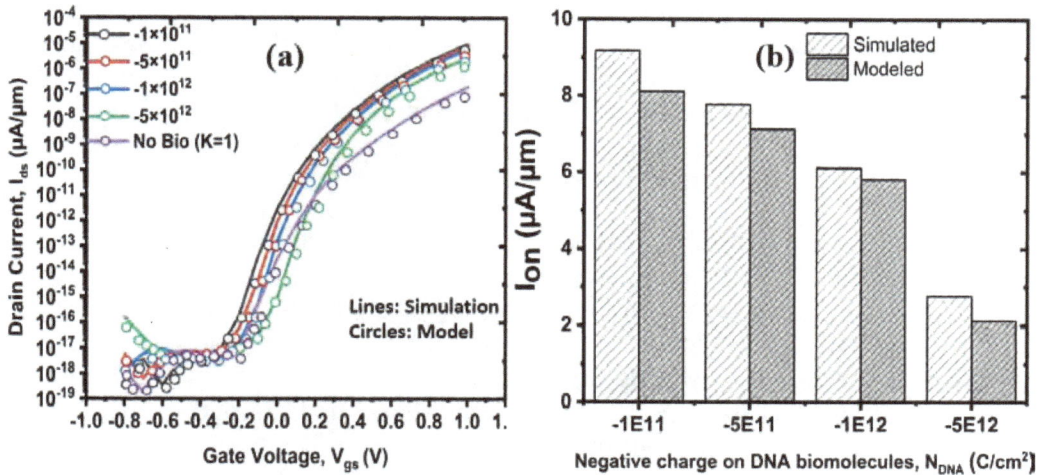

Fig. (3). (a) Transfer characteristics for modeled and simulated drain current for DNA biomolecules; (b) comparison of ON-current values for modeled and simulated equations.

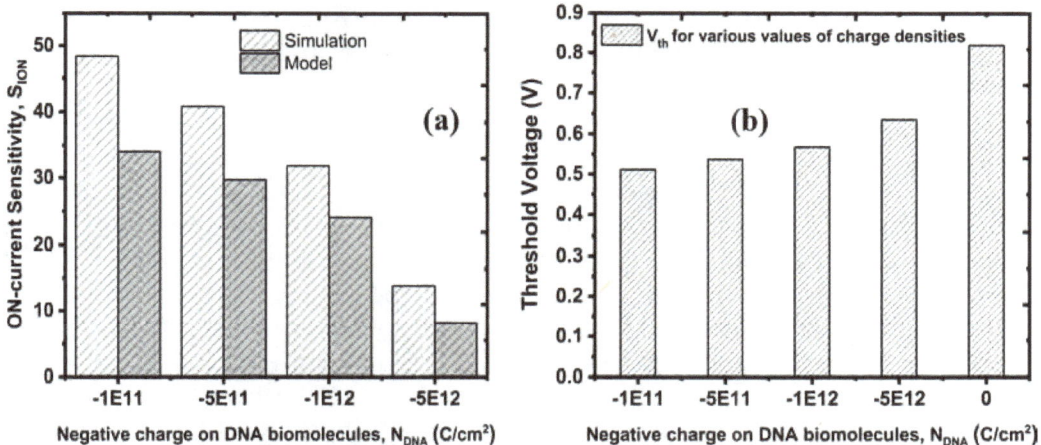

Fig. (4). (a) ON-current sensitivity for modeled and simulated values for DNA biomolecules; (b) threshold voltage variation for different charge densities of DNA biomolecules.

RADIATION-INDUCED EFFECTS ON TFET

Radiation-induced effects on field-effect transistors can significantly impact the performance and reliability of electronic devices [28 - 31]. When exposed to ionizing radiation, such as gamma rays or energetic particles, FETs may experience various phenomena, such as threshold voltage shift, charge trapping in semiconductor materials, degradation of gate oxide, *etc*. The defects and trap sites within the oxide layer may arise due to the formation of oxygen vacancies or it can also happen due to the displacement of atoms caused by the radiation's energy deposition. Trapped charge at these defect sites can alter the electrical properties, affecting the insulating capabilities and leading to increased leakage currents. The presence of trapped charge near the channel region of the transistor can modulate the V_{th} and affect the device's switching characteristics. As a result, these effects can degrade device performance, reduce switching speed, and compromise reliability, particularly in radiation-sensitive applications.

Radiation exposure not only creates trapped charges in the oxide layer, but also induces additional interface states at the silicon/silicon dioxide boundary. This boundary, despite maintaining lattice continuity, experiences bond ruptures due to the differing lattice constants of silicon and silicon dioxide. Radiation catalyzes this bond breakage, disrupting the regular potential field at the interface. Consequently, a new, fixed potential field emerges, with charged centers located at the band gap midpoint, defining the interface state [32]. These states tend to change the characteristics of the device.

Sometimes, the high-energy radiation particles may cause transient or permanent disruptions in field-effect devices that can compromise the integrity of the complete electronic systems. These effects are comprised of Single-event Effects (SEES), including single-event upsets, single-event transients, and single-event latch-ups [33, 34]. In one of the studies [34], a Junctionless Tunnel FET (JLTFET) was specifically engineered and tested for its resilience to radiation by modifying the work function of its auxiliary gate. To identify the most susceptible area of the device, heavy ion radiation with varying Linear Energy Transfer (LET) values is directed at all regions, revealing the channel as the most vulnerable. Consequently, the radiation investigation focuses solely on this region. Parameters, such as collected charge, transient peak current, and bipolar gain reciprocated minimal sensitivity, when exposed to higher doses of LET values, show the efficiency of JLTFET in a radiated environment. Prolonged exposure to ionizing radiation can result in Total Ionizing Dose (TID) effects. This cumulative radiation damage gradually degrades the electrical properties of FETs, leading to shifts in device parameters, increased leakage currents, and reduced reliability over time.

Due to its low power consumption, TFET emerges as a promising option for space or medical applications. Therefore, exploring TFET behavior in a radiation environment is crucial. Wang*et al.*. [35] conducted a simulation-based investigation on the single-event transient effect in L-shaped TFETs, highlighting the vulnerable areas to heavy ion hits that could induce soft errors in integrated circuits. Additionally, Ding*et al.*. conducted experimental research on the effect of 10 keV X-ray irradiation on silicon material TFETs, revealing significant changes in radiation-affected parameters [36, 37]. Their findings underscore the importance of considering electrical stress effects in total ionizing dose tests. The impact of radiation is mentioned in Table **1**.

IMPACT OF X-RAYS ON THE SENSITIVITY OF TFET BIOSENSOR

The impact of X-rays on TFET (Tunnel Field-effect Transistor) devices is notable across various aspects of their electrical behavior and performance. When exposed to X-rays, TFETs undergo significant changes due to the ionization of atoms within their semiconductor material. This ionization process generates electron-hole pairs, contributing to increased device conductivity. One notable effect is the alteration of the TFET's threshold voltage. X-ray-induced modifications in the density of trapped charges within the gate oxide or at the silicon-insulator interface layer can alter the threshold voltage, affecting the TFET's switching behavior. Additionally, X-ray exposure may induce higher leakage currents in TFETs, as radiation-induced defects or trap states within the semiconductor material act as recombination centers for charge carriers. This can lead to degraded device performance, especially in the off-state. Furthermore, X-rays can influence the subthreshold swing of TFETs, measuring their efficiency in switching between on and off states. Changes in trap states or interface properties induced by radiation can affect the subthreshold swing, impacting the TFET's power consumption and switching speed. Prolonged or high-dose X-ray exposure may also cause threshold voltage instability in TFETs, stemming from radiation-induced charge trapping or material degradation processes, leading to variations in device performance over time or under different operating situations. Contemplating the effects of these radiations is crucial for designing TFET-based electronics for applications where X-ray exposure is prevalent, such as medical imaging, radiation detection, aerospace electronics, and biosensing devices.

A similar effect is expected to be seen in TFET-based biosensors. The TID effects can be included in semiconductor devices by the addition of interface and oxide traps that are formed when rays of particular intensity hit the device. Table **2** shows the defect density for SiO_2 material when X-rays of different intensities hit the semiconductor device containing SiO_2 as a dielectric material.

The impact of X-rays on the essential parameters of the biosensor is demonstrated in this section. As discussed in the analytical modeling section, SiO_2 serves as the dielectric material. The biosensor presented in Fig. (**5**) contains an n^+ pocket, which is doped to attain maximum sensitivity of the biosensor towards biomolecular detection. A description of the biosensor's operation and its sensitivity equations is already made in the preceding sections of this chapter. The transfer characteristics for an empty cavity are drawn in Fig. (**6**) considering the data in Table **2**. It can be noticed that the presence of oxide and interface traps due to X-rays impacts the tunneling in OFF as well as ON state operation of the device. The presence of these intermediate states not only increases the requirement of biasing voltage, but also lowers the overall drain current for each biomolecule insertion in the affected biosensor. Raising the overall biasing voltage results in the elevation of the V_{th} of the device. The repercussions of X-rays thus affect the sensing metrics of biosensors. Since the trapped charges possess a positive charge, they facilitate carrier tunneling in both the ON and OFF states.

Table 1. Impact of radiation on TFET devices.

Device	Nature of Radiation	Maximum Radiation Dose	Impact on Device Parameters
SOI TFET [38]	Gamma (γ)	100 Mrad (SiO$_2$)/s	V_{th} decreases Interface trap charge increases The drain current remains stable in the ON state, but changes significantly below the threshold region The I_{ON}/I_{OFF} ratio increases
Ferroelectric TFET [38]	Gamma (γ)	10^{11} rads (Si)/s	Reduced photocurrent is observed Subthreshold Swing (SS) experiences degradation I_{ON}/I_{OFF} decreases.
N$^+$ Pocketed TFET [39]	^{60}Co γ-rays	1 Mrad (Si)/s	Deep state trap increases OFF state leakage The interface trap charge increases the SS
SOI TFET [40]	Gamma (γ)	100 Mrad (SiO$_2$)/s	Less sensitivity to gamma radiations TFET also exhibits a good I_{ON}/I_{OFF} current ratio at higher radiation doses
SOI TFET [41]	Gamma (γ)	500 krad (SiO$_2$)/s	Subthreshold Swing (SS) experiences degradation I_{ON}/I_{OFF} decreases
Vertical DG-FeTFET and SOI-FeTFET [42]	Gamma (γ)	10^{11} rads (Si)/s	VDG-FeTFET had fewer trapped charges and higher SS and I_{OFF} ratios than planar VDG-FeTFET had reduced SS, increased I_{ON}/I_{OFF} ratio, and minimal curve shift after TID radiation, suggesting better protection from TID, hence offering better performance

Table 2. Variation in SiO$_2$ defect density with different X-ray radiation doses [43].

Defect Density	Radiation Dose (krad)				
	10	30	50	70	100
Oxide × 10^{12} cm^{-2}	0.86	1.2	1.48	1.55	1.6
Interface × 10^{12} cm^{-2} eV^{-1} (Si/SiO$_2$)	0.55	0.71	0.85	0.92	1.45

A similar effect can be observed in Fig. (7) for keratin biomolecule. When the cavity contains a biomolecule, the key factors influencing tunneling are the gate and drain voltages, the trapped charges at the interface and oxide in the semiconductor device, and the dielectric constant of the biological molecules. All these parameters support the tunneling of carriers in both ON and OFF states.

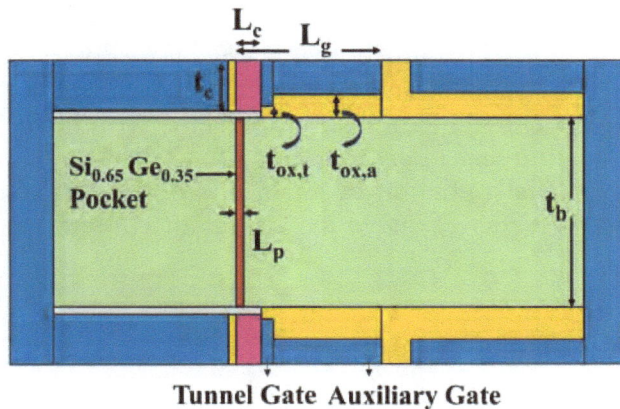

Fig. (5). Schematic of a double-gate Si $_{(0.65)}$ and Ge $_{(0.35)}$ pocketed dopingless TFET biosensor (t$_b$=10 nm).

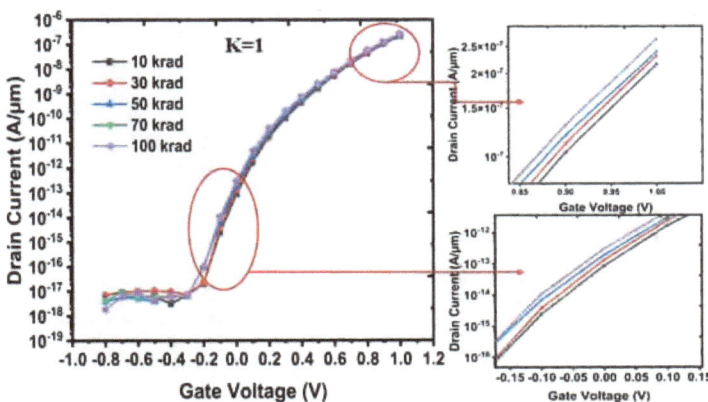

Fig. (6). Transfer characteristics for different X-ray irradiated doses at K=1 for the presented biosensor.

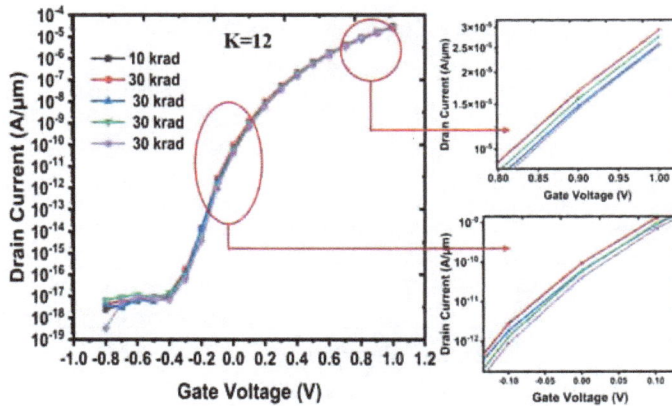

Fig. (7). Transfer characteristics of the designed biosensor for different X-ray irradiated doses at K=12.

The electric field is another parameter that is crucial for any device operation. The electric field along the channel and maximum electric field curves are shown in Fig. (**8a-d**) for an empty and biomolecule-filled cavity of the respected biosensor. In Fig. (**8a**), it is notable that in the ON state for K=1, the electric field improves with the intensity of the radiation dose. In addition, the leakage at the drain-channel interface also increases due to the availability of intermediate states.

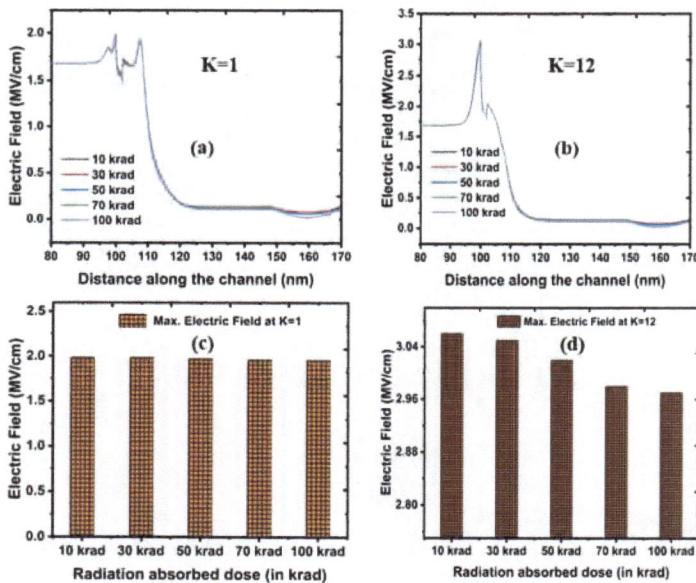

Fig. (8). Electric field along the channel in designed biosensor for different X-ray irradiated doses at (**a**) K=1 and (**b**) K=12 and maximum electric field along the channel in designed biosensor for different X-ray irradiated doses at (**c**) K=1 and (**d**) K=12.

Fig. (**8b**) shows a similar trend as Fig. (**8a**); however, a single peak of the electric field near the source-channel interface in Fig. (**8b**) is caused due to the strong impact of keratin biomolecule on the electric field at this interface. Fig. (**8c**) shows that the maximum electric field is almost constant (slightly decreasing) with the increase in radiation dose for K=1. This happens due to the slight decrease in net tunneling in the device, showing that the TFET is mostly unaffected by trap charges. The maximum electric field decreases with the increase in radiation dose, as shown in Fig. (**8d**). This happens due to the decrease in net tunneling in the presence of trap states at the interface and oxide.

Fig. (**9**) illustrates the variation in on-current and threshold voltage sensitivity for different doses of X-ray radiation. Specifically, a 100 krad radiation dose leads to a 5.1% reduction in V_{th} sensitivity and a 19.2% reduction in on-current sensitivity. Table **3** presents further details on the impact of radiation dose on the sensitivity of the TFET biosensor under X-ray irradiation.

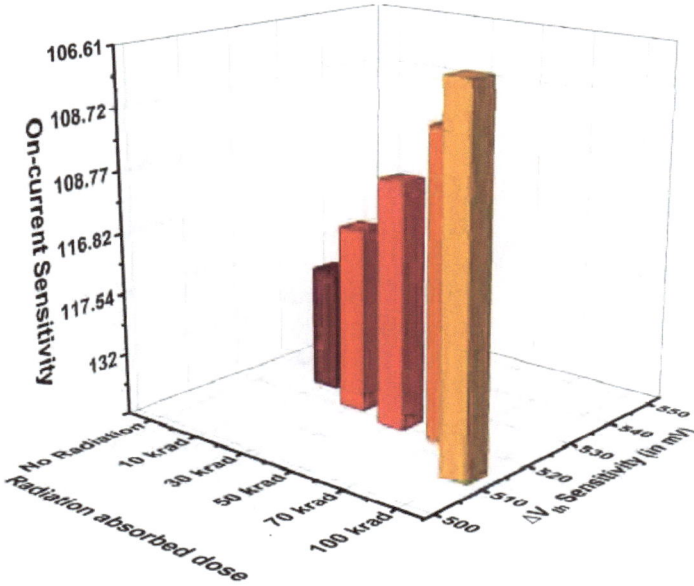

Fig. (9). Sensitivity *vs.* radiation absorbed doses.

Table 3. Impact of radiation dose absorbed on the sensitivity of TFET biosensor under X-ray irradiation.

Radiation Dose in krad (SiO$_2$)	ΔV_{th} Sensitivity (mV)	On-current Sensitivity
No radiation	539.5	132.00
10	527.4	117.50
30	523.1	116.80

(Table 3) cont.....

50	520.9	108.77
70	520.6	108.72
100	512.0	106.61

CONCLUSION

In conclusion, this chapter has delved into the multifaceted realm of TFET biosensor development, encompassing design optimization, analytical modeling, and radiation considerations. Through an exploration of design optimization methodologies, the chapter has underscored the importance of tailoring TFET biosensors to achieve optimal performance characteristics, including sensitivity, selectivity, and reliability. Analytical modeling techniques provide insights into the underlying physics of TFET operation, enabling predictive capabilities essential for informed design decisions and performance optimization. Furthermore, the chapter has delved into radiation considerations, emphasizing the importance of fortifying TFET biosensors against radiation-induced effects. By synthesizing the key elements, the chapter offers a comprehensive roadmap for researchers and engineers engaged in TFET biosensor development, guiding them toward the realization of high-performance biosensing platforms. Through a multidimensional approach encompassing design optimization, analytical modeling, and radiation considerations, the chapter has strived to advance the state-of-the-art in biosensing technology, fostering innovations with transformative implications across diverse applications. As the field of TFET biosensors continues to evolve, this chapter serves as a foundational resource, driving progress and inspiring future advancements in biosensing technology.

KEY POINTS

- This chapter discusses the TFET biosensors and their significance in biomedical sensing.
- Practical considerations for consistent sensor operation across diverse application scenarios are discussed along with insights into the complexities of TFET biosensor design and their impact on sensor performance.
- Various design strategies focusing on material selection, device architecture, and surface functionalization to optimize sensor performance are explained.
- Analytical modeling techniques for DNA detection, including electrostatics modeling and charge transport simulations, are presented as well as validated with experimental data.
- Research on radiation-induced phenomena, like gate oxide damage, and their implications for sensor functionality is proposed and an investigation of X-ray radiation effects on TFET biosensor performance is analyzed.

- Discussion on strategies to mitigate radiation effects and enhance TFET biosensor resilience has been made.

ACKNOWLEDGEMENTS

The authors would like to thank the Department of Electronics and Communication Engineering, NIT Hamirpur, India, for providing TCAD software(s) to carry out the research presented in this chapter.

REFERENCES

[1]   G. Dewey, "Fabrication, characterization, and physics of III–V heterojunction tunneling field effect transistors (H-TFET) for steep sub-threshold swing", In: *in IEDM Tech. Dig.* Washington, DC, USA, 2011, p. 33.6.1–33.6.4.
[http://dx.doi.org/10.1109/IEDM.2011.6131666]

[2]   Qin Zhang, Wei Zhao, and A. Seabaugh, "Low-subthreshold-swing tunnel transistors", *IEEE Electron Device Lett.,* vol. 27, no. 4, pp. 297-300, 2006.
[http://dx.doi.org/10.1109/LED.2006.871855]

[3]   Byung-Gook Park, B-G. Park, J.D. Lee, and T-J.K. Liu, "Tunneling field-effect transistors (TFETs) with subthreshold swing (SS) less than 60 mV/dec", *IEEE Electron Device Lett.,* vol. 28, no. 8, pp. 743-745, 2007.
[http://dx.doi.org/10.1109/LED.2007.901273]

[4]   U.E. Avci, D.H. Morris, and I.A. Young, "Tunnel field-effect transistors: prospects and challenges", *IEEE J. Electron Devices Soc.,* vol. 3, no. 3, pp. 88-95, 2015.
[http://dx.doi.org/10.1109/JEDS.2015.2390591]

[5]   A.C. Seabaugh, and Q. Zhang, "Low-voltage tunnel transistors for beyond CMOS logic", *Proc. IEEE,* vol. 98, no. 12, pp. 2095-2110, 2010.
[http://dx.doi.org/10.1109/JPROC.2010.2070470]

[6]   J. Zhu, Y. Zhao, Q. Huang, C. Chen, C. Wu, R. Jia, and R. Huang, "Design and simulation of a novel graded-channel heterojunction tunnel FET with high ION/IOFF ratio and steep swing", *IEEE Electron Device Lett.,* vol. 38, no. 9, pp. 1200-1203, 2017.
[http://dx.doi.org/10.1109/LED.2017.2734679]

[7]   K.K. Bhuwalka, M. Born, M. Schindler, and I. Eisele, "Scaling rules for tunnel field-effect transistors", *2005 International Semiconductor Device Research Symposium,* 2005, pp. 13-14 Bethesda, MD, USA
[http://dx.doi.org/10.1109/ISDRS.2005.1595952]

[8]   K.K. Bhuwalka, J. Schulze, and I. Eisele, "Scaling the vertical tunnel FET with tunnel bandgap modulation and gate workfunction engineering", *IEEE Trans. Electron Dev.,* vol. 52, no. 5, pp. 909-917, 2005.
[http://dx.doi.org/10.1109/TED.2005.846318]

[9]   A. Theja, and M. Panchore, "Performance investigation of GaSb/Si heterojunction based gate underlap and overlap vertical TFET biosensor", *IEEE Trans. Nanobiosci.,* vol. 22, no. 2, pp. 284-291, 2023.
[http://dx.doi.org/10.1109/TNB.2022.3183934] [PMID: 35709121]

[10] J. Chowdhury, A. Sarkar, K. Mahapatra, and J.K. Das, "Analytical modeling of dielectrically modulated broken-gate tunnel FET biosensor considering partial hybridization effect", *Comput. Electr. Eng.,* vol. 99, p. 107859, 2022.
[http://dx.doi.org/10.1016/j.compeleceng.2022.107859]

[11] A. Bhattacharyya, D. De, and M. Chanda, "Ovarian-cancer biomarker (HE4) recognition in serum using hetero TFET biosensor", *IEEE Trans. Nanotechnol.,* vol. 22, pp. 238-244, 2023.

[http://dx.doi.org/10.1109/TNANO.2023.3272926]

[12] A. Anam, S. Anand, and S.I. Amin, "Design and performance analysis of tunnel field effect transistor with buried strained Si1-xGex Source structure based biosensor for sensitivity enhancement", *IEEE Sens. J.,* vol. 20, no. 22, pp. 13178-13185, 2020.
[http://dx.doi.org/10.1109/JSEN.2020.3004050]

[13] S. Tiwari, and R. Saha, "Sensitivity analysis of i-shape TFET biosensor considering repulsive steric and trap effects", *IEEE Trans. Nanotechnol.,* vol. 22, pp. 518-524, 2023.
[http://dx.doi.org/10.1109/TNANO.2023.3309411]

[14] P.K. Bera, R.R. Sahoo, R. Kar, and D. Mandal, "Design and sensitivity investigation of dielectric modulated and electrolyte-based pH sensing of vertical TFET biosensor", *IEEE Trans. Nanotechnol.,* vol. 22, pp. 537-544, 2023.
[http://dx.doi.org/10.1109/TNANO.2023.3307632]

[15] Gupta, R., Beg, S., & Singh, S. "Evaluation of design and performance of biosensor utilizing ferroelectric vertical tunnel field-effect transistor (V-TFET)", *Silicon,* pp. 1-12, 2024.
[http://dx.doi.org/10.1016/j.mseb.2023.116841]

[16] S. Kumar, K. Singh, S. Chander, E. Goel, P.K. Singh, K. Baral, B. Singh, and S. Jit, "2-D analytical drain current model of double-gate heterojunction TFETs with a SiO2/HfO2 stacked gate-oxide structure", *IEEE Trans. Electron Dev.,* vol. 65, no. 1, pp. 331-338, 2018.
[http://dx.doi.org/10.1109/TED.2017.2773560]

[17] P. Goma, and A. K. Rana, "Investigation of scaling on the sensitivity and performance of tunnel FET biosensor", *IEEE Sensors Letters,* vol. 7, no. 8, pp. 1-4, 2023.
[http://dx.doi.org/10.1109/LSENS.2023.3296346]

[18] P. Goma, and A.K. Rana, "Investigation of the effects of trap-assisted tunneling and temperature variations on the sensitivity metrics of a tunnel FET biosensor", *in Micro and Nanostructures,* vol. 184, p. 207697, 2023.
[http://dx.doi.org/10.1016/j.micrna.2023.207697]

[19] K.N. Priyadarshani, and S. Singh, "Ultra sensitive label-free detection of biomolecules using vertically extended drain double gate $Si_{0.5}Ge_{0.5}$ Source Tunnel FET", *IEEE Trans. Nanobiosci.,* vol. 20, no. 4, pp. 480-487, 2021.
[http://dx.doi.org/10.1109/TNB.2021.3106333] [PMID: 34424845]

[20] R. Priyanka, L. Chandrasekar, R. R. Shaik, and K. P. Pradhan, "Label free DNA detection techniques using dielectric modulated FET: inversion or tunneling?", *in IEEE Sensors Journal,* vol. 21, no. 2, pp. 2316-2323, 2021.
[http://dx.doi.org/10.1109/JSEN.2020.3019103]

[21] B. Das, and B. Bhowmick, "Dielectrically modulated ferroelectric-TFET (Ferro-TFET) based biosensors", *Mater. Sci. Eng. B,* vol. 298, p. 116841, 2023.
[http://dx.doi.org/10.1016/j.mseb.2023.116841]

[22] R. Narang, K.V.S. Reddy, M. Saxena, R.S. Gupta, and M. Gupta, "A dielectric-modulated tunnel-FET-based biosensor for label-free detection: analytical modeling study and sensitivity analysis", *IEEE Trans. Electron Dev.,* vol. 59, no. 10, pp. 2809-2817, 2012.
[http://dx.doi.org/10.1109/TED.2012.2208115]

[23] S.H. Kim, H. Kam, C. Hu, and T.J.K. Liu, "Germanium-source tunnel field-effect transistors with record high ION/IOFF", *in Proc. Int. Symp. VLSI Technol.,* pp. 178-179, 2009.

[24] J.K. Mamidala, R. Vishnoi, and P. Panday, *Tunnel field-effect transistors (TFET): modelling and simulation.* John Wiley & sons, 2017.

[25] A.S. Verhulst, W.G. Vandenberghe, K. Maex, and G. Groeseneken, "Tunnel field-effect transistor without gate-drain overlap", *Appl. Phys. Lett.,* vol. 91, no. 5, p. 053102, 2007.
[http://dx.doi.org/10.1063/1.2757593]

[26] M.G. Bardon, H.P. Neves, R. Puers, and C. Van Hoof, "Pseudo-two-dimensional model for double-gate tunnel FETs considering the junctions depletion regions", *IEEE Trans. Electron Dev.,* vol. 57, no. 4, pp. 827-834, 2010.
[http://dx.doi.org/10.1109/TED.2010.2040661]

[27] P. Venkatesh, K. Nigam, S. Pandey, D. Sharma, and P.N. Kondekar, "A dielectrically modulated electrically doped tunnel FET for application of label free biosensor", *Superlattices Microstruct.,* vol. 109, pp. 470-479, 2017.
[http://dx.doi.org/10.1016/j.spmi.2017.05.035]

[28] M.M. Pejovic, A. Jaksic, and M. Momcilo, "Pejovic Contribution of fixed oxide traps to the sensitivity of pMOS dosimeters during gamma ray irradiation and annealing at room and elevated temperature",

[29] A. Holmes-Siedle, and L. Adams, "RADFET: a review of the use of metal-oxide-silicon devices as integrating dosimeters", *Radiat. Phys. Chem.,* vol. 28, pp. 235-244, 1986.

[30] E.G. Moreno, R. Picos, E. Isern, M. Roca, S. Bota, and K. Suenaga, Picos, R. Isern, E. Roca, M. Bota, K. Suenaga, "Radiation sensor compatible with standard CMOS technology", *IEEE Trans. Nucl. Sci.,* vol. 56, no. 5, pp. 2910-2915, 2009.
[http://dx.doi.org/10.1109/TNS.2008.2011804]

[31] A. Dubey, A. Singh, R. Narang, M. Saxena, and M. Gupta, R. Narang, M. Saxena, M. Gupta, "Modeling and simulation of junctionless double gate radiation sensitive FET (RADFET) dosimeter", *IEEE Trans. Nanotechnol.,* vol. 17, no. 1, pp. 49-55, 2018.
[http://dx.doi.org/10.1109/TNANO.2017.2719286]

[32] C. Chong, H. Liu, S. Wang, and X. Wu, "Research on total ionizing dose effect and reinforcement of SOI-TFET", *in Micromachines,* vol. 12, no. 10, p. 1232, 2021.
[http://dx.doi.org/10.3390/mi12101232]

[33] H.R. Yaghobi, K. Eyvazi, and M.A. Karami, "Investigation of single-event-transient effects on n+ pocket double-gate tunnel FET", *Radiat. Phys. Chem.,* vol. 212, p. 111094, 2023.
[http://dx.doi.org/10.1016/j.radphyschem.2023.111094]

[34] K. Aishwarya, and B. Lakshmi, "Investigation of heavy ion radiation and temperature on junctionless tunnel field effect transistor", *J. Nanopart. Res.,* vol. 25, no. 7, p. 137, 2023.
[http://dx.doi.org/10.1007/s11051-023-05793-4]

[35] Q. Wang, H. Liu, S. Wang, and S. Chen, "TCAD simulation of single-event-transient effects in L-shaped channel tunneling field-effect transistors", *IEEE Trans. Nucl. Sci.,* vol. 65, no. 8, pp. 2250-2259, 2018.
[http://dx.doi.org/10.1109/TNS.2018.2851366]

[36] L. Ding, E. Gnani, S. Gerardin, M. Bagatin, F. Driussi, P. Palestri, L. Selmi, C.L. Royer, and A. Paccagnella, "Effects of electrical stress and ionizing radiation on Si-based TFETs", *2015 Joint International EUROSOI Workshop and International Conference on Ultimate Integration on Silicon,* 2015pp. 137-140
[http://dx.doi.org/10.1109/ULIS.2015.7063792]

[37] L. Ding, E. Gnani, S. Gerardin, M. Bagatin, F. Driussi, P. Palestri, L. Selmi, C.L. Royer, and A. Paccagnella, "Total ionizing dose effects in Si-based Tunnel FETs", *IEEE Trans. Nucl. Sci.,* vol. 61, no. 6, pp. 2874-2880, 2014.
[http://dx.doi.org/10.1109/TNS.2014.2367548]

[38] K. Aishwarya, and B. Lakshmi, "Radiation study of TFET and JLFET-based devices and circuits: a comprehensive review on the device structure and sensitivity", *Radiat. Eff. Defects Solids,* vol. 178, no. 3-4, pp. 229-257, 2023.
[http://dx.doi.org/10.1080/10420150.2022.2133708]

[39] K. Xi, J. Bi, J. Chu, G. Xu, B. Li, H. Wang, M. Liu, and M. Sandip, "Total ionization dose effects of N-type tunnel field effect transistor (TFET) with ultra-shallow pocket junction", *Appl. Phys., A Mater.*

Sci. Process., vol. 126, no. 6, p. 440, 2020.
[http://dx.doi.org/10.1007/s00339-020-03622-2]

[40] A. Dubey, R. Narang, M. Saxena, and M. Gupta, "Investigation of total ionizing dose effect on SOI tunnel FET", *Superlattices Microstruct.,* vol. 133, p. 106186, 2019.
[http://dx.doi.org/10.1016/j.spmi.2019.106186]

[41] C. Chong, H. Liu, S. Wang, and X. Wu, "Research on total ionizing dose effect and reinforcement of SOI-TFET", *Micromachines (Basel),* vol. 12, no. 10, p. 1232, 2021.
[http://dx.doi.org/10.3390/mi12101232] [PMID: 34683283]

[42] G. Yan, "Accumulative total ionizing dose (TID) and transient dose rate (TDR) effects on planar and vertical ferroelectric tunneling-field-effect-transistors (TFET)", *Microelectron.,* vol. 114, 2020.
[http://dx.doi.org/10.1016/j.microrel.2020.113855]

[43] H. Huang, S. Wei, J. Pan, W. Xu, C-C. Chen, Q. Mei, J. Chen, L. Geng, Z. Zhang, and Y. Du, "Threshold voltage model of total ionizing irradiated short-channel FD-SOI MOSFETs with gaussian doping profile", *IEEE Trans. Nucl. Sci.,* vol. 65, no. 10, pp. 2679-2690, 2018.
[http://dx.doi.org/10.1109/TNS.2018.2864977]

CHAPTER 10

A New Paradigm Shift in the Semiconductor Industry for 6G Technology: A Review

Karabi Baruah[1] and **Prachi Gupta**[1,*]

[1] SOET, CMR University, Lakeside Campus, Bangalore, India

Abstract: Sixth-generation (6G) wireless communication networks are expected to combine terrestrial, maritime, and aerial communications into a scalable, fast, and resilient network that can support a lot of devices with very low latency requirements. 6G semiconductor materials need to have particular properties in order to satisfy the goals of substantially faster data speeds, reduced latency, and enhanced device connection over earlier generations. Novel semiconductor materials are being discovered, and current ones are being optimized to satisfy the ever-changing needs of 6G technology. With an emphasis on wide bandgap semiconductors, like GaN and SiC, which offer improved efficiency and performance, this overview examines significant developments in semiconductor materials. To satisfy the particular requirements of the next-generation wireless networks, the semiconductor industry will probably witness breakthroughs and advances in these and other components as 6G technology develops. It is anticipated that the advancement of 6G technology will present novel demands and obstacles for semiconductor components. For 6G networks, the semiconductor industry is seeing major paradigm developments. In order to support higher frequencies and data rates, this shift places an emphasis on the integration and shrinking of components. For 6G devices to be widely adopted in a sustainable manner, advances in energy efficiency are essential. The 6G network dimensions with air interface and related prospective technologies are thoroughly outlined in this article. With regards to the 6G network, we primarily focus on a variety of semiconductor materials and components, as well as Key Performance Indicators (KPI), like high thermal conductivity, low noise, and wide bandgap.

Keywords: III-V compound semiconductors, 6G technology, Millimeter wave, Semiconductor device, Terahertz frequency.

INTRODUCTION

While 5G mobile network standards are still in their early stages, the basis for the future 6G wireless network standards is already beginning to take shape. It is acceptable to start thinking about the possible features of 6G technology, even

---

* **Corresponding author Prachi Gupta:** SOET, CMR University, Lakeside Campus, Bangalore, India; E-mail: prachig048@gmail.com

Ashish Raman, Prabhat Singh, Naveen Kumar & Ravi Ranjan (Eds.)

though it might not be accessible for a while [1 - 4]. 6G networks will be equipped with more supporting applications than what consumers usually use on their phones because they are anticipated to be more diversified than previous mobile networks. The Internet of Everything (IoE) has led to the development of numerous new vertical services and applications that improve industrial processes, society, and our daily lives [5, 6]. These include mass production, smart housing, smart transit systems, and Virtual Reality (VR) technologies. These applications have a wide range of effects, and there are numerous Quality of Service (QoS) criteria. However, the capabilities of fifth-generation (5G) networks are now short of what is required to enable new vertical applications. As a result, researchers are looking into novel approaches to building and managing 6G networks.

6G networks will replace 5G cellular technology in the near future, as 6G networks will have far greater capacity, lower latency, and the ability to operate at higher frequencies than the 5G network. With latency as low as one microsecond, the 6G network would aim to facilitate communication between the users. Compared to a millisecond's worth of throughput, it would be 1000 times faster. Unlike 5G, which uses mmWave to operate in the microwave frequency range, 6G will use even shorter wavelengths to operate in the Terahertz (THz) region, spanning from 100 GHz to 3 THz. Radio Access Networks (RAN) are greatly impacted by 5G, but 6G networks will have a much greater impact because of a large frequency increase that will reduce the need for antennas nearly everywhere [7].

The Internet of Things (IoT), massive Machine-type Communications (MTC), and other technologies have grown, making 5G far more than just standard cellular networks. It can deliver gigabits per second instead of megabits per second and achieve a 1000x increase in capacity over 4G networks, among other Key Performance Indicators (KPIs) [8]. Target peak rates are stated to be 20 Gbps for the downlink and 10 Gbps for the uplink in a document on 5G scenarios and requirements for access technologies released by the European Telecommunication Standards Institute (ETSI) (3GPP TR 38.913). Table **1** provides further information about the different KPIs. We have also made some educated guesses about how the needs for potential 6G networks would differ from those for 5G [8, 9].

This chapter aims to offer an expert opinion on the most innovative and popular research trajectories that could influence 6G mobile technologies. Though the development of 6G is still in its early stages and some concepts may not become clear for some time, this visionary piece adopts a bold stance in speculating on potential enabling technologies and revolutionary 6G elements, outlining features that go beyond what 5G can offer. First, we provide an overview of the evolution of 6G technology in this chapter. After that, a summary of the 6G market overview and current state is provided. Next, we talk about the 6G technology's uses and semiconductor parts. We have covered new paradigm developments in the semiconductor sector for 6G technology in the section that follows. After that, we discuss the characteristics needed for improved semiconductor materials in 6G. We have gone into great detail about new semiconductor materials for 6G technologies. We finally wrap up this chapter and outline some potential future study topics. The next section covers new paradigm advances in the semiconductor industry for 6G technology. The features required for enhanced semiconductor materials in 6G are then covered. New semiconductor materials for 6G technologies have been covered in great depth. In the end, we discuss several possible areas of future research as we wind up this chapter.

Table 1. KPIs for 5G *vs.* 6G.

Features	6th Gen	5th Gen
Frequency of operation	Up to 1 THz	3 − 300 GHz
Individual data rate	100 Gbps	1 Gbps
U-plane latency	< 0.1 ms	0.5 ms1 [9]
Download data rate	> 1 Tbps	20 Gbps
C-plane latency	< 1 ms	10 ms
Mobility	Up to 1000 km/hr	Up to 500 km/h
Spectral efficiency	100 bps/Hz	30 bps/Hz

EVOLUTION OF 1G TO 6G

Significant technological advancements and conceptual shifts have characterised the remarkable development of mobile telecommunications from 1G to 6G [6, 8, 10 - 13]. Below is a summary of each generation, along with its salient features:

First Generation (1G)

1G represents first-generation wireless mobile communication, which first appeared in the late 1970s and early 1980s. It made use of voice services *via* the Advanced Mobile Phone System (AMPS) and analog technology. With a maximum speed of 2.4 kbps, it used FDMA technology with a channel capacity of 30 kHz, operating in the 824–894 MHz frequency band. Spectrum widening allowed AMPS to increase its capacity by 10 MHz in 1988. In 1982, the US implemented 1G AMPS [10].

Second Generation (2G)

With the advent of 2G wireless mobile communication in the early 1990s, a shift towards digital technology was indicated. By supporting both voice and basic data services, it improved over 1G. Two of the technologies employed by 2G networks were the Global System for Mobile Communications (GSM) and Code Division Multiple Access (CDMA). Text messaging, limited data capabilities, and improved audio quality were made possible by these networks. With average data transfer rates of 9.6 kbps, basic internet access and applications, like SMS, were made possible. More sophisticated mobile communication technologies were made possible by the 2G era [10].

Third Generation (3G)

Introduced in the early 2000s, 3G wireless mobile communication represented a major improvement over 2G. Faster data rates were made available, allowing for more comprehensive multimedia capabilities, video calling, and internet access. Data rates were enhanced by technologies, like CDMA2000 and UMTS (Universal Mobile Telecommunications System), which provided speeds of several megabits per second. During this period, cell phones became widely used, and the amount of time spent on mobile internet increased. Three-generation (3G) networks enabled the development of mobile data services and the shift to higher-end wireless communication generations.

Fourth Generation (4G)

Compared to 3G, 4G wireless mobile communication offered significant improvements and was first deployed in the late 2000s. It offered noticeably quicker data speeds, enabling better internet browsing, online gaming, and streaming of high-quality videos. Tens of megabits per second of data were made possible by technologies, like WiMAX and LTE (Long-Term Evolution). 4G networks drove the smartphone revolution and made it possible for a more connected digital lifestyle by enabling the emergence of mobile apps, video sharing, and connected gadgets. A major advancement toward seamless high-speed mobile internet experiences was made during this era.

Fifth Generation (5G)

The emergence of 5G wireless mobile communication in the late 2010s has served as the first tangible example of this significant leap in connectivity. It has provided incredibly low latency, lightning-fast communication speeds, and the ability to connect a large number of devices at once. 5G has made possible the Internet of Things (IoT), driverless cars, Augmented Reality (AR), and virtual reality usage. 5G networks, which offer multi-gigabit speeds and nearly instantaneous connectivity by utilizing technologies, like massive MIMO and millimeter wave (mmWave) frequencies, are transforming the way people interact with technology, paving the way for innovative uses and discoveries.

Sixth Generation (6G)

With 6G, users will obtain benefits from services, like Massive-URLLC, Massive-eMBB, and eMBB. Massive URLLC includes techniques, like assertion multiple access, NOMA, and OMA. As the number of devices increases, the amount of bandwidth needed may grow linearly. To get the greatest results possible, trade-offs among latency, dependability, and scalability, and alternative multiple access mechanisms may be used. Energy efficiency will be eMBB's primary goal. Large-scale data transmission/processing, cross-talk, and handover are all expected to be major benefits of this proposed service class for mobile communication networks. It is also important to understand the underlying security and privacy concerns. A high link density would be required for massive-eMBB to gather tactile impressions and convert them to digital data. Massive eMBB is going to be a major issue in the 6G future to improve large-scale IoT operations and capabilities. It will allow personnel, sensors, and actuators to communicate with one another in large quantities at low latency. Additional services, like uHSLLC (Ultra-high Speed with Low Latency), UHDD (Ultra-high Data Density), uHRUx (Ultra-high Reliability and User experience), Uhs (Ultra-high security), *etc.*, will also be included in 6G. One microsecond latency would be provided by 6G. Compared to the throughput in a millisecond, this would be 1000 times faster. Compared to 5G, which employs mmWave in the microwave frequency range, 6G will use even shorter wavelengths in the Terahertz (THz) band, spanning 100 GHz to 3 THz [10, 11]. Each generation builds upon the advancements of its predecessors, offering faster speeds, lower latency, and new capabilities, ultimately driving innovation in communication technology and enabling new applications and services. A compariso the generations from 1G to 6G in terms of year, speed, technology, services, and many more, is shown in Table **2**. The evaluation of mobile technology from 1st Generation to 6th Generation is shown in Fig. (**1**).

Table 2. Comparison of all generations from 1G to 6G.

Generation→/Features↓	1st Gen	2nd Gen	3rd Gen	4th Gen	5th Gen	6th Gen
Year	1970-1980	1980-1990	1990-2000	2000-2010s	2015	2030
Technology	Analog	Digital	Broadband CDMA, IP	Unified IP and seamless combination of broadband LAN, WAN, WLAN, PAN	4G+WWWW	5G+satellite
Speed	2.4Kbps	64Kbps	2Mbps	200Mbps to 1Gbps	1Gbps and higher	10 to 11Gbps
Multiplexing	FDMA	TDMA, CDMA	CDMA	CDMA	CDMA	CDMA
Standard	AMPS	GSM, PDC, IS-95, IS-136, EDGE, GPRS	CDMA2000, UMTS, TD-SCD MA, WCDMA	LTE, WiMAX	LAS-CDMA, OFDM, MC-CDMA, UWB, Network-LMDS, IPv6	GPS, COMPASS, GLONASS, Galileo systems
Core network	PSTN	PSTN and packet network	Packet network	Internet	Internet	Internet
Handoff	Horizontal	Horizontal	Horizontal and vertical	Horizontal and vertical	Horizontal and vertical	Horizontal and vertical
Switching	Circuit	Circuit and packet	Packet except for circuit for air interface	Packet	Packet	Packet
Services	Voice only	Digital voice and short messaging, packetized data	Integrated high-quality audio, video, and data	Dynamic information access, wearable devices	Dynamic information access, wearable devices with AI capabilities	Ultra-fast internet access

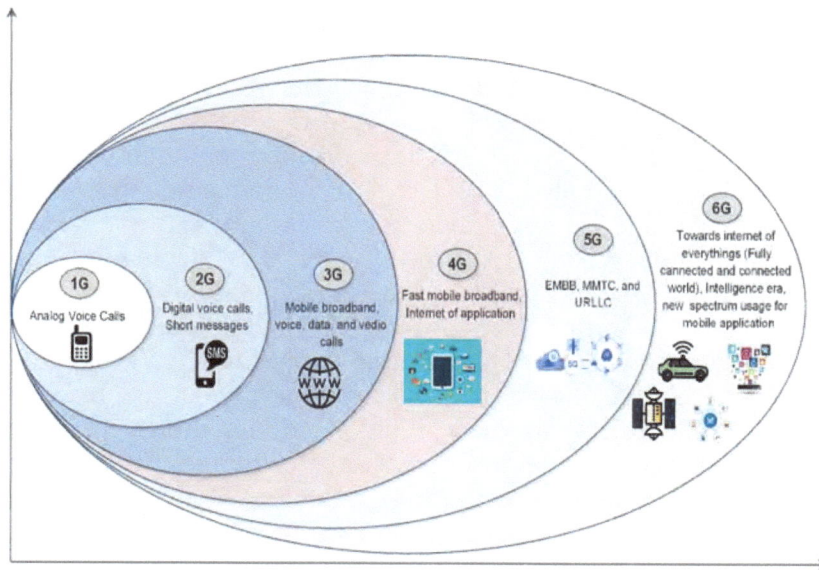

Fig. (1). Evolution of 1G to 6G [14].

OUTLINE OF THE 6G MARKET AND PRESENT SITUATION

It is expected that the 6G network sector will lead to the development of new ideas in location awareness, present technology, and imaging. 6G is anticipated to launch by 2030. The COVID-19 pandemic has had a complicated effect on the sector and presented a number of obstacles to the global 6G market's expansion. Many companies, including Apple Inc., AT&T Inc., Nokia Corporation, Verizon Communications, Inc., Intel Corporation, Google LLC, Samsung Electronics, Inc. Co. Ltd., LG Corporation, Cisco Systems Inc., and Sony Corporation, are primarily focused on developing 6G networks [15 - 17]. China and the US are working hard on 6G networks, the next advancement in the field of fast communication, while the rest of the world waits for the new 5G technology to take off. One of the main reasons the industry is progressing is the extensive development and application of 6G technology worldwide. China asserted that by January 2022, 6G mobile technology would have advanced. This was reported by the South China Morning Post. Verizon has focused on developing 6G in order to keep its attention on digital platforms, cloud computing, and 6G development. Telecom specialists from Japan and Finland have proposed collaborating to advance 6G technology in addition to what has previously been mentioned. Many companies, governments, and enterprises prioritize the development of 6G technology. Such significant technological advancements are anticipated to boost the overall growth of the global 6G market. As a result of several companies and government initiatives, it is projected that North America will hold the largest market share for 6G globally. The US is lagging behind other countries in advancing the next wave of communication technologies. Leading the charge to position North America as an explorer in 6G networks for the US market are executives from Ericsson and AT&T. The Next G Alliance was announced to be formed in October 2020 by the Alliance for Telecommunications Industry Solutions. The alliance included AT&T, Ericsson, Samsung, Verizon, Microsoft, and other businesses. The alliance aimed to solidify North America's leadership in mobile communications globally over the next ten years. China launched the first test satellite into orbit, sparking a wave of interest in 6G. For example, China launched a system rocket, a test satellite for 6G networks, and roughly twelve other satellites into orbit on 6th Nov., 2020. Additionally, India's Minister of Communication declared that 6G technology would be in use by 2024. On the other hand, 6G networks can offer high-speed network access due to the construction of a high-frequency spectrum; furthermore, higher-throughput satellite services can be employed to provide extensive coverage in isolated and remote areas (Fig. **2**) [17, 18].

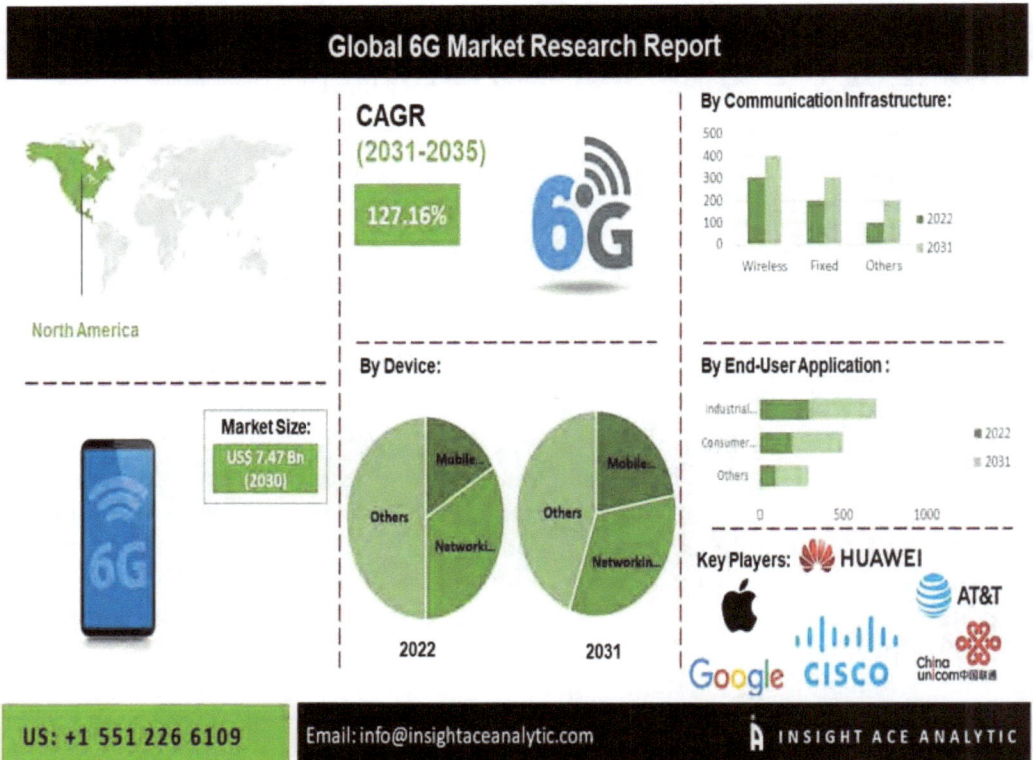

Fig. (2). 6G market overview [18].

The following sections identify and discuss innovative service types for the 6G network [19 - 21]:

Massive Ultra-reliable Low Latency Communication (URLLC)

For vital applications, like the Industrial IoT and robotic surgery, URLLC stands for communications with enhanced availability, decreased latency, and dependability. Massive-URLLC is a novel type of service that combines traditional mMTC with 5G-URLLC. It will emerge due to the necessity for 6G to greatly expand the 5G URLLC service. Autonomous Intelligent Driving (AID) is one application of massive URLLC that some wish to implement. AID requires balancing multiple priorities at once, such as motion planning, obstacle detection, and automated driving. Many access strategies, including OMA (Orthogonal Multiple Access), NOMA (Non-orthogonal Multiple Access), and assertion of multiple access are intriguing substitutes for massive-URLLC. With OMA approaches, such as massive-URLLC, the bandwidth needed for 6G could increase exponentially with the number of devices added. To get the optimal trade-off among scalability, reliability, and latency, alternative multiple-access approaches, like NOMA, can be employed. Massive-URLLC requires a large number of little data packets to be delivered in order to guarantee low latency and good resource efficiency for "time-critical" 6G applications [20].

eMBB (Enhanced Mobile Broadband)

High dependability, low latency, and quick transmission speeds are frequently needed for applications, such as AR (Augmented Reality), VR (Virtual Reality), and holographic meetings. Furthermore, these requirements must be fulfilled in circumstances requiring a high degree of mobility, such as air and sea travel. Consequently, an improved mobile broadband URLLC has been proposed for the next new service class for 6G. Energy efficiency is a top concern for this category of services. This new network class is expected to be far more adept at enhancing handover, interference, massive data transfer, and processing in mobile communications networks than the URLLC and eMBB in 5G networks. It is also necessary to consider security and privacy issues with the upgraded mobile broadband URLLC communication service.

Massive eMBB

In Industry 4.0-based scenarios, a high link frequency is required to capture tactile perceptions and translate them into digital data. Consequently, massive eMBB will be a popular issue in the 6G communication as a means of enhancing large-scale Industrial IoT operations and capabilities by permitting the wide-ranging, low-latency connection among an actuator, worker, and sensor.

Innovations in technology and intelligent autonomous driving will give rise to new service-oriented programs that will define the network of the future. Peak data rates for 6G are anticipated to be 50 times higher than those for 5G, and users should see at least a 10-fold increase in data speed compared to 5G networks. The requirement of 6G can be better understood by Fig. (**3**).

The following list of benefits of 6G networks over 5G networks is provided (Fig. **4**) [19 - 24]:

1. **Frequency Bands**: 5G offers sub-5GHz frequency bands and millimeter wave (mmWave), and fixed access. While 6G would enable millimeter waves and sub-6 GHz frequency bands, THz bands, non-RF bands, and other bands for mobile communication.
2. **Data Rate**: 5G offers data rates of 10 Gbps uplink and 20 Gbps downlink. However, 6G would provide 1 Tbps for both the downlink and the uplink.
3. **Latency:** 6G aims to achieve a latency of less than 1 microsecond, while 5G delivers a delay of approximately 1 millisecond.
4. **Architectural Style**: Dense sub-6 GHz smaller Base Stations (BSs) with umbrella macro BSs are combined with Mmwave microcells, which have a range of about 100 meters. While testing using tiny THz cells, transient hotspots created by BSs mounted on drones, and cell-free smart surfaces functioning at higher frequencies will all be a part of the 6G design.
5. **Device Type**: 5G encompasses gadgets, such as smartphones, drones, and sensors. 6G will be made up of DLT (Distributed Ledger Technology) devices, as well as smart implants, CRAS, XR, and BCI technologies.
6. **Reliability:** 5G is 10–5 reliable, while 6G would be 10–9.
7. **Localization Accuracy:** 10 cm in 2D is for 5G networks, while 1 cm in 3D will be for 6G-based networks.
8. **Customer Engagement**: On a 5G network, 50 Mbps 2D is available everywhere, while on a 6G network, 10 Gbps 3D will be accessed everywhere.

APPLICATIONS OF 6G IN VARIOUS ASPECTS

The 6th Generation of wireless technology, or 6G, is anticipated to significantly outperform its predecessors in terms of data throughput, latency, dependability, and energy efficiency [25 - 28]. Here are mentioned a few possible uses for 6G in a variety of contexts:

Security

For threat identification and crime prevention, sophisticated surveillance systems with high-resolution cameras and real-time analytics will be utilised. Emergency response systems would be able to react more quickly because of enhanced communication and location monitoring features.

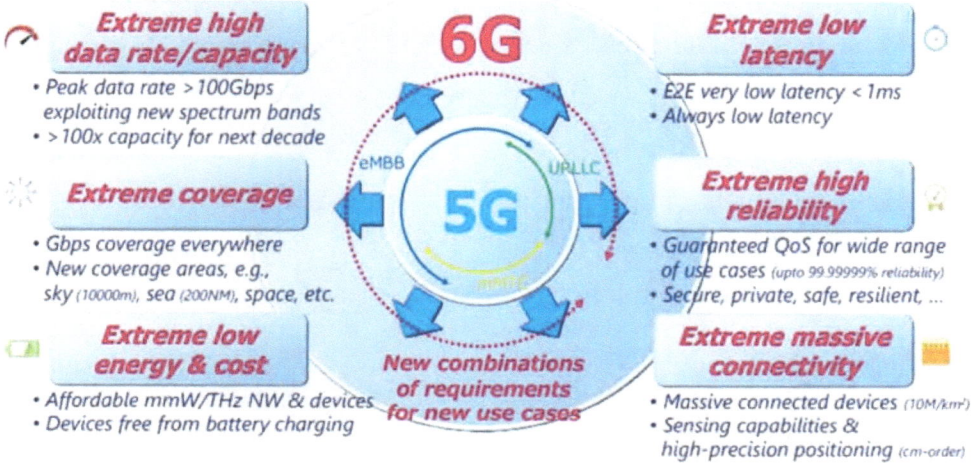

Fig. (3). Requirements for 6G technology [19].

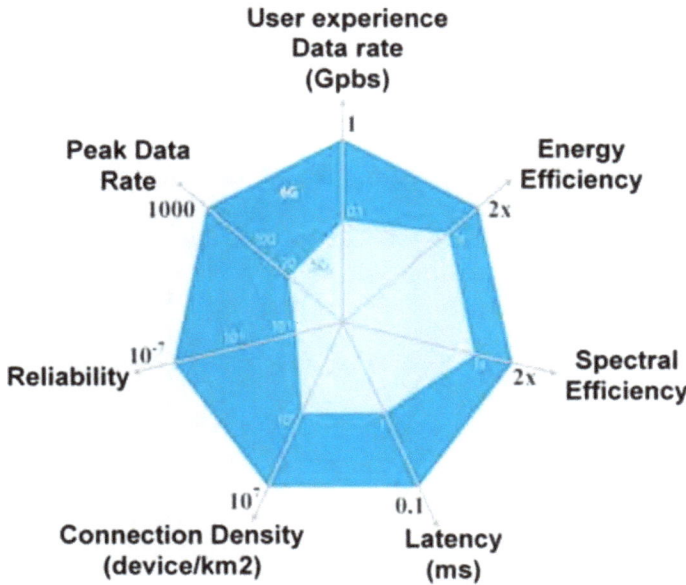

Fig. (4). Notional 6G performance improvements compared to 5G [16].

Telecommunication

With 6G, communication between devices and networks will be possible over incredibly fast and dependable networks. With less latency, it might enable high-definition video streaming, virtual reality experiences, real-time gaming, and enormous Internet of Things (IoT) connectivity, making it possible for a wide range of devices to interact effectively and support smart cities, industrial automation, and healthcare monitoring.

Manufacturing Industry

Real-time communication and data interchange between machines and systems will enable industry projects with advanced automation and robotics.

Artificial Intelligence (AI) along with machine learning algorithms will be used by predictive maintenance systems to foresee equipment faults and optimise maintenance schedules.

Environmental Performance

Real-time monitoring of crop health, soil conditions, and environmental elements would be made possible by precision agriculture using drones and sensors.

Climate monitoring systems will be in place for effective response planning and early natural disaster identification.

Healthcare

With the help of haptic feedback and remote surgery, doctors may operate with extreme precision even when they are far away from the patient.

Real-time health monitoring will be made possible by sensors and wearable technology, which will further allow for ongoing vital sign tracking and the early identification of health problems.

Smart Cities

Smart grids, effective waste management, and environmental monitoring would be greatly enhanced, which are examples of intelligent and integrated urban infrastructure. Improved public services will include parking control, public transportation, intelligent street lighting, *etc*.

Entertainment

Interactive content, augmented reality overlays, and 3D holographic projections would improve multimedia experiences. Lag-free multiplayer, high-fidelity graphics, and cloud gaming services will all combine to create immersive gaming experiences.

Education and Transportation

Interactive simulations and virtual field trips will be made possible by immersive learning experiences powered by Virtual Reality (VR) and Augmented Reality (AR) technologies. High-quality video streaming and collaborative tools will enable remote learning, enabling learners to access education from any location. Ultra-reliable autonomous car communication would allow cars to interact with infrastructure and other vehicles in real-time, improving traffic control and safety. Sophisticated traffic control systems would help manage congestion and optimise routing by using real-time data.

The various applications of 6G technology are summarised in Fig. (**5**).

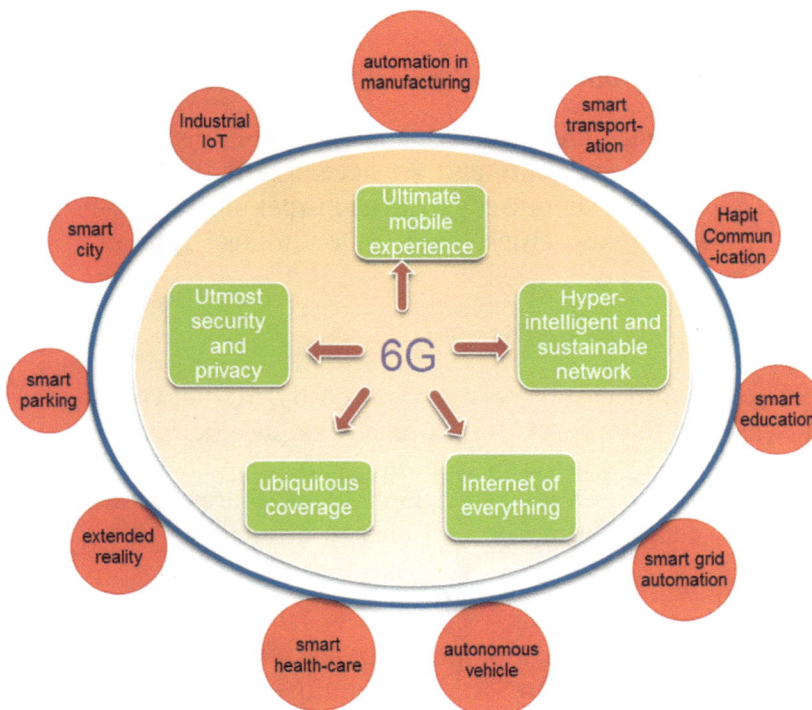

Fig. (5). Various applications of 6G technology [28].

Semiconductor Components in 6G

It is anticipated that the advancement of 6G technology will present novel demands and obstacles for semiconductor components. Here are some essential semiconductor parts that are probably going to be very important in 6G networks; however, specifics may change as the technology advances [29, 30]:

Millimeter-wave and Terahertz Transceivers

It is anticipated that 6G will communicate in higher frequency bands, which may include terahertz and millimeter-wave frequencies. Transceivers and other semiconductor parts will need to be made to function well at these higher frequencies in order to support large data rates and low-latency communication.

Advanced Semiconductor Components for Signal Processing and Modulation

Semiconductor components for signal processing and modulation will need to be more sophisticated in order to reach the high data rates planned by 6G. Among these are signal-processing circuits, modulators, and demodulators that can effectively process enormous volumes of data and handle intricate modulation schemes.

Edge Processing Units and AI Accelerators

As Artificial Intelligence (AI) will be included in 6G networks, semiconductor parts, like edge processing units and AI accelerators, will be essential. These elements will facilitate network-edge decision-making in real-time, supporting augmented reality, smart cities, and autonomous systems, among other applications.

Power-efficient Processors and Memory

With the increased processing power required by 6G networks, semiconductor makers must concentrate on creating processors and memory solutions that are low-power. This includes memory technology that can handle the increased data throughput, as well as CPUs with sophisticated architectures that strike a compromise between performance and energy efficiency.

Antennas and Beamforming Components

In order to enhance communication performance, 6G networks are anticipated to make use of sophisticated beamforming techniques. Phased-array antennas and beamforming circuits are two examples of semiconductor components that are essential for guiding and shaping signals to improve coverage, capacity, and dependability.

Components of Quantum Communication

In order to improve security, quantum communication is expected to be incorporated into the 6G environment. To offer secure communication channels, semiconductors associated with quantum cryptography, quantum entanglement, and Quantum Key Distribution (QKD) may be incorporated into the network architecture.

Sensors and Sensor Fusion Chips

As 6G networks will enable the widespread use of sensors for IoT and smart city sensing, semiconductor components for data gathering and sensor fusion will be needed. To enable sophisticated functionality, these components will be involved in the collection and processing of data from several sensors.

Security Hardware Modules

Semiconductor components devoted to hardware-based security modules will be crucial, given the vital role of security in 6G networks. To guard against cyberattacks, these could contain parts for encryption and decryption, safe key storage, and authentication.

Photonic Components for Optical Communication

With high-speed and low-latency connections in particular, optical communication is anticipated to be important in 6G. Optical interconnects, modulators, detectors, and other semiconductor components for integrated photonics may become more common.

Personalized System-on-Chip (SoC) Solutions

Semiconductor producers may provide highly integrated and personalized System-on-Chip (SoC) solutions to meet the various needs of 6G applications. By combining various features and functionalities designed for particular use cases, these SoCs might maximize power efficiency and performance.

The semiconductor industry will probably witness breakthroughs and advances in these and other components with the development of 6G technology to satisfy the particular requirements of the next-generation wireless networks. It is important to keep up with the latest findings and developments in the field in order to obtain the most reliable information.

Impact of 6G Technology on Semiconductor Industry

The semiconductor sector is anticipated to be significantly impacted by the development of 6G technology [30 - 33]. While specific details may evolve as the technology develops, here are some potential impacts:

Increased Demand for Advanced Chips

Higher data rates, reduced latency, and improved connectivity are anticipated with 6G technology. Because of this, there will be a greater need for sophisticated semiconductor chips that can meet the demands of increased connectivity. There might be more demand for companies making memory, communication semiconductors, and high-performance processors.

Innovation in Semiconductor Design

Innovations in semiconductor design may be required to meet the demands of 6G networks in order to integrate specialised functionalities, increase energy efficiency, and improve performance. This may result in advancements in packaging, interconnects, and transistor technology.

Advanced Materials Development

Silicon carbide (SiC), gallium nitride (GaN), and possibly other novel materials with special qualities will likely be developed and adopted as a result of the need for higher frequencies and better efficiency.

Higher Frequency Components

It is expected that 6G would function in higher frequency bands, possibly reaching the terahertz range. Because of this, new materials, device architectures, and manufacturing techniques will be needed in order to produce semiconductor components that can handle these higher frequencies.

Integration of AI and Machine Learning

It is anticipated that 6G networks will make use of Machine Learning (ML) and Artificial Intelligence (AI) for a range of purposes, including intelligent communication and network optimization. For AI and ML processing, the semiconductor industry might need to design specialized hardware accelerators in order to support these features.

Security and Privacy Enhancements

Security and privacy are becoming more and more important issues as communication networks get more complicated. To solve these issues and guarantee the integrity of 6G communications, the semiconductor industry could need to concentrate on creating secure hardware solutions.

Global Economic Impact

The global economy benefits greatly from the semiconductor industry. Adoption of 6G technology may improve the economy overall by promoting investment in related organisations, job creation, and economic growth.

Supply Chain Adjustments

A change in the semiconductor supply chain might be required as a result of the adoption of 6G technology. Companies that manufacture cutting-edge semiconductor machinery, materials, and manufacturing techniques might have to change to keep up with the ever-changing demands of the 6G internet.

Research and Development Investments

Research and development spending in the semiconductor sector is expected to grow as 6G technology advances and becomes more standardised. Organisations may spend money on staying on the cutting edge of technology and helping to develop components that are compatible with 6G.

It is critical to take into account that the actual impact on the semiconductor sector will vary depending on the speed at which 6G technology develops, the state of the regulations, and how rapidly the industry will adapt to the new demands.

Required Properties of Advanced Semiconductor Materials in 6G

The development of 6G technology is greatly influenced by the characteristics of advanced semiconductor materials. 6G semiconductor materials need to have particular properties in order to satisfy the requirements of substantially faster data speeds, reduced latency, and enhanced device connection over earlier generations. Regarding 6G communications, some essential characteristics of advanced semiconductor materials are as follows [31 - 35]:

High Electron Mobility

For rapid electron movement to be possible within the semiconductor material, high electron mobility is essential. This feature supports the higher data rates anticipated in 6G networks by enabling high-speed and high-frequency operation in electronic devices.

Heterogeneous Integration

Combining several semiconductor technologies and materials on a single chip is known as heterogeneous integration. Because of this characteristic, it is possible to design multipurpose devices with a wide range of functions, allowing different functionality to be integrated into the limited 6G device space.

Wide Bandgap

Higher energy is needed to transfer electrons from the valence band to the conduction band in materials having a large bandgap. For high-frequency and high-power applications, wide bandgap semiconductors are useful because they reduce energy losses and boost 6G device efficiency.

Low Noise

In high-frequency communication systems, in particular, low noise characteristics are crucial for preserving signal integrity. Improved signal-to-noise ratios and the dependable transfer of data in 6G networks are made possible by advanced semiconductor materials with low noise characteristics.

Quantum Properties

Applications in secure communication for 6G networks may arise for semiconductor materials possessing quantum characteristics. In 6G networks, quantum communication technologies, like quantum key distribution, can improve data transmission security and privacy.

High Thermal Conductivity

High-frequency and high-power electronic devices demand efficient heat dissipation. High thermal conductivity semiconductors aid in efficient heat management, minimising overheating and preserving the stability of equipment used in 6G applications under demanding working conditions.

Low Power Consumption

One of the most important factors in the development of energy-efficient 6G devices is reduced power consumption. The overall energy efficiency of communication systems and connected devices is enhanced by advanced semiconductor materials that provide low-power operation.

Flexibility and Stretchability

Designing conformal and wearable devices for 6G networks requires semiconductor materials to be flexible and stretchy. It is possible to create unique form factors and incorporate communication technologies into unconventional devices with the help of flexible materials.

Photonics Integration

For optical communication systems in 6G, photonics integration with semiconductor materials is crucial. High-speed optical communication components are developed with the use of materials that provide effective light modulation, switching, and detection.

Compatibility with Fabrication Processes

Advanced semiconductor materials should be compatible with state-of-the-art manufacturing processes to ensure scalable and cost-effective production of 6G devices.

It is noteworthy that the continuing advancements in research and development are revealing new semiconductor materials and refining the ones that already exist to satisfy the changing demands of 6G technology. The combination of these features would improve the overall performance and usefulness of 6G communication networks.

Emerging Semiconductor Materials in 6G Technology

Here are provided some semiconductor materials and their key properties [34 - 38]:

Gallium Nitride (GaN)

Excellent properties for high-frequency and high-power applications include high thermal conductivity, broad bandgap, and significant electron mobility. GaN-based devices could be crucial for implementing power amplifiers and switches in 6G base stations and devices.

Silicon Carbide (SiC)

High thermal conductivity, wide bandgap, and resistance to high temperatures make it ideal for high-power and high-frequency applications. It could be employed in power electronics and RF devices for 6G systems, enabling higher efficiency and power handling.

2D Materials

They are atomically thin and have distinct electronic properties, with the potential for high-speed electronic devices. Besides graphene, other two-dimensional materials, like hexagonal boron nitride (h-BN) and Transition Metal Dichalcogenides (TMDs), are being investigated for their potential role in 6G technology, particularly in nanoelectronics and optoelectronics [39, 40].

Graphene

It is a single layer of carbon atoms with the ability to create high-frequency transistors and interconnects, and has high electron mobility and superior thermal conductivity. It could be utilized in various components of 6G systems, such as high-speed transistors, flexible electronics, and terahertz communication devices.

III-V Compound Semiconductors [e.g., Gallium Arsenide (GaAs), Indium Phosphide (InP)]

Optoelectronics and high-speed devices employ high mobility of electrons, which is appropriate for high-frequency applications. These materials could find applications in 6G communication devices for amplifiers, high-speed switches, and photonic components [41].

Silicon Germanium (SiGe)

Silicon and germanium alloy, which is appropriate for high-frequency applications, is utilised to improve integrated circuit performance. SiGe technology is ideally suited for constructing high-frequency Radio Frequency (RF) circuits, including oscillators, mixers, power amplifiers, and Low Noise Amplifiers (LNAs).

Nanowires and Nanotubes

These possess one-dimensional structures, distinct electrical characteristics, and the possibility of nanoscale electronics and interconnects. Using nanowires and nanotubes, electrical components for 6G communication systems, such as high-speed transistors, can be made. For terahertz communication systems in 6G networks, antennas, waveguides, and other parts can be made of nanowires and nanotubes.

Quantum Dots

Quantum confinement phenomena and nanoscale semiconductor particles are utilised in quantum sensing and communication applications. In 6G technology, quantum dots hold promise for various applications, like high-efficiency photovoltaics, Quantum dot LEDs (QLEDs), quantum dot lasers, quantum dot photodetectors, *etc*.

Diamond Semiconductors

High thermal conductivity, high electron mobility, radiation hardness, and wide bandgap offer the possibility of high-power and high-frequency electronic devices. Diamond semiconductors possess a high breakdown field strength and thermal conductivity, making them suitable for high-power, high-frequency electronic devices, high-frequency RF devices, laser diodes, high-power laser diodes, quantum computing and sensing, *etc*.

Organic Semiconductors

These are carbon-based materials that have flexibility and the possibility for organic and flexible electronics. Since organic semiconductors are naturally flexible and of low-cost, solution-based processing methods can be used to handle them. This qualifies them for the fabrication of flexible electronics for wearable electronics and 6G communication systems, including flexible displays and flexible sensors. With flexible electronics, conformal devices that fit seamlessly into clothes, accessories, or even the human body can be developed, enabling seamless communication and interaction

Perovskite

Perovskite materials have drawn interest because of their exceptional optoelectronic qualities and tunability, which may find use in photovoltaics, LEDs, and photodetectors. Perovskite-based devices are being investigated by researchers for usage as inexpensive and effective photonic components, like modulators and photodetectors, in 6G networks.

These materials are intriguing prospects for advanced semiconductor applications, especially those related to 6G technology, due to their diverse characteristics. As the area develops, more research and development will likely find additional candidates and further refine our understanding of these materials. In the context of 6G communications, it is imperative to refer to more recent literature for the most recent advancements in semiconductor materials.

CONCLUSION

The semiconductor industry is witnessing a profound paradigm shift to meet the demands of 6G technology, characterized by the emergence of new materials, such as gallium-nitride (GaN), silicon-germanium (SiGe), diamond, and organic semiconductors. Since regulations take time to develop and technology advances swiftly, we have always moved aggressively toward 6G. The idea of 6G demonstrates the speed at which technology advances. 6G is the logical next step towards a faster, more dependable wireless connection, considering how quickly we have transitioned from 1G to 5G. To provide a higher data rate, every spectrum will be thoroughly investigated, including sub-6 GHz, mmWave, THz, and optical frequency bands. This shift is driven by the need to operate at terahertz frequencies, requiring novel semiconductor devices capable of ultra-high-speed data transmission and processing. Heterogeneous integration, advanced packaging solutions, and the integration of Artificial Intelligence (AI) and Machine Learning (ML) capabilities are becoming essential for optimizing performance and enabling multifunctional systems. Additionally, there is a heightened focus on security, trustworthiness, and environmental sustainability throughout the semiconductor lifecycle, reflecting the industry's commitment to addressing evolving challenges in 6G technology.

This book chapter has gone into extensive detail about the numerous aspects of 6G technology in the semiconductor trade and its current market dynamics. This chapter has discussed the integration of 6G technology in advanced semiconductors, such as silicon-germanium (SiGe) and gallium-nitride (GaN), and diamond semiconductors, due to the growing demand for significantly improved spectral/cost/energy efficiency, higher rates of data (Tbps), tenfold lower-latency, a hundred-fold greater connection density, more intelligence for full automation, and better frequency handling capabilities. Additionally, to facilitate the enhanced functionalities that 6G promises, edge computing and Artificial Intelligence (AI) integration into semiconductor designs has become indispensable, affecting the industry's emphasis on compact, low-power System-on-Chip (SoC) solutions. We have, herein, mainly emphasized various semiconductor materials and components, together with their attributes and applications, with respect to the 6G network.

KEY POINTS

- This chapter explores the revolutionary changes taking place in the semiconductor industry in advance of the introduction of 6G technology.
- It discusses how semiconductor technological developments are essential to delivering the ultra-fast data transmission, minimal latency, and huge connection that 6G networks offer.
- Potential topics for discussion in this chapter include emerging semiconductor materials and their properties, semiconductor components in 6G, and their impacts on the semiconductor industry, which have the potential to completely

transform the market and make it easier to build 6G networks.

- The chapter also looks at the benefits and difficulties that come with this paradigm change, such as the requirement for creative solutions to problems with integration, power consumption, and reliability.
- In addition, a summary of the evolution of 6G technology is provided, and subsequently, an overview of the 6G market's current status is given.
- The chapter further discusses the semiconductor components and applications of 6G technology, as well as the key parameters of 6G technology.

ACKNOWLEDGEMENTS

With great appreciation, the authors would like to thank everyone who has helped them finish this book chapter. They extend their sincere gratitude to Bentham Science Publishers for giving them the chance and venue to share their findings and perspectives. They also express their gratitude to CMR University (Lakeside Campus), Bangalore, for its institutional and financial support, which has allowed them to continue their research and development.

REFERENCES

[1]   N. Cahoon, P. Srinivasan, and F. Guarin, "6G roadmap for semiconductor technologies: Challenges and advances", *2022 IEEE International Reliability Physics Symposium (IRPS)*, 2022.
[http://dx.doi.org/10.1109/IRPS48227.2022.9764582]

[2]   I.F. Akyildiz, A. Kak, and S. Nie, "6G and beyond: The future of wireless communications systems", *IEEE Access*, vol. 8, pp. 133995-134030, 2020.
[http://dx.doi.org/10.1109/ACCESS.2020.3010896]

[3]   M. Božanić, and S. Sinha, *Device technologies and circuits for 5G and 6G*, pp. 99-154, 2021.

[4]   P. Yang, Y. Xiao, M. Xiao, and S. Li, "6G wireless communications: Vision and potential techniques", *IEEE Netw.*, vol. 33, no. 4, pp. 70-75, 2019.
[http://dx.doi.org/10.1109/MNET.2019.1800418]

[5]   P. Padhi, and F. Charrua-Santos, "6G enabled industrial Internet of Everything: Towards a theoretical framework", *Appl. Syst. Innov.*, vol. 4, no. 1, p. 11, 2021.
[http://dx.doi.org/10.3390/asi4010011]

[6]   Y. Shahzad, H. Javed, H. Farman, J. Ahmad, B. Jan, and M. Zubair, "Internet of energy: Opportunities, applications, architectures and challenges in smart industries", *Comput. Electr. Eng.*, vol. 86, no. 106739, p. 106739, 2020.
[http://dx.doi.org/10.1016/j.compeleceng.2020.106739]

[7]   A. Shahraki, M. Abbasi, M.J. Piran, and A. Taherkordi, "A comprehensive survey on 6G networks: Applications, core services, enabling technologies, and future challenges", *arXiv preprint arXiv*, p. 12475.

[8]   M. Fuentes, J.L. Carcel, C. Dietrich, L. Yu, E. Garro, V. Pauli, F.I. Lazarakis, O. Grøndalen, O. Bulakci, J. Yu, W. Mohr, and D. Gomez-Barquero, "5G new radio evaluation against IMT-2020 key performance indicators", *IEEE Access*, vol. 8, pp. 110880-110896, 2020.
[http://dx.doi.org/10.1109/ACCESS.2020.3001641]

[9]   J.G. Andrews, S. Buzzi, W. Choi, S.V. Hanly, A. Lozano, A.C.K. Soong, and J.C. Zhang, "What will 5G be?", *IEEE J. Sel. Areas Comm.*, vol. 32, no. 6, pp. 1065-1082, 2014.
[http://dx.doi.org/10.1109/JSAC.2014.2328098]

[10]   A.U. Gawas, "An overview on evolution of mobile wireless communication networks: 1G-6G", *Int. J. Recent Innov. Trends Comput. Commun.,* vol. 3, no. 5, pp. 3130-3133, 2015.

[11]   P. Gupta, "Performance improvement of millimeter wave antennas (review)", *Radioelectron. Commun. Syst.,* vol. 65, no. 9, pp. 447-463, 2022.
[http://dx.doi.org/10.3103/S0735272722100016]

[12]   P. Gupta, and V. Gupta, "Linear 1× 4 microstrip antenna array using slotted circular patch for 5G communication applications", *Wirel. Pers. Commun.,* vol. 127, no. 4, pp. 2709-2725, 2022.
[http://dx.doi.org/10.1007/s11277-022-09896-4]

[13]   Prachi and T. K. Mandal, "Dual frequency millimeter-wave perturbed ring patch antenna array for 5G applications", *J. Inst. Electron. Telecommun. Eng.,* pp. 1-11, 2021.

[14]   R. Dangi, "6G Mobile Networks: Key Technologies, Directions, and Advances", In: *in Telecom. MDPI* vol. 4. , 2023.

[15]   J.R. Bhat, and S.A. Alqahtani, "6G ecosystem: Current status and future perspective", *IEEE Access,* vol. 9, pp. 43134-43167, 2021.
[http://dx.doi.org/10.1109/ACCESS.2021.3054833]

[16]   Y. Lu, and X. Zheng, "6G: A survey on technologies, scenarios, challenges, and the related issues", *J. Ind. Inf. Integr.,* vol. 19, no. 100158, p. 100158, 2020.
[http://dx.doi.org/10.1016/j.jii.2020.100158]

[17]   B. Ji, Y. Wang, K. Song, C. Li, H. Wen, V.G. Menon, and S. Mumtaz, "A survey of computational intelligence for 6G: Key technologies, applications and trends", *IEEE Trans. Industr. Inform.,* vol. 17, no. 10, pp. 7145-7154, 2021.
[http://dx.doi.org/10.1109/TII.2021.3052531]

[18]   InsightAce   Analytic,   "6G   Market"   [Online].   Available   from: https://www.insightaceanalytic.com/report/6g-market/1820

[19]   Tele.net, "Exploring 6G: Industry begins to look into the new technology" [Online]. Available from: https://tele.net.in/exploring-6g-industry-begins-to-look-into-the-new-technology/

[20]   A. Dogra, R.K. Jha, and S. Jain, "A survey on beyond 5G network with the advent of 6G: Architecture and emerging technologies", *IEEE Access,* vol. 9, pp. 67512-67547, 2021.
[http://dx.doi.org/10.1109/ACCESS.2020.3031234]

[21]   A.I. Salameh, and M. El Tarhuni, "From 5G to 6G—challenges, technologies, and applications", *Future Internet,* vol. 14, no. 4, p. 117, 2022.
[http://dx.doi.org/10.3390/fi14040117]

[22]   F. Salahdine, T. Han, and N. Zhang, "5G, 6G, and Beyond: Recent advances and future challenges", *Ann. Telecommun.,* vol. 78, no. 9-10, pp. 525-549, 2023.
[http://dx.doi.org/10.1007/s12243-022-00938-3]

[23]   Samsung Research, *6G Vision - The Next Hyper-Connected Experience for All,* 2020.

[24]   Z. Zhang, Y. Xiao, Z. Ma, M. Xiao, Z. Ding, X. Lei, G.K. Karagiannidis, and P. Fan, "6G wireless networks: Vision, requirements, architecture, and key technologies", *IEEE Veh. Technol. Mag.,* vol. 14, no. 3, pp. 28-41, 2019.
[http://dx.doi.org/10.1109/MVT.2019.2921208]

[25]   M.Z. Chowdhury, M. Shahjalal, S. Ahmed, and Y.M. Jang, "6G wireless communication systems: Applications, requirements, technologies, challenges, and research directions", *IEEE Open J. Commun. Soc.,* vol. 1, pp. 957-975, 2020.
[http://dx.doi.org/10.1109/OJCOMS.2020.3010270]

[26]   W. Saad, M. Bennis, and M. Chen, "A vision of 6G wireless systems: Applications, trends, technologies, and open research problems", *IEEE Netw.,* vol. 34, no. 3, pp. 134-142, 2020.
[http://dx.doi.org/10.1109/MNET.001.1900287]

[27] H.H.H. Mahmoud, A.A. Amer, and T. Ismail, "6G: A comprehensive survey on technologies, applications, challenges, and research problems", *Trans. Emerg. Telecommun. Technol.,* vol. 32, no. 4, p. e4233, 2021.
[http://dx.doi.org/10.1002/ett.4233]

[28] S. Rajoria, and K. Mishra, "A brief survey on 6G communications", *Wirel. Netw.,* vol. 28, no. 7, pp. 2901-2911, 2022.
[http://dx.doi.org/10.1007/s11276-022-03007-8]

[29] T. Maiwald, "A review of integrated systems and components for 6g wireless communication in the D-Band", *Proceedings of the IEEE,* 2023.
[http://dx.doi.org/10.1109/JPROC.2023.3240127]

[30] J. Lee, M. Nouwens, and K. L. Tay, *Strategic Settings for 6G: Pathways for China and the US,* 2022.

[31] Z. Li, J. Pan, H. Hu, and H. Zhu, "Recent advances in new materials for 6G communications", *Adv. Electron. Mater.,* vol. 8, no. 3, p. 2100978, 2022.
[http://dx.doi.org/10.1002/aelm.202100978]

[32] W. Jiang, and H.D. Schotten, "The KICK-OFF of 6G research worldwide: An overview", *in 2021 7th International Conference on Computer and Communications (ICCC),* 2021.
[http://dx.doi.org/10.1109/ICCC54389.2021.9674614]

[33] Q. Bi, "Ten trends in the cellular industry and an outlook on 6G", *IEEE Commun. Mag.,* vol. 57, no. 12, pp. 31-36, 2019.
[http://dx.doi.org/10.1109/MCOM.001.1900315]

[34] Y. Zhao, "A survey of 6G wireless communications: Emerging technologies", In: *in Advances in Intelligent Systems and Computing.* Springer International Publishing: Cham, 2021, pp. 150-170.

[35] N. Collaert, A.R. Alian, A. Banerjee, V. Chauhan, R.Y. ElKashlan, B. Hsu, M. Ingels, A. Khaled, K.V. Kodandarama, B. Kunert, Y. Mols, U. Peralagu, V. Putcha, R. Rodriguez, A. Sibaja-Hernandez, E. Simoen, A. Vais, A. Walke, L. Witters, S. Yadav, H. Yu, M. Zhao, P. Wambacq, B. Parvais, and N. Waldron, "(plenary) the revival of compound semiconductors and how they will change the world in a 5G/6G era", *ECS Trans.,* vol. 98, no. 5, pp. 15-25, 2020.
[http://dx.doi.org/10.1149/09805.0015ecst]

[36] K. Baruah, R.G. Debnath, and S. Baishya, "Technology computer-aided design simulation of G e-source double-gate S i-tunnel Field Effect Transistor: Radio frequency and linearity analysis", *Int. J. RF Microw. Comput.-Aided Eng.,* vol. 32, no. 10, 2022.
[http://dx.doi.org/10.1002/mmce.23316]

[37] K. Baruah, and S. Baishya, "Electrical noise in Ge-source double-gate PNPN tunnel field effect transistor", *Proceedings of the Indian Association for the Cultivation of Science,* 2022.

[38] E.C. Strinati, "The hardware foundation of 6G: The NEW-6G approach", *in 2022 Joint European Conference on Networks and Communications & 6G Summit (EuCNC/6G Summit),* 2022.
[http://dx.doi.org/10.1109/EuCNC/6GSummit54941.2022.9815700]

[39] W. Cao, J. Kang, D. Sarkar, W. Liu, and K. Banerjee, "2D semiconductor FETs—Projections and design for sub-10 nm VLSI", *IEEE Trans. Electron Dev.,* vol. 62, no. 11, pp. 3459-3469, 2015.
[http://dx.doi.org/10.1109/TED.2015.2443039]

[40] D. Chi, *2D semiconductor materials and devices.* Elsevier, 2019.

[41] M.P. Mikhailova, K.D. Moiseev, and Y.P. Yakovlev, "Discovery of III–V semiconductors: Physical properties and application", *Semiconductors,* vol. 53, no. 3, pp. 273-290, 2019.
[http://dx.doi.org/10.1134/S1063782619030126]

Semiconductor Nanoscale Devices, 2025, 311-336   311

CHAPTER 11

Exploring the Depths of Sigma-Delta Analog-to Digital Converters: A Comprehensive Review

Ravita[1,*], **Ashish Raman**[1] and **Ramesh K Sunkaria**[1]

¹ Dr. B. R. Ambedkar National Institute of Technology, Jalandhar, Punjab, India

Abstract: Sigma-delta Analog-to-Digital Converters (ADCs) have arisen as critical components in modern electronic systems due to their capability to achieve high-resolution conversions with minimal power consumption. This chapter provides a comprehensive review of sigma-delta ADC architectures, operating principles, design considerations, and applications. It begins by outlining the sigma-delta modulation technique and proceeds to explore various architectures, including first-order, higher-order, and multi-bit sigma-delta converters. The discussion extends to analyzing noise shaping, quantization noise, and dynamic range, which are pivotal in shaping the performance of sigma-delta ADCs. The chapter meticulously addresses design challenges, such as stability concerns, handling non-idealities, and optimizing circuit implementation techniques to achieve peak performance. Furthermore, it explores the evolution of sigma-delta ADC technology, highlighting recent advancements and emerging trends. This includes advancements in oversampling techniques, digital decimation filters, and calibration methods aimed at further enhancing the efficiency and accuracy of sigma-delta ADCs. Finally, the chapter concludes with an overview of the diverse applications of sigma-delta ADCs across various domains. These include their use in communications, sensor interfaces, audio processing, and medical instrumentation, underscoring their versatility and importance in modern electronics. Each application domain benefits uniquely from the precision and efficiency offered by sigma-delta ADCs, making them indispensable in today's technology landscape.

Keywords: ADCs (analog-to-digital converters), $\sum \Delta$ADCs (sigma-delta analog-to-digital converters), Modulation techniques, Noise shaping, Quantization noise.

INTRODUCTION

ADCs are essential parts of contemporary electronic systems, enabling the transformation of continuous analog signals into digital formats for processing, storage, and transmission [1]. Among the plethora of ADC architectures available,

---

*** Corresponding author Ravita:** Dr. B. R. Ambedkar National Institute of Technology, Jalandhar, Punjab, India; E-mail: ravita.ec.23@nitj.ac.in

Ashish Raman, Prabhat Singh, Naveen Kumar & Ravi Ranjan (Eds.)

$\sum\Delta$ADCs have attracted much interest since they can provide high-resolution conversions, while consuming relatively low power. This chapter comprehensively explores $\sum\Delta$ADCs, elucidating their fundamental principles, architectures, design considerations, performance analysis, recent advancements, and applications.

The modulation technique forms the cornerstone of $\sum\Delta$ADCs, leveraging oversampling and noise shaping to attain high-resolution conversion [2, 3]. By exploiting the inherent trade-off between sampling rate and resolution, sigma-delta ADCs can achieve resolutions exceeding those of traditional ADC architectures without necessitating high-precision analog components.

Various architectures of sigma-delta ADCs, including first-order, higher-order, and multi-bit converters, are examined in detail, highlighting their advantages, limitations, and design considerations. The intricacies of stability analysis, non-idealities, and circuit implementation techniques are also discussed, offering insights into achieving optimal performance in practical implementations.

Performance analysis constitutes a crucial aspect of sigma-delta ADC evaluation, encompassing noise shaping, dynamic range, resolution, accuracy, speed, and power consumption. By comprehensively assessing these factors, designers can tailor $\sum\Delta$ADCs to meet the requirements of diverse applications spanning communications, sensor interfaces, audio processing, and medical instrumentation.

Furthermore, this chapter delves into recent advancements and emerging trends in sigma-delta ADC technology, including novel oversampling techniques, digital decimation filters, and advanced calibration methods, to enhance performance and efficiency.

In conclusion, this chapter is a valued resource for researchers, engineers, and students seeking a deeper understanding of $\sum\Delta$ADCs. By providing a comprehensive overview of architectures, design considerations, performance analysis, recent advancements, and applications, it aims to foster continued innovation and advancement in analog-to-digital conversion.

OVERVIEW OF ANALOG-TO-DIGITAL CONVERTERS (ADCs)

ADCs are essential components in modern electronics [4]. They allow continuous analog signals to be converted into digital representations, making them suitable for storage, processing, and transmission within digital systems. They play a critical role in bridging the analog and digital domains, allowing for the integration of analog signals into digital processing chains.

The quality of the digital output, including resolution, accuracy, and speed, is paramount in determining the overall performance of the ADC. Different ADC architectures offer unique characteristics and trade-offs. The selection of an ADC architecture is determined by the particular requirements of an application, including the required speed, power consumption, resolution, and financial concerns. Some common types include the following:

Successive Approximation ADCs

These ADCs employ a binary search algorithm to establish the digital representation of the incoming signal. They are known for their moderate speed and reasonable resolution.

Flash ADCs

Flash ADCs use an array of voltage comparators to convert an analog input into a digital output directly. They offer high-speed conversion, but are limited in resolution and power efficiency.

Sigma-Delta ADCs

Oversampling and noise-shaping methods are used by sigma-delta ADCs to attain high-resolution conversions, while keeping power consumption relatively low. They are particularly suited for high-resolution applications with low to moderate bandwidth.

Pipeline ADCs

Pipeline ADCs divide the conversion process into multiple stages, each contributing to the ADC's overall resolution. They balance speed and resolution and are commonly used in high-speed applications.

In addition to the architecture, other factors, such as sampling rate, input range, and signal conditioning, influence the performance of ADCs. Noise, linearity, and dynamic range are key factors in guaranteeing the precise and reliable conversion of analog signals into digital format. The first block in Fig. (1) is the Anti-alias Filter (AAF), which reduces the input signal's bandwidth to less than half of the sampling frequency. This filter prevents higher-frequency signals from folding back into the frequency range of interest during sampling, thereby avoiding aliasing. Depending on the sampling rate and signal bandwidth relationship, ADCs are typically categorized as Nyquist rate ADCs or oversampled ADCs, such as sigma-delta ADCs. As the sampling rate increases, the transition width of the anti-alias filter also increases proportionally.

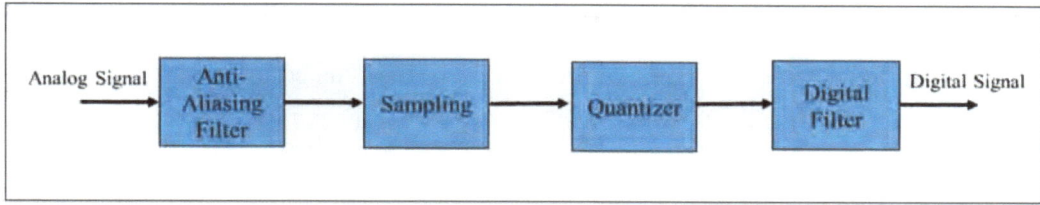

Fig. (1). Block diagram of ADC.

NYQUIST RATE ANALOG-TO-DIGITAL CONVERTERS

The sampling theorem, which states that in order to permit reliable reconstruction, the input signal must be sampled at a frequency at least twice that of the signal bandwidth, is followed by Nyquist rate converters. These ADCs are relatively fast, but face resolution constraints that are typically limited to 10-12 bits due to the need for precise component matching and circuit imperfections [5]. Anti-aliasing filters with very sharp cut-offs are essential to avoid aliasing. Flash ADCs, pipelined ADCs, and successive approximation ADCs fall under Nyquist converters.

An ideal impulse train with uniform spacing at nT, (T) is the reciprocal of sampling frequency fs, which is multiplied by the continuous-time band-limited signal x(t) to produce the sampled signals, such that $x_s(t)$ is a kind of sampling. One way to think of this sampled signal, $x_s(t)$, is as a continuous-time signal.

In the frequency domain, the process is like convolving the input spectrum with an impulse train and generating the image of the input spectrum centered at integer multiples of such sampling frequency.

All things considered, ADCs act as the link between the analog and digital realms, allowing analog signals to be seamlessly integrated into digital systems for a wide range of domains where applications are in variety for industrial purposes, such as audio interference analysis processing, sensor interfaces, instrumentation, and telecommunications.

ADC CHARACTERISTICS

In ADCs, there are two primary categories of characteristics to consider: static and dynamic. Static features that are essential for evaluating the accuracy of the ADC are offset, gain error, Differential Non-linearity (DNL), and Integral Non-linearity (INL). On the other hand, dynamic properties, like Spurious-free Dynamic Range (SFDR), signal-to-noise ratio, and dynamic range, are critical for evaluating the performance of oversampled converters, which operate differently from other types of ADCs. The following are crucial dynamic parameters used to measure and compare the effectiveness of these converters [6, 7].

Signal-to-noise ratio: The SNR of an ADC is defined as equation 1.

$$SNR = \frac{\text{Input signal power}}{\text{Power noise at the output (ADC)}} \tag{1}$$

Peak SNR (SNRp) is maximum SNR of an ADC. The combination of the quantization and internal noise of the circuit is called noise [8].

Signal-to-Noise and Distortion Ratio (SNDR): This characteristic of an ADC is defined as equation 2.

$$SNDR = \frac{\text{Input signal power}}{\text{Noise and distortion components measured power at the ADC output}} \tag{2}$$

Dynamic Range (DR): The DR is defined as equation 3.

$$DR = \frac{\text{Most input powers applied to the input of the ADC}}{\text{Minimum detectable input power}} \tag{3}$$

A major performance decline in an ADC occurs when the SNR falls roughly 3 dB below its maximum value. The input power that yields an SNR of 0 dB in an ADC is referred to as the smallest measurable input signal.

Overload level: The relative input amplitude at which the SNR decreases by 3 dB below its highest value is known as the Overload Level (OL) [9].

Spurious free dynamic range: The SFDR is described as equation 4.

$$SFDR = \frac{\text{Signal power}}{\text{Highest harmonic or spurious noise component power}} \tag{4}$$

The SFDR shows the minimum signal level at which noise or distortion can occur.

Total harmonic distortion: This characteristic is described as equation 5.

$$THD = \frac{\text{Total harmonic distortion power}}{\text{Power of fundamental frequencies of the signal}} \tag{5}$$

Typically, the THD is computed for harmonics for specific numbers [10].

Effective Number of Bits (ENOB): This property measures an ADC's associated circuitry's Dynamic Range (DR). The amount of bits utilized to digitally represent the analog value determines the resolution of an ADC [11].

Figure of Merit (FOM): This characteristic is a valuable tool for comparing the conversion efficiency of ADCs. In engineering, FOM is often described for specific materials or devices to find their relative utility for an application. A good figure of merit should accurately reflect the merits of the ADC in the context and for the purpose for which FOM is used. The figure of merit used is Schreier FOM presented in equation 6.

$$FOMS = DR + 10\log(BW/Power) \tag{6}$$

IMPORTANCE OF SIGMA-DELTA ADCs

Sigma-delta ADCs have gained significant importance in various applications due to their unique advantages over other ADC architectures. These advantages include high resolution, low power consumption, and suitability for applications requiring high precision and low-to-moderate bandwidth.

These ADCs can handle input signal bandwidth in the MHz range due to advanced semiconductor technology and design methods. Second, these ADCs are appropriate for low-power design due to their oversampling feature, which significantly minimizes the performance requirement of the antialiasing filters, where AAF is the most power-hungry block in the wireless receivers. Third, $\sum\Delta$ADCs ADCs have a noise shaping feature, which provides shaping to almost all analog circuit errors that go away from the signal band to obtain higher accuracy in the band of interest [12 - 15].

Furthermore, sigma-delta ADCs exhibit inherent linearity and stability, making them well-suited for applications requiring high accuracy and dynamic range. Their digital-intensive nature enables easy integration with Digital Signal Processing (DSP) algorithms, facilitating additional signal conditioning and processing within the ADC. Because of their low power requirements, sigma-delta ADCs can be used in energy-efficient and battery-powered applications. By leveraging oversampling techniques and low-power circuit design, sigma-delta ADCs can achieve high-resolution conversions, while consuming minimal power, making them ideal for portable devices, sensor nodes, and other low-power electronics.

Moreover, $\sum\Delta$ADCs are widely used in applications, such as audio processing, sensor interfaces, communications, and medical instrumentation. In audio applications, sigma-delta ADCs are prized for their ability to faithfully capture subtle nuances in audio signals, making them essential components in high-fidelity audio systems and professional audio equipment. In sensor interfaces, sigma-delta ADCs excel at digitizing small analog signals from sensors, enabling precise measurements in applications ranging from temperature sensing to industrial process control. Sigma-delta ADCs are essential because they can deliver high-resolution, low-power, and high-accuracy analog-to-digital conversion, making them indispensable components in various applications across various harmonics industries.

SIGMA-DELTA MODULATION TECHNIQUE

In order to provide high-resolution conversion, sigma-delta modulation is essential to $\sum\Delta$ADCs. This technique involves noise shaping to remove quantization noise from the signal band after oversampling the input signal at a frequency far higher than the Nyquist rate. The oversampled signal is quantized with a one-bit quantizer, producing a high-density digital data stream that carries the desired information encoded in its high-frequency components. This approach allows sigma-delta ADCs to achieve high-resolution conversion with relatively low oversampling ratios and simple analog components [16]. Integrators (sigma) and subtractors (delta) are used in designing, also called sigma-delta or delta-sigma ADCs, as shown in Fig. (**2**).

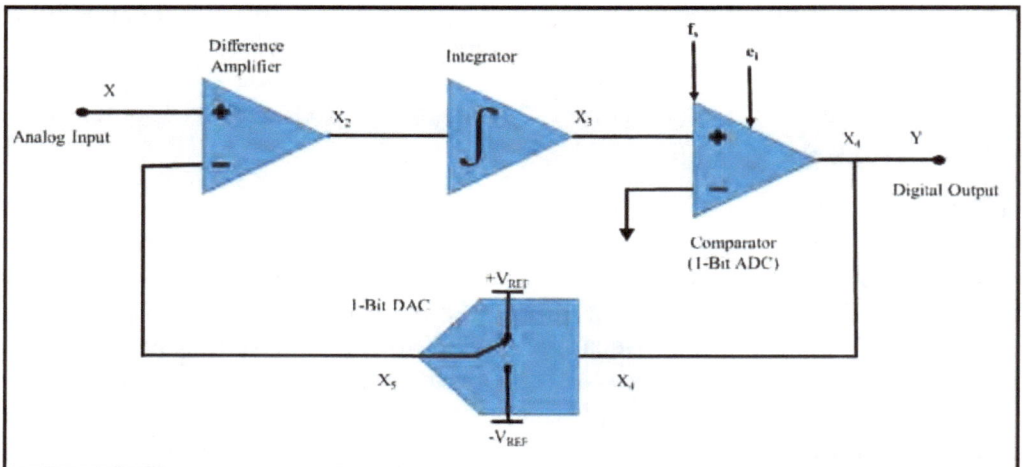

Fig. (2). Block diagram of sigma-delta ADC.

BASIC PRINCIPLE

The principles of noise shaping and oversampling are the foundation of sigma-delta modulation. Sigma-delta ADCs can remove quantization noise from the signal band by oversampling the incoming signal at a frequency that is significantly greater than the Nyquist rate [17, 18]. A one-bit quantizer is used to quantize the difference between the input and feedback signals in a feedback loop that processes the oversampled signal. In order to create a high-density digital data stream with high-frequency components carrying the appropriate signal information, the quantization error that results known as the delta is then fed back and deducted from the input signal [19].

The noise shaping process ensures that most quantization noise is concentrated at higher frequencies, effectively reducing its impact on the desired signal. This approach enables sigma-delta ADCs to achieve high-resolution conversion with relatively low oversampling ratios and simple analog components [20].

Sigma-delta ADCs do not require precise component matching, making them ideal for applications demanding resolutions of 20 bits and higher. Nyquist rate ADCs rely on accurate quantization of the input signal for each digital word, whereas sigma-delta ADCs produce output through coarse quantization of input samples. Since the majority of the conversion process occurs in the digital domain, these converters enable the construction of high-speed and high-density digital circuits. This contrasts with Nyquist rate ADCs, which depend on precise analog components for accurate conversion.

Sigma-delta converters use relatively simple and compact analog components, contrasting their Nyquist rate counterparts. When employing high sampling rates, anti-aliasing filters are unnecessary or only need a broad roll-off filter, eliminating the need for sharp cut-off filters. These ADCs achieve high precision through two main techniques: oversampling and noise shaping.

QUANTIZATION

There are two main processes involved in converting an analog continuous-time signal to a digital signal. The analog signal is first sampled and held for a pre-agreed amount of time. Next, it is quantized, meaning the signal is mapped to fixed quantization levels based on the resolution of the ADC. This quantization is carried out by a component called a quantizer, which assigns the incoming signal to the closest quantization level [21].

Quantization introduces a form of error or noise known as quantization error (or noise), denoted as e_q, computed as the difference between the quantizer output (y) and the analog input (x). The magnitude of this error decreases as the converter's resolution increases because of the corresponding decrease in the size of the least significant bit (LSB) or step size Δ.

Quantization error is an innate part of the procedure for converting analog to digital, and it is typically distributed uniformly between $\pm\Delta/2$. Its Power Spectral Density (PSD), shown in Fig. (3), is often considered flat within the Nyquist band of $[-f_s/2, f_s/2]$, where $_{fs}$ is the sampling frequency. The relationship between the quantization step size and noise power depends on the converter's resolution, commonly called quantization noise.

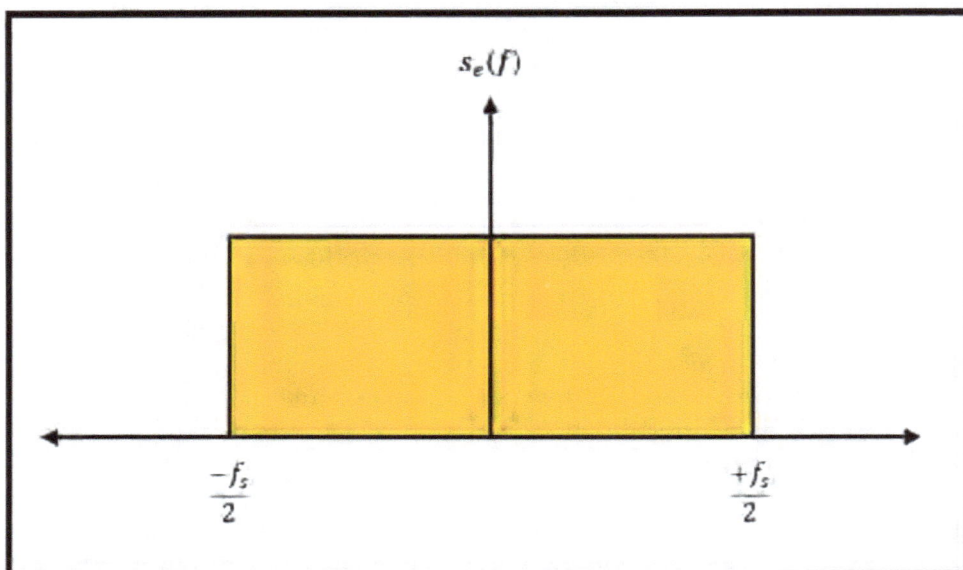

Fig. (3). Power spectral density of quantization noise.

In sigma-delta ADCs, the quantizer is represented as an additional noise source. Nonetheless, sigma-delta ADCs use noise-shaping and oversampling techniques to convert data at high resolution. These ADCs rearrange quantization noise in the frequency domain, shifting most of the noise energy to higher frequencies and away from the signal band [22]. This leads to high-resolution conversion using relatively simple analog components and low-resolution quantizers.

The power spectral density amplitude in sigma-delta ADCs often exhibits a high-pass characteristic, concentrating most of the noise energy at higher frequencies as shown in Fig. (4). The sigma-delta modulator's loop filter transfer function and feedback loop architecture determine this shaping of the quantization noise spectrum.

The quantization noise Power Spectral Density (PSD) in sigma-delta ADCs can be derived from the loop filter's transfer function and the quantizer characteristics. This PSD is influenced by factors, such as the sampling frequency, quantizer resolution, and the design of the loop filter. The quantization noise PSD in sigma-delta ADCs can be described using the Least Significant Bit (LSB) and the frequency response of the loop filter, allowing the noise to be pushed out of the signal band and concentrated at higher frequencies [23].

The integration of quantization and noise-shaping methods in sigma-delta ADCs allows for high-precision conversion, while reducing the effect of quantization noise within the signal band.

OVERSAMPLING AND NOISE SHAPING

In order to accomplish high-resolution ADC in sigma-delta modulation, two essential techniques are oversampling and noise shaping. The ratio of the sampling rate to the Nyquist rate is known as the oversampling ratio. Oversampled converters have the widest resolution *versus* bandwidth. In order to oversample, one must sample the incoming analog signal x(t) at a frequency that is significantly more than the Nyquist rate (fs = 2 * fmax), where fmax is the input signal's highest frequency component [24, 25]. The oversampled signal $x_0(t)$ is then quantized using a one-bit quantizer, resulting in a sequence of high-density digital samples (equation 7).

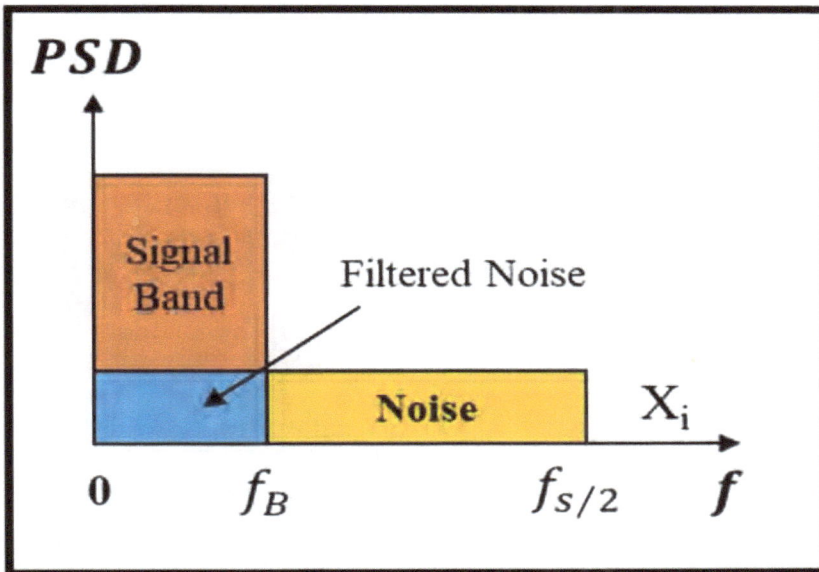

Fig. (4). Power density spectrum of quantization noise.

$$X_0[n] = Quantizer(x(t_n)) \tag{7}$$

Where, $X_0[n]$ represents the digital output of the quantizer at time t_n.

Noise shaping is achieved by introducing a feedback loop in the sigma-delta modulator, which modifies the quantization noise spectrum to push it away from the signal band [26]. The error signal, e[n], is attained by subtracting the quantization noise, eq[n], from the input signal at the quantizer's output (equation 8). This signal is then integrated and fed back into the input:

$$e[n] = x[n] - x_0[n] \tag{8}$$

The feedback loop introduces a transfer function *H(z)*, which shapes the quantization noise spectrum. The overall loop transfer function *T(z)* can be expressed as equation 9:

$$T(z) = \frac{1}{1+H(z)} \tag{9}$$

The transfer function from quantization noise to the output is represented by the noise transfer function N(z), as presented in equation 10:

$$N(z) = \frac{H(z)}{1+H(z)} \tag{10}$$

By appropriately designing the loop filter transfer function *H(z)*, the quantization noise can be shaped and pushed towards higher frequencies, effectively reducing its energy within the signal bandwidth.

With relatively low-resolution ADCs and straightforward analog circuitry, sigma-delta ADCs can accomplish high-resolution conversion thanks to overwhelming noise shaping [21]. These techniques are essential for their performance, allowing them to achieve high levels of linearity, dynamic range, and resolution.

ARCHITECTURES OF SIGMA-DELTA ADCs

A sigma-delta modulator reaches high resolution by using noise shaping and oversampling to function as an ADC. The order, structure, and quantity of quantizer bits in these modulators can all differ [27, 28]. Higher-order modulators provide better noise shaping and accuracy, whereas the modulator's order determines the number of integrator stages in the loop filter. Structures, such as Cascaded Integrator-Feedforward (CIFF), Cascaded Resonator-Feedforward (CRFF), Cascaded Integrator-Feedback (CIFB), and Cascaded Resonator-

Feedback (CRFB) determine the arrangement and feedback paths within the modulator, affecting performance and stability. Additionally, modulators can be designed with either one-bit or multi-bit quantizers. One-bit quantizers offer high linearity, but they can be more sensitive to noise, while multi-bit quantizers improve signal-to-noise ratio and reduce quantization noise. Overall, higher-order modulators with complex structures and multi-bit quantizers provide superior performance, increase design complexity, and may require careful calibration.

First-order Sigma-delta Converters

First-order $\sum\Delta$ converters are the simplest form of $\sum\Delta$ADCs. They typically consist of a single integrator in the feedback loop, resulting in a noise-shaping characteristic. These converters are often used in low-cost and low-power applications where moderate performance is sufficient. In the feedback loop of a first-order $\sum\Delta$ converter, there is usually only one integrator. The quantization error is produced by deducting the incoming signal from the feedback signal and integrating the result. To conclude the loop, this quantization error is then sent back and deducted from the input signal. The single integrator in the loop limits the first-order sigma-delta converter's noise-shaping performance. The noise shaping effect pushes quantization noise to higher frequencies, but is less effective than higher-order sigma-delta architectures. First-order sigma-delta converters typically offer moderate resolution, which may be sufficient for many applications. The oversampling ratio, quantizer, resolution, and loop filter design determine the resolution. When mediocre performance is acceptable and the application is low-cost and low-power, sigma-delta converters are frequently utilized. Examples include low-resolution audio ADCs, sensor interfaces, and battery-powered devices. While first-order sigma-delta converters offer simplicity and low cost, they may provide a lower resolution or performance than higher-order architectures. When selecting the appropriate ADC architecture, designers must balance performance requirements with cost and power constraints.

First-order sigma-delta converters balance performance, cost, and complexity for many applications, making them popular in various consumer electronics and industrial applications.

Higher-order Sigma-delta Converters

Higher-order sigma-delta converters utilize multiple integrators in the feedback loop to achieve higher noise shaping and resolution levels. By cascading several integrators, the noise shaping performance can be improved, allowing higher resolution conversion. These converters, such as audio and instrumentation systems, are commonly employed in applications requiring higher performance and resolution.

Higher-order sigma-delta converters utilize multiple integrators (typically second-order shown in Fig. (**5**) or higher shown in Fig. (**6**) in cascade within the feedback loop. Each integrator introduces additional poles in the loop transfer function, increasing noise shaping capability and resolution. With multiple integrators in the loop, higher-order sigma-delta converters offer superior noise-shaping performance compared to first-order converters. The increased number of poles in the loop transfer function allows for more aggressive shaping of the quantization noise spectrum, pushing noise energy further away from the signal band. Higher-order sigma-delta converters can achieve upper resolution than first-order converters. The increased noise shaping capability, combined with higher oversampling ratios and finer quantization, enables these converters to achieve higher effective resolutions. Higher-order sigma-delta converters are more complex and require additional circuitry and computational requirements. As a result, they may consume more power and require more silicon area compared to first-order architectures. Superior quality applications that require greater performance and resolution, like high-fidelity music, instrumentation, and precision measurement systems, frequently use sigma-delta converters. They work especially effectively in applications that need accurate acquisition of low-level signals in noisy environments. Higher-order sigma-delta converters offer superior resolution and noise shaping, but come with increased complexity and cost. Designers must carefully balance performance requirements with practical constraints, such as power consumption, price, and silicon area.

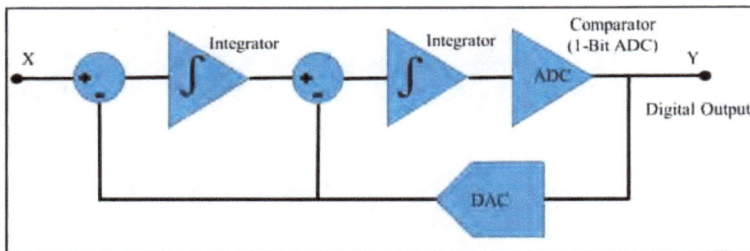

Fig. (5). Second-order sigma-delta ADC.

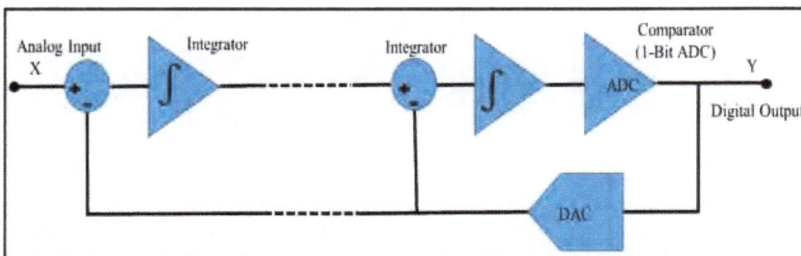

Fig. (6). Higher-order sigma-delta ADC.

Multi-bit Sigma-delta Converters

Multi-bit sigma-delta converters use multi-level quantizers (typically two or more bits) in the response loop instead of a one-bit quantizer. By employing multi-level quantization, these converters can achieve higher resolution with reduced quantization noise. Multi-bit sigma-delta converters, such as high-fidelity audio systems and precision measurement equipment, are often used in high-resolution, high-performance applications [29]. Each architecture has advantages and trade-offs in performance, complexity, power consumption, and cost. Designers choose the appropriate architecture based on the application's specific requirements. The quantization error is represented by a multi-bit binary word, allowing for finer quantization and increased resolution compared to single-bit converters. A multi-bit sigma-delta converter's quantization levels determine its resolution. Higher-resolution converters use more bits in the quantizer, providing finer quantization and higher effective resolution. Noise-shaping techniques are still used in multi-bit sigma-delta converters to shift quantization noise out of the signal band. The quantizer's properties and the loop filter's design determine the noise-shaping performance [1]. Multi-bit $\sum\Delta$ converters can achieve greater resolution than single-bit converters, especially at lower oversampling ratios. The increased quantization levels allow for a more precise representation of the input signal, leading to higher effective resolution. These converters typically offer an improved dynamic range compared to single-bit converters. The additional quantization levels reduce quantization noise and increase the signal-to-noise ratio, allowing for accurate conversion of low-level and high-amplitude signals. These are commonly used in high-resolution and high-dynamic range conversion applications, such as high-fidelity audio, medical imaging, and scientific instrumentation. They are particularly well-suited for applications where precise measurement and accurate signal reproduction are critical. While multi-bit sigma-delta converters offer superior resolution and dynamic range, they may require more complex circuitry and calibration than single-bit converters. When selecting the appropriate ADC architecture, designers must consider trade-offs between resolution, complexity, power consumption, and cost.

In general, multi-bit sigma-delta converters offer a practical way to convert analog to digital data at high resolution for applications requiring accurate signal reproduction and exact measurement [30].

DESIGN CONSIDERATIONS

Design considerations are crucial in developing sigma-delta ADCs to ensure optimal performance, stability, and efficiency. Here are mentioned some key design considerations:

Stability Analysis

Stability analysis ensures that the sigma-delta modulator remains stable under all operating conditions. This includes analyzing the loop stability, stability margins, and stability criteria, such as the Nyquist criterion. Stability analysis helps prevent instability, oscillations, and limit cycles in the feedback loop.

Non-idealities and Error Sources

Understanding and mitigating non-idealities, such as capacitor mismatch, amplifier offset, and finite amplifier gain, is critical. Error sources, such as quantization noise, clock jitter, and thermal noise, should be minimized through careful design and calibration. Calibration techniques may be employed to compensate for non-idealities and improve overall performance.

Circuit Implementation Techniques

Choosing appropriate circuit topologies, transistor sizing, and biasing schemes is essential for achieving desired performance metrics. Circuit implementation techniques, such as switched-capacitor circuits, chopper stabilization, and low-noise amplifiers, may be employed to improve performance and reduce noise.

Linearity and Distortion

Ensuring high linearity and low distortion is critical for accurate signal reproduction. Techniques, such as dynamic element matching, digital calibration, and dynamic range enhancement, may improve linearity and reduce distortion [31].

Power Consumption

Minimizing power consumption is essential, especially in battery-powered and portable applications. Low-power design techniques, such as power gating, voltage scaling, and dynamic power management, may be employed to reduce power consumption while maintaining performance.

Clocking and Timing

Proper clocking and timing are essential for the operation of sigma-delta ADCs. Clock jitter, phase noise, and clock distribution should be carefully considered to ensure accurate sampling and conversion.

Layout and Parasitic Aspects

Layout considerations, such as minimizing parasitic capacitances, inductances, and resistances, are crucial for maintaining signal integrity and minimizing noise. Proper grounding, shielding, and isolation techniques should be employed to reduce crosstalk and interference.

Technology Scaling

With advancements in semiconductor technology, scaling sigma-delta ADC designs to smaller process nodes may offer benefits, such as improved performance, reduced power consumption, and increased integration.

By addressing these design considerations, designers can develop sigma-delta ADCs that meet the requirements of their target applications while achieving high performance and efficiency.

OVERSAMPLING TECHNIQUES

Advancements in oversampling techniques have enabled the development of sigma-delta ADCs with higher resolution and improved dynamic range. By increasing the oversampling ratio and optimizing digital filtering algorithms, designers can achieve superior noise shaping and suppression of quantization noise. Oversampling techniques have been a cornerstone of sigma-delta ADC development, enabling higher resolution and improved dynamic range. Sigma-delta ADCs improve resolution by increasing the number of quantization levels by sampling the input data at a rate that is much greater than the Nyquist rate. By sampling at a higher rate, these ADCs can achieve finer quantization of the input signal, leading to higher resolution and improvement. By using noise shaping techniques, oversampling enables sigma-delta ADCs to push quantization noise outside of the signal band and into higher frequencies. This reduces noise within the desired frequency range, enlightening the overall signal-to-noise ratio and dynamic range.

The optimization of the trade-offs between resolution, speed, and power consumption has been the main focus of recent developments in oversampling approaches [32]. Techniques, such as noise shaping, multi-bit quantization, and digital filtering, are employed to achieve superior performance [33]. Oversampling techniques are used in various domains, including audio processing, telecommunications, and instrumentation. In audio applications, oversampling ADCs capture subtle nuances in music and voice signals, resulting in high-fidelity audio [34].

Researchers and engineers continue to push the boundaries of sigma-delta ADC performance by continuously improving oversampling techniques, enabling applications that demand high resolution, accuracy, and dynamic range.

DIGITAL DECIMATION FILTERS

Digital decimation filters play a crucial role in sigma-delta ADCs by reducing the sampling rate of the oversampled signal, while preserving signal integrity. Recent advancements in digital filter design have led to more efficient and compact implementations, allowing for higher-order decimation with reduced power consumption and latency.

Digital decimation filters play a crucial role in sigma-delta ADCs by reducing the sampling rate of the oversampled signal, while preserving signal integrity. These filters work to downsample the high-rate signal following the modulator step and are usually implemented in the digital domain [21].

Decimation is essential for sigma-delta ADCs to achieve the desired output data rate, while maintaining high resolution and dynamic range. Decimation filters enable efficient data processing and reduce the burden on downstream digital circuitry by lowering the sampling rate.

The design of digital decimation filters involves several considerations, including stopband attenuation, passband ripple, phase response, and computational complexity. In order to achieve performance requirements while limiting resource use, advanced filter design techniques are used, such as Finite Impulse Response (FIR) and Infinite Impulse Response (IIR) designs.

Recent advancements in digital filter design have focused on improving filter performance, efficiency, and flexibility. Optimized coefficient quantization, distributed arithmetic, and polyphase structures are utilized to achieve high-performance decimation with reduced hardware complexity and power consumption [35 - 37].

Digital decimation filters find applications in various domains, including wireless communication, audio processing, and sensor networks. In wireless communication systems, decimation filters extract baseband signals from the Intermediate Frequency (IF) or Radio Frequency (RF) band [30], enabling demodulation and further processing.

By advancing digital decimation filter design, researchers and engineers are enhancing the overall performance and efficiency of sigma-delta ADCs, enabling high-speed and high-resolution analog-to-digital conversion in various applications [37, 38].

APPLICATIONS OF SIGMA-DELTA ADCs

Because they offer high-precision, high-resolution analog-to-digital conversion at low power consumption and decreased sensitivity to analog flaws, sigma-delta ADCs are widely employed in a variety of areas.

Communications

Sigma-delta ADCs are commonly employed in communication systems for wireless transceivers' baseband and Intermediate Frequency (IF) signal processing. They are used in applications, such as digital radio, Software-defined Radio (SDR), cellular networks, and broadband communication systems. Sigma-delta ADCs' high resolution and low noise characteristics make them suitable for capturing weak signals in noisy environments.

Sigma-delta ADCs are essential parts of many communication systems because they make it easier to digitize analog data for processing and sending. Their high resolution, low noise, and efficient oversampling techniques make them well-suited for applications in both wired and wireless communications.

Wireless Transceivers

In wireless communication systems, sigma-delta ADCs are commonly used in transmitter and receiver paths. They play a crucial role in digitizing baseband signals for modulation and demodulation. In transmitters, sigma-delta ADCs convert digital baseband signals into analog signals for transmission, while in receivers, they convert received analog signals back into digital form for further processing [39].

Software-defined Radio (SDR)

Sigma-delta ADCs are critical components in Software-defined Radio (SDR) platforms, enabling flexible and reconfigurable radio systems. SDR systems rely on digital signal processing techniques to perform modulation, demodulation, and other communication functions in software. Sigma-delta ADCs digitize the antenna's RF signals, allowing software-based processing and demodulation.

Cellular Networks

Sigma-delta ADCs are utilized in cellular networks in base stations, mobile devices, and other network equipment. They digitize voice, data, and control signals in uplink and downlink channels. Sigma-delta ADCs enable efficient signal processing and modulation techniques, contributing to cellular networks' overall performance and capacity.

Broadband Communication Systems

Sigma-delta ADCs are employed in broadband communication systems, such as DSL, cable, and fiber-optic transceivers. They digitize analog signals at high speeds and with high resolution, enabling the transmission of voice, video, and data over long distances.

By providing accurate and efficient analog-to-digital conversion, sigma-delta ADCs play a vital role in enabling reliable, high-performance communication systems across various applications.

Sensor Interfaces

In environmental monitoring applications, sigma-delta ADCs are used in sensor interfaces to measure humidity, air quality, radiation, and pollution levels. They convert analog sensor signals into digital form, allowing for continuous monitoring and analysis of environmental conditions. $\sum\Delta$ADCs offer high resolution and low power consumption, making them suitable for remote and battery-powered sensor networks.

Medical Devices

Sigma-delta ADCs are vital in medical sensor interfaces for patient monitoring, diagnostic equipment, and medical imaging. They digitize signals from sensors, such as Electrocardiogram (ECG) electrodes, blood glucose monitors, temperature sensors, and imaging sensors. Sigma-delta ADCs provide accurate and reliable measurement of physiological signals, enabling healthcare professionals to make informed decisions and diagnoses.

By enabling precise and reliable digitization of sensor signals, sigma-delta ADCs contribute to the advancement of sensor technology and its applications in various industries.

Audio Processing

Sigma-delta ADCs are widely utilized in audio processing applications, including professional audio equipment, consumer audio devices, and digital audio recording systems, because of their capacity to provide high-fidelity audio reproduction with little noise and distortion. Sigma-delta ADCs offer advantages, such as high dynamic range, low noise, and superior linearity, making them suitable for capturing the subtle nuances of music and voice.

Professional Audio Equipment

Sigma-delta ADCs are used in high-end audio interfaces and digital audio workstations in professional audio applications, such as recording studios. They digitize analog audio signals with exceptional accuracy and resolution, preserving the nuances and dynamics of the original sound source. Sigma-delta ADCs offer high dynamic range, low distortion, and superior linearity, making them ideal for capturing the subtle details of music and voice recordings.

Consumer Audio Devices

In consumer electronic devices, such as smartphones, tablets, and digital audio players, sigma-delta ADCs are integral components for audio input and output functionalities. They are used to digitize analog audio signals from microphones, line inputs, and instrument inputs, and generate digital audio outputs for headphones, speakers, and external audio systems. Sigma-delta ADCs enable high-quality audio recording, playback, and processing in compact, energy-efficient devices.

Audio Interfaces and Sound Cards

Sigma-delta ADCs are also employed in audio interfaces, sound cards, and audio processing units for professional and consumer applications. They serve as the interface between analog audio sources and digital audio processing systems, allowing for real-time audio recording, mixing, and editing. Sigma-delta ADCs offer low latency, high-resolution conversion, and low noise performance, ensuring pristine audio quality in demanding production environments.

High-resolution Audio

The need for high-resolution audio formats, like Direct Stream Digital (DSD) and FLAC (Free Lossless Audio Codec), is growing, and sigma-delta ADCs are essential for recording and maintaining the audio signal's complete quality. They enable the digitization of analog audio signals with sampling rates and bit depths exceeding those of standard CD audio, resulting in a more immersive and realistic listening experience.

By delivering high-performance analog-to-digital conversion, sigma-delta ADCs contribute to the advancement of audio technology and the enjoyment of high-quality audio content across various platforms and devices.

Medical Instrumentation

In medical instrumentation, sigma-delta ADCs are employed for various purposes, including medical imaging, patient monitoring, and diagnostic equipment. They are used in ultrasound machines, MRI systems, ECG, and blood glucose monitors. The high-resolution and low-noise performance of sigma-delta ADCs enables accurate measurement and analysis of physiological signals.

By catering to diverse applications, sigma-delta ADCs enable advanced signal processing and data acquisition in modern electronic systems.

Sigma-delta ADCs play a vital role in various medical instrumentation applications, contributing to the accurate measurement and analysis of physiological signals in healthcare settings.

Medical Imaging

When using medical imaging tools, like Computed Tomography (CT), Magnetic Resonance Imaging (MRI), and ultrasound, sigma-delta ADCs digitize analog signals from sensors, such as transducers and detectors. They enable converting raw sensor data into digital images for visualization and diagnosis. Sigma-delta ADCs offer high resolution, low noise, and wide dynamic range, making them suitable for capturing fine details and subtle abnormalities in medical images.

Patient Monitoring

The ECG, EEG, EMG, and blood pressure are among the physiological data that sigma-delta ADCs digitize for use in patient monitoring systems found in clinics, hospitals, and home healthcare settings. They enable continuous monitoring of vital signs and patient parameters, allowing healthcare professionals to assess patient health and detect abnormalities in real-time. $\sum\Delta$ADCs offer high precision, low power consumption, and immunity to common-mode interference, ensuring reliable and accurate measurement of physiological signals.

Diagnostic Equipment

To measure and analyze patient samples and parameters, $\sum\Delta$ADCs are used in various diagnostic equipment, such as blood glucose meters, spirometers, and pulse oximeters. They make it possible to transform analog data from transducers and sensors into digital data for analysis and diagnosis. Sigma-delta ADCs are appropriate for applications demanding quick and precise measurements because of their high resolution, low noise, and quick response times.

Research Instruments

In research laboratories and academic institutions, sigma-delta ADCs are utilized in scientific instruments and experimental setups to acquire and process analog signals in biomedical research. They enable researchers to collect data from sensors, probes, and experimental setups with high precision and fidelity. Sigma-delta ADCs offer flexibility, programmability, and compatibility with various sensors and transducers, making them versatile tools for biomedical research applications.

By providing accurate and reliable analog-to-digital conversion, sigma-delta ADCs contribute to advancing medical technology and improving patient care in healthcare settings.

CONCLUSION

In conclusion, sigma-delta ADCs have emerged as essential components in modern electronic systems, offering high-resolution, low-noise analog-to-digital conversion across various applications. Throughout this review, we have explored the fundamental principles, architectures, design considerations, and performance characteristics of sigma-delta ADCs.

We have begun by providing an overview of ADCs and then highlighted the importance of sigma-delta ADCs in achieving high resolution and dynamic range. We have discussed the basic principles of $\sum\Delta$ modulation, including oversampling and noise shaping techniques, which are instrumental in improving ADC performance.

Furthermore, we have examined various architectures of $\sum\Delta$ADCs, including first-order, higher-order, and multi-bit converters, each offering unique advantages and trade-offs. Design considerations, such as stability analysis, non-idealities, and circuit implementation techniques, have been discussed to ensure robust and reliable ADC operation.

Performance analysis has been focused on noise shaping, dynamic range, resolution, accuracy, speed, and power consumption, emphasizing the trade-offs in optimizing ADC performance for specific applications. Recent advancements and trends in oversampling techniques, digital decimation filters, and calibration methods have been highlighted, indicating the continuous evolution of sigma-delta ADC technology.

We have also explored these ADCs' applications in communications, sensor interfaces, audio processing, and medical instrumentation, demonstrating their versatility and widespread adoption across various industries.

In conclusion, sigma-delta ADCs continue to drive innovation and enable advancements in electronic systems, offering high-performance analog-to-digital conversion for a diverse range of applications. As technology progresses, further research and development efforts are expected to enhance the capabilities and expand the applications of sigma-delta ADCs in the future.

Through their contributions to signal processing, data acquisition, and measurement systems, sigma-delta ADCs play a pivotal role in advancing technology and shaping the future of electronic engineering.

KEY POINTS

1. Sigma-delta ADCs employ oversampling and noise-shaping methods to attain high resolution and dynamic range.
2. Sigma-delta ADCs can be implemented in various architectures, including first-order, higher-order, and multi-bit converters, each with advantages and trade-offs.
3. Stability analysis, non-idealities, and circuit implementation techniques are crucial for ensuring reliable ADC operation.
4. With trade-offs between these parameters, sigma-delta ADC performance is evaluated based on noise shaping, dynamic range, resolution, accuracy, speed, and power consumption.
5. Advancements in oversampling techniques, digital decimation filters, and calibration methods have improved sigma-delta ADC performance and efficiency.
6. Sigma-delta ADCs find applications in communications, sensor interfaces, audio processing, and medical instrumentation, enabling accurate signal processing and data acquisition across various industries. Sigma-delta ADCs play a vital role in modern electronic systems, offering high-performance analog-to-digital conversion for multiple applications.

ACKNOWLEDGEMENTS

The authors would like to thank the Department of Electronics and Communication Engineering, Dr. B.R Ambedkar National Institute of Technology, Jalandhar, Punjab, for providing valuable support to carry out this study.

REFERENCES

[1]   S. Pavan, R. Schreier, and G.C. Temes, "Understanding delta-sigma data converters", 2017. [Online]. Available from: http://www.mathworks.com/matlabcentral/fileexchange/

[2]   W. Kester, "ADC architectures III: sigma-delta ADC basics MT-022 tutorial ADC architectures III: sigma-delta   ADC   basics",   [Online]   2011.   Available   from: https://www.researchgate.net/publication/239569030

[3]   "Implementation Considerations for ADCs | part of understanding delta-sigma data converters | wiley-IEEE Press books | IEEE xplore", Accessed: Apr. 23, 2024. [Online]. Available from: https://ieeexplore.ieee.org/document/5264510

[4]   E. T. King, A. Eshraghi, I. Galton, and T. S. Fiez, "A Nyquist-rate delta-sigma A/D converter", *in IEEE Journal of Solid-State Circuits,* vol. 33, no. 1, pp. 45-52, 1998.
[http://dx.doi.org/10.1109/4.654936]

[5]   K. Kang, "Simulation, and overload and stability analysis of continuous simulation, and overload and stability analysis of continuous time sigma delta modulator time sigma delta modulator", 2014.
[http://dx.doi.org/10.34917/7048593]

[6]   A. Gupta, "Design and implementation of efficient low power sigma delta ADC", *Ph.D. dissertation,* 2020.

[7]   A. Nowacki, "Design of sigma-delta modulators for a-to-digital conversion intensively using passive circuits", *Ph.D. dissertation,* 2016.

[8]   S.-S. C. C. S. IEEE, and and E. E. I. of E., "1999 IEEE international solid-state circuits conference", *IEEE,* 1999.

[9]   B. Hoefflinger, "ITRS: The international technology roadmap for semiconductors", In: *Springer,* 2011.Berlin, Heidelberg

[10]   E. Troels, and E. E. Kolding, "Review of RF CMOS performance and future process innovations", *Unpublished,* 1998.

[11]   S. Tsukamoto, W. G. Schofield, and T. Endo, "A CMOS 6-b, 400-M sample/s ADC with error correction", *Unpublished,* 1998.

[12]   Galton, "Delta-sigma data conversion in wireless transceivers", In: *Unpublished,* 2002.

[13]   S. Pavan, "Alias rejection of continuous-time $\Delta\Sigma$ modulators with switched-capacitor feedback DACs", *IEEE Transactions on Circuits and Systems I: Regular Papers,* vol. 58, no. 2, pp. 233-243, 2011.
[http://dx.doi.org/10.1109/TCSI.2010.2071930]

[14]   S. Pavan, R. Schreier, and G. C. Temes, "Understanding delta-sigma data converters", In: *Unpublished,* 2017.

[15]   M. de la Rosa, and R. del Rio, "CMOS sigma-delta converters: Practical design guide", In: *Wiley-IEEE Press,* 2013.

[16]   A. Applications Journal, and B. Baker, A. Applications Journal and B. Baker, "How delta-sigma ADCs work, Part 1", 2011. [Online]. Available from: www.ti.com/aaj

[17]   A. Cherry, and W.M. Snelgrove, *Snelgrove, Continuous-Time Delta-Sigma Modulators for High-Speed A/D Conversion*. 1st ed. Springer New York: NY, 2006.

[18]   E. Jackson, "Optimal design of discrete-time delta sigma modulators", *Master's thesis, University of Nevada, Las Vegas, 2009*.
[http://dx.doi.org/10.34917/1354531]

[19]   R. Schreier, and G.C. Temes, "Implementation considerations for ADCs", *Understanding Delta-Sigma Data Converters*, no. Feb, 2010.
[http://dx.doi.org/10.1109/9780470546772.ch6]

[20]   Park Sangil, "Motorola digital signal processors principles of sigma-delta modulation for analog-to-digital converters", *Motorola*, 1993.

[21]   *Pelgrom, Analog-to-Digital Conversion*. 3rd ed. Springer Cham, 2016.

[22]   B. Farahani, and M. Ismail, "Ismail Adaptive digital techniques to suppress quantization noise of $\Sigma\Delta$ analog to digital converters", *Proceedings of the 15th ACM Great Lakes symposium on VLSI - GLSVSLI 05 GLSVSLI 05,*, 2005.

[23]   L. Yao, M. Steyaert, and W.M.C. Sansen, *Low power low voltage sigma delta modulators in nanometer CMOS*. vol. 868. Springer Science & Business Media, 2006.

[24]   M. Ortmanns, F. Gerfers, and Y. Manoli, "Compensation of finite gain-bandwidth induced errors in continuous-time sigma-delta modulators", *IEEE Trans. Circ. Syst. I Fundam. Theory Appl.*, vol. 51, no. 6, pp. 1088-1099, 2004.
[http://dx.doi.org/10.1109/TCSI.2004.829234]

[25]   S.S. Panda, "A 15 Bit third order power optimized continuous time sigma delta modulator for audio applications." [Online] 99-103, 2014. Available from: http://www.ijret.org

[26]   C.-W. Lo, and H. C. Luong, "A 1.5-V 900-MHz monolithic cmos fast-switching frequency synthesizer for wireless applications", 2002.

[27]   F. Maloberti, *Data Converters*. 1st ed. Springer New York: NY, 2007.

[28]   T. Caldwell, D. Alldred, and Z. Li, *A Reconfigurable $\Delta\Sigma$ Modulator with up to 100MHz Bandwidth using Flash Reference Shuffling*. IEEE, 2013.

[29]   S. Nadeem, C. G. Sodini, and H.-S. Lee, "6-channel oversampled analog-to-digital converter", 1994.
[http://dx.doi.org/10.1109/4.309903]

[30]   A. Rusu, B. R. J, M. Ismail, and H. Tenhunen,, "Linearity enhancement in a configurable sigma-delta modulator", In: *in IEEE NEWCAS 2005, Quebec, Canada*, 2004, pp. 673-676.

[31]   J.M. Rosa, B. Pérez-Verdú, and A. Rodríguez-Vázquez, *Systematic Design of CMOS Switched-Current Bandpass Sigma-Delta Modulators for Digital Communication Chips*. 1st ed. Springer New York: NY, 2007.

[32]   A. Rodríguez-Vázquez, F. Medeiro, E. Janssens, Ed., *CMOS Telecom Data Converters*. 1st ed. Springer New York: NY, 2010.

[33]   J.M. de la Rosa, R. Schreier, K.P. Pun, and S. Pavan, "Next-generation delta-sigma converters: trends and perspectives", *IEEE J. Emerg. Sel. Top. Circuits Syst.*, vol. 5, no. 4, pp. 484-499, 2015.
[http://dx.doi.org/10.1109/JETCAS.2015.2502164]

[34]   S. Pavan, "Continuous-time Delta-Sigma modulator design using the method of moments", *IEEE Trans. Circuits Syst. I Regul. Pap.*, vol. 61, no. 6, pp. 1629-1637, 2014.
[http://dx.doi.org/10.1109/TCSI.2013.2290846]

[35]   J.M. de la Rosa, "Sigma-delta modulators: Tutorial overview, design guide, and state-of-the-art survey", *IEEE Trans. Circuits Syst. I Regul. Pap.*, vol. 58, no. 1, pp. 1-21, 2011.
[http://dx.doi.org/10.1109/TCSI.2010.2097652]

[36]   P. Singh, and D.S. Yadav, "Performance analysis of ITCs on analog/RF, linearity and reliability performance metrics of tunnel FET with ultra-thin source region", *Appl. Phys., A Mater. Sci. Process.,* vol. 128, no. 7, p. 612, 2022.
[http://dx.doi.org/10.1007/s00339-022-05741-4]

[37]   B. Leung, "Brief papers BiCMOS current cell and switch for digital-to-analog converters", 1993.

[38]   B. Murmann, "ADC performance survey 1997-2010", *Online (Bergh.),* 2010.

[39]   X. Chen, A wideband low-power continuous-time delta-sigma modulator for next generation wireless applications, 2007.Ph.D. dissertation, 2007.

Semiconductor Nanoscale Devices, 2025, 337-359

Photovoltaic Performance Estimation of Thin Film Lateral Pn-Junction Solar Devices and Comprehensive Consideration of Performances of Various Homo- and Hetero-Junction Structures

Yasuhisa Omura[1,*]

[1] *ORDIST, Kansai University, Suita, Osaka 564-8680, Japan*

Abstract: This chapter provides a practical theoretical foundation and perspective on the performance of various thin-film lateral pn-junction solar devices under illumination. It focuses on Si- and Ge-based homo-junctions, as well as ZnO/Si (Type-I) and GaN/Si (Type-II) based hetero-junctions. Theoretical models assume poly-crystalline or amorphous semiconductor materials. The study demonstrates that highly-doped Si- and Ge-based homo-junction architectures show great promise for high-performance solar devices. By utilizing published experimental results, the predicted performances of homo-junction and hetero-junction solar devices are primarily compared at room temperature. Additionally, the chapter addresses the behaviors of these devices at low and high temperatures, considering various applications. The results reveal the superiority of Si- and Ge-based homo-junctions. The chapter delves into the theoretical aspects, providing a robust understanding of the principles governing the performance of these solar devices. It evaluates the advantages and challenges associated with each type of junction, offering a comprehensive analysis of their operational efficiencies. Through detailed comparisons and analysis, the study underscores the potential of Si- and Ge-based homo-junctions in advancing solar technology. This investigation into the practical and theoretical aspects of thin-film lateral pn-junction solar devices serves as a valuable resource for understanding their performance under different conditions. It highlights the critical role of material selection and doping strategies in optimizing device efficiency, paving the way for future research and development of high-performance solar technologies.

Keywords: - Lateral pn junction, Ge, GaN, Photovoltaic performance, Si, Thin film, ZnO.

---

* **Corresponding author Yasuhisa Omura:** ORDIST, Kansai University, Suita, Osaka 564-8680, Japan; E-mail: omuray@kansai-u.ac.jp

INTRODUCTION

A pn-junction-based optical rotor operating under external illumination was proposed [1]. Though the device was proposed as a mechanical part for MEMS (Micro Electro Mechanical System) applications [2], its full potentiality has yet to be examined in experiments. Since many MEMS applications [3] must have very low-power operation, because most must work with very small batteries, we need advanced mobile battery-powered devices, not AC power supplies. One possible way is to make a built-in battery that is charged by an external energy source, like optical illumination.

Combinations of various flexible electronics technologies and solar batteries are now being investigated in order to support new advanced applications [4 - 7] because they have high potential for future medical implantation applications [8] and sensor-network applications for the IoT society [9], as well as space applications [10]. Since low-temperature deposition techniques, like mist CVD method [11 - 14], have recently been developed, the above device applications are seen as real targets for the electronics industry. The potential of lateral pin diodes has recently been investigated in order to attain high-performance photodiodes [15 - 18], and the influence of temperature and device thickness is being studied. However, their future potential has yet to be discussed.

In this paper, the author reconsiders the theoretical base for photovoltaic performance estimation of thin film lateral pn-junction solar devices and comprehensively discusses the performances of homo- and hetero-junction devices because it is anticipated that such lateral pn-junction solar devices are much more robust against local damage than the conventional vertical junction structure. In addition, the author reveals by theoretical predictions that the carrier diffusion behaviors of sub-100-nm-thick Si films are quite insensitive to carrier lifetime values [2]. Such characteristic is needed in future various flexible electronics. Though the primary theoretical base was given in an earlier work [19], the physical conditions implicit in the discussion were not deeply considered. Accordingly, this paper revisits the theoretical formulation in order to widen the consideration [20, 21], and some additional theoretical concepts are introduced for the hetero-junction structures and numerically evaluated [20]. Here, various calculations are performed for Si- and Ge-based homo-junction devices and ZnO/Si (Type-I) and GaN/Si (Type-II) based hetero-junction devices [6, 7, 22 - 25] from the viewpoint of future applications. In addition, device design issues are also discussed.

DEVICE STRUCTURES ASSUMED

Though some advanced proposals and experiments have recently been published [5 - 7, 23 - 27], some of them are based on chalcogenide materials because of their promising future potential [5, 20]. On the other hand, this study has paid attention to the ubiquitous society of solar battery devices with their low costs and robustness from the viewpoint of industrial application. So, only well-known materials are assumed in simulations.

A simplified physical image of the thin film lateral pn junction solar device is shown in Fig. (**1**), where the top view of the device is given in Fig. (**1a**), the cross-sectional view of Si- or Ge-based homo-junction device is shown in Fig. (**1b**), that of the n-ZnO/p-Si (or n-GaN/p-Si) hetero-junction device is shown in Fig. (**1c**), and that of the multi-stack architecture is shown in Fig. (**1d**); it is assumed that green light, with a wavelength of 550 nm, directly illuminates to the sheet's top surface [1]. Here, we assume the coordinate system, as shown in Fig. (**1a**).

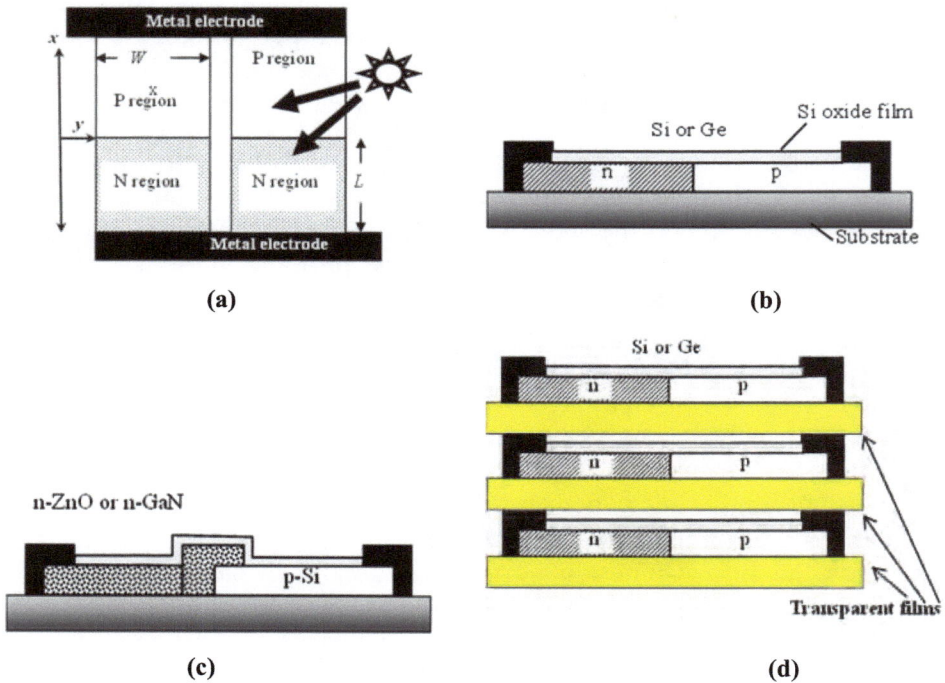

Fig. (1). Schematic representation of the device structure. (**a**) Top view, (**b**) cross-sectional view of Si pn-junction or Ge pn-junction, (**c**) n-ZnO/p-Si or n-GaN/p-Si hetero-pn-junction, (**d**) schematic view of a multi-stack structure.

This paper assumes dimensions of devices, as summarized in Table **1**, and physical parameters are shown in Tables **2-4**, where very short carrier lifetimes are assumed because future potential applications must consider the poly-crystalline or amorphous quality of the film [10]. The following mathematical analyses are based on Fermi-Dirac statistics because high-doping levels are assumed in some devices at room temperature and low temperatures.

Table 1. Device parameters of the control device.

Parameters	Values	Units
Length of n-region and p-region (L)	10.0	μm
Width of n-region and p-region (W)	10.0	μm
Thickness (t_s)	100.0	nm
Doping level of n-region (N_D)	$1.0 \times 10^{19} \sim 10^{20}$	cm^{-3}
Doping level of p-region (N_A)	$1.0 \times 10^{19} \sim 10^{20}$	cm^{-3}

Table 2. Physical parameters of the control device (Si and Ge).

Parameters	Values	Unit
Lifetime of electrons (τ_{n0})	10.0	ns (@300K)
Lifetime of holes (τ_{p0})	10.0	ns (@300K)
Electron mobility (μ_n)	-	-
Si	140	cm^2/Vs (@300K)
Ge	390	cm^2/Vs (@300K)
Hole mobility (μ_p)	-	-
Si	50	cm^2/Vs (@300K)
Ge	190	cm^2/Vs (@300K)
Coefficient of temperature dependence of mobility	-	-
γ_n (Si)	2.42	--
γ_p (Si)	2.20	--
γ_n (Ge)	2.30	--
γ_p (Ge)	1.60	--
Illumination source power density (green light)	100.0	mW/cm^2
Wavelength of light (λc)	0.55	μm
Reflectance (R)	0.5	--
Bandgap parameters	-	-
α (Si)	4.73×10^{-4}	--
β (Si)	636.0	K

(Table 2) cont.....

α (Ge)	4.774x10⁻⁴	--
β (Ge)	235.0	K

Table 3. Physical parameters of ZnO films.

Parameters	Values	Unit
Electron mobility	-	-
ZnO	20	cm²/Vs(@300K)
Optical reflectance	0.75	-
Absorption length	(λ=0.55 μm)	-
	5.0	μm
Bandgap parameters	-	-
α	2.0x10⁻⁴	-
β	325.0	K

Table 4. Physical parameters of GaN films.

Parameters	Values	Unit
Electron mobility	-	-
GaN	200	cm²/Vs(@300K)
Optical reflectance	0.80	-
Absorption length	(λ=0.55 μm)	-
	0.95	μm
Bandgap parameters	-	-
α	9.39x10⁻⁴	-
β	772.0	K

THEORETICAL BASE

Preparing the Theoretical Procedure

For Si- and Ge-based homo-junctions, lateral one-dimensional transport is assumed due to the thickness of tens nanometers, where it is assumed that the parameter values are uniform over the cross-sectional area, and it is also assumed that the top and bottom surfaces of the film are well passivated; the surface recombination effect is not assumed here for simplicity. Thus, the hole generation and recombination process in the p-type region under illumination generally follows the continuity equation shown below:

$$\frac{\partial p_p(x,t,T)}{\partial t} = G_p - \frac{p_p(x,t,T) - p_{p0}(T)}{\tau_p} - p_p(x,t,T)\mu_p \frac{\partial F}{\partial x} - \mu_p F \frac{\partial p_p(x,t,T)}{\partial x} + D_p \frac{\partial^2 p_p(x,t,T)}{\partial x^2} \quad (1)$$

Where, $p_{p0}(T)$ denotes the hole concentration at thermal equilibrium, G_p denotes the generation rate of holes, τ_p denotes the lifetime of holes, F denotes the local electric field, μ_p denotes the hole mobility, and D_p denotes the diffusion constant of holes. In this formulation, it is assumed that holes generated in the n-type region are injected into the p-type region; that is, for the Si homo-pn-junction, the excess hole density $P_J(0,T)$ at the junction interface exhibits the following relation:

$$P_J(0,T) = p_n(0,T) - p_{n0}(T) - G_p \tau_p \quad (2)$$

Where, $p_n(0,T)$ is the hole density value at the junction interface. This is the sum of the hole density generated in the n-type region under illumination and its intrinsic value in the n-type region, where $p_{n0}(0,T)$ is the hole density in the n-type region at thermal equilibrium without illumination. A similar equation is made for the electron density in the n-type region as:

$$\frac{\partial n_n(x,t,T)}{\partial t} = G_n - \frac{n_n(x,t,T) - n_{n0}(T)}{\tau_n} - n_n(x,t,T)\mu_n \frac{\partial F}{\partial x} - \mu_n F \frac{\partial n_n(x,t,T)}{\partial x} + D_n \frac{\partial^2 n_n(x,t,T)}{\partial x^2} \quad (3)$$

Where, every notation has a counterpart for electrons.

In this paper, equation (3) is solved using the same mathematical techniques as applied to equation (1) because it has the same form as eq. (1). Since the electron density in the p-type region is very low, Poisson's equation in the p-type region can be approximated as:

$$\frac{\partial F(x,T)}{\partial x} \cong \frac{q}{\varepsilon_S}\left[p_p(x,T) - p_{p0}(T)\right] \quad (4)$$

Where, q denotes the elemental charge and ε_s is the permittivity of the semiconductor material. When this form is not inserted into eq. (1) as an approximation, equation (1) cannot be solved. So, this approximation is needed if equation (1) is to be solved.

Previous papers [19 - 21] have assumed the film to have bulk crystalline quality with a low-doping density of around 10^{15} cm^{-3}, where only the local electric field in the p-type Si region was considered. This yielded the approximate form of

$$F(x,T) \cong \frac{qL_{Dp}}{\varepsilon_S}\left[p_p(x,T)-p_{p0}(T)\right] \tag{5}$$

Where, L_{Dp} is the Debye length of holes in the p-type region. The theoretical formulation under this condition has already been presented earlier [20]. Consequently, the performances of Si-based and Ge-based lateral solar panels were not so good. However, this approximation is not reasonable for practical situations because it is considered that the final solution of the excess hole density takes the exponential function with the diffusion length of the film. Therefore, in this paper, it is assumed that the local electric field in the p-type Si region, for example, takes the approximate form of:

$$F(x,T) \cong \frac{qL_p}{\varepsilon_S}\left[p_p(x,T)-p_{p0}(T)\right] \tag{6}$$

Where, L_p is the diffusion length of holes in the p-type region. When the device is fabricated on a flexible film, we have to assume that the Si film, for example, has poly-crystalline or amorphous electronic properties, so a very short carrier lifetime and small carrier mobility must be assumed. This could result in smaller diffusion lengths of carriers. Even in this case, it is expected that the diffusion process would be dominant, so equation (6) is valid for a theoretical discussion.

When equation (1) was analytically solved using eq. (6), the semiconductor film was assumed to have infinite length for simplicity in the previous papers [19 - 21]. At the steady state, equation (1) is thus rewritten as:

$$D_p \frac{\partial^2 p_p(x,T)}{\partial x^2}-\left(\frac{eL_p\mu_p}{\varepsilon_S}\right)\frac{\partial p_p(x,T)\left[p_p(x,T)-p_{p0}(T)\right]}{\partial x}-\frac{p_p(x,T)-p_{p0}(T)-G_p\tau_p}{\tau_p}=0, \tag{7}$$

Here, we have introduced the following parameter change in the p-type region.

$$P(x,T)=p_p(x,T)-p_{p0}(T)-G_p\tau_p, \tag{8}$$

Equation (7) takes the form of:

$$\frac{\partial^2 P(x,T)}{\partial x^2}-\left(\frac{2eL_p\mu_p}{D_p\varepsilon_S}\right)P(x,T)\frac{\partial P(x,T)}{\partial x}-\left(\frac{eL_p\mu_p}{D_p\varepsilon_S}\right)\left(p_{p0}(T)+2G_p\tau_p\right)\frac{\partial P(x,T)}{\partial x}-\frac{P(x,T)}{D_p\tau_p}=0, \tag{9}$$

This equation can be represented as:

$$\frac{\partial^2 P(x,T)}{\partial x^2} - A_p P(x,T) \frac{\partial P(x,T)}{\partial x} - B_p \frac{\partial P(x,T)}{\partial x} - C_p P(x,T) = 0, \tag{10}$$

$$A_P = \frac{2eL_p\mu_p}{D_p\varepsilon_S}, \tag{11}$$

$$B_P = \left(\frac{eL_p\mu_p}{D_p\varepsilon_S}\right)\left(p_{p0}(T) + 2G_p\tau_p\right), \tag{12}$$

$$C_P = \frac{1}{D_p\tau_p}. \tag{13}$$

In order to attain a practical solution, a slight approximation is introduced here. When the doping level is high or the optical illumination is strong, the following approximation can be introduced (***Appendixes A to C***).

$$A_p P(x,T) + B_p \rightarrow A_p P(x,T), \tag{14}$$

Therefore, equation (10) can be rewritten as:

$$\frac{\partial^2 P(x,T)}{\partial x^2} - A_p P(x,T) \frac{\partial P(x,T)}{\partial x} - C_p P(x,T) = 0, \tag{15}$$

Details of the mathematical technique used to solve eq. (15) are described in ***Appendixes A*** and ***B***.

Under steady-state light illumination, the condition of $P_{00} \ll N_A$ can be assumed for P_{00}, as given in ***Appendix A***. Equation (A10) can be reduced to:

$$\int \frac{dP}{P^2(x,T) \pm 2\frac{\sqrt{2C_P}}{A_P}P(x,T)} = \int \frac{A_P}{2} dx, \tag{16}$$

Then, we have for $x>0$ (Fig. **1**).

$$P(x,T) = \frac{2\frac{\sqrt{2C_P}}{A_P}\dfrac{P(0,T)}{P(0,T)+2\dfrac{\sqrt{2C_P}}{A_P}}\exp\left(-\sqrt{2C_P}\,x\right)}{1-\dfrac{P(0,T)}{P(0,T)+2\dfrac{\sqrt{2C_P}}{A_P}}\exp\left(-\sqrt{2C_P}\,x\right)} \qquad (17)$$

In a similar way, we have the solution of the excess electron density $N(x,T)$ for $x<0$ as:

$$N(x,T) = \frac{2\frac{\sqrt{2C_N}}{A_N}\dfrac{N(0,T)}{N(0,T)+2\dfrac{\sqrt{2C_N}}{A_N}}\exp\left(\sqrt{2C_N}\,x\right)}{1-\dfrac{N(0,T)}{N(0,T)+2\dfrac{\sqrt{2C_N}}{A_N}}\exp\left(\sqrt{2C_N}\,x\right)} \qquad (18)$$

Additional Model Option to Consider the Hetero-junction

Though we can obtain appropriate solutions for Si- and Ge-film pn junctions (Fig. **1b**) by solving the conventional technique with possible approximations [19], similar solutions for the device shown in Fig. (**1c**) are unavailable because we must take into account the conduction-band notch at the ZnO/Si (type-I) hetero junction (Fig. **2a**) [21]. In Fig. (**2a**), the schematic energy band diagram of the n-ZnO/p-Si junction is shown, where non-degenerate semiconductors are assumed for simplicity, ΔEc is the energy level difference between the conduction band bottom of ZnO and that of Si, and ΔEv is the energy level difference between the valence band top of ZnO and that of Si. The energy band diagram of the n-GaN/-Si (type-II) hetero-pn-junction is given in Fig. (**2b**) [28], where the non-degenerate semiconductor is assumed for simplicity.

Since degenerate semiconductors or highly doped materials are assumed in the present case, the previous technique of the simplified Boltzmann's approximation for the Poisson equation [19] does not give a precise solution. Since equation (6) gives a very good approximation of the electric field under this condition, it is used in the following calculations for highly doped semiconductor materials.

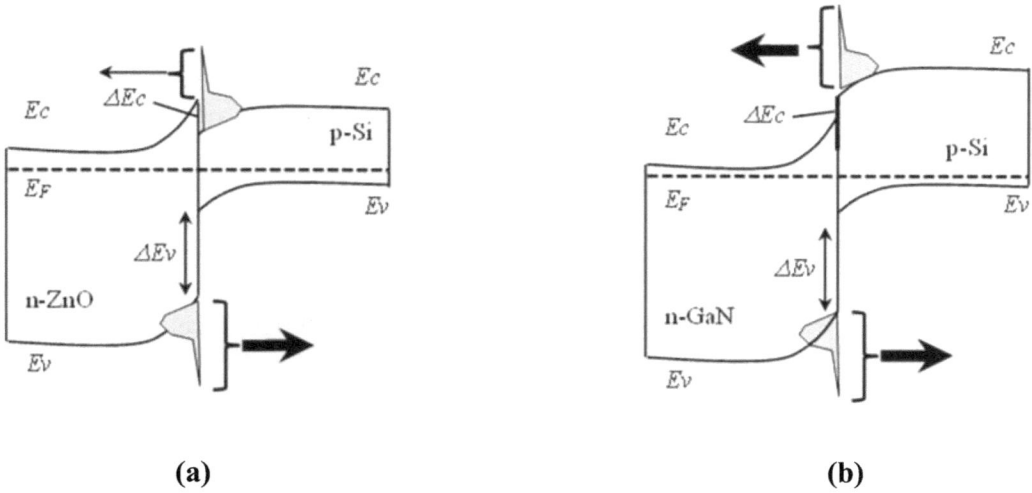

(a)   **(b)**

Fig. (2). Schematic energy band diagram for hetero-junction. (**a**) n-ZnO/p-Si hetero-pn-junction, (**b**) n-GaN/p-Si hetero-pn-junction.

For example, for the n-ZnO film, we have to use the solution for the electron density distribution function in the n-ZnO region due to the type-I hetero-junction. Using eqs. (3) and (10), we have a similar function form for the excess electron density, $N(x,T)$. At steady state, for example, the counterpart solution for the excess electron density, $N(x)$, for eq. (17) in the ZnO region ($x<0$) is expressed as:

$$N(x,T) = \dfrac{2\dfrac{\sqrt{2C_N}}{A_{Nh}}\dfrac{N(0,T)}{N(0,T)+2\dfrac{\sqrt{2C_N}}{A_{Nh}}}\exp\left(\sqrt{2C_N}\,x\right)}{1-\dfrac{N(0,T)}{N(0,T)+2\dfrac{\sqrt{2C_N}}{A_{Nh}}}\exp\left(\sqrt{2C_N}\,x\right)} \tag{19}$$

$$N(0,T) = n_{n,\mathrm{ZnO}}(0,T) - n_{n0,\mathrm{ZnO}}(T) - G_n\tau_n \tag{20}$$

$$A_{Nh} = \dfrac{2qL_n\mu_n}{D_n\varepsilon_{ZnO}} \tag{21}$$

$$C_N = \dfrac{1}{D_n\tau_n} \tag{22}$$

Where, ε_{ZnO} is ZnO film permittivity. Since the notch at the ZnO/Si interface is the obstacle to electron injection from the p-Si region, its form must be theoretically estimated. Its form is approximately given below (*Appendix C*).

$$n_{n,ZnO}\left(0,T\right) \cong n_{p,Si}\left(0,T\right)\sqrt{\frac{\Delta E_C}{k_B T}}\exp\left(-\frac{\Delta E_C}{k_B T}\right) \tag{23}$$

Where, $n_{p,Si}(0,T)$ is the electron density in the p-Si region. For the case of n-GaN/p-Si (type-II) hetero-pn-junction, eq. (23) is not needed because effectively, there is no energy barrier to electron transfer at the junction interface.

Temperature Dependences of Physical Parameters

In calculating the excess electron density and the excess hole density, the temperature dependence of various parameters must be introduced. The temperature dependence of carrier mobility for Si, Ge, GaN, and ZnO films is expressed as [6, 29 - 33]

$$\mu_n = \mu_{n0}\left(\frac{300}{T}\right)^{\gamma_n}, \tag{24}$$

$$\mu_p = \mu_{p0}\left(\frac{300}{T}\right)^{\gamma_p}. \tag{25}$$

Parameter values (μ_{n0}, μ_{p0}, γ_n, and γ_p) are given in Tables **2-4**.

In addition, the temperature dependence of the energy band gap is expressed as [29, 34, 35].

$$E_G\left(T\right) = E_G\left(0\right) - \frac{\alpha T^2}{T+\beta}. \tag{26}$$

Parameter values (α, β) are also given in Tables **2-4**. The temperature dependence of lifetime for Si, Ge, ZnO, and GaN is expressed as [29, 36, 37].

$$\tau_n = \tau_{n0}\exp\left[-0.25\frac{E_G\left(T\right)}{k_B}\left(\frac{1}{300}-\frac{1}{T}\right)\right], \tag{27}$$

$$\tau_p = \tau_{p0}\exp\left[-0.25\frac{E_G\left(T\right)}{k_B}\left(\frac{1}{300}-\frac{1}{T}\right)\right]. \tag{28}$$

Where, parameter values (τ_{n0}, τ_{p0}) are also given in Tables **2-4**.

CALCULATION RESULTS AND DISCUSSION

Calculation Results of Carrier Diffusion from the Metallurgical Junction in Homo-junction Devices

It has been assumed in all calculations that the film has carrier mobility values one-tenth those of bulk values (Table **2**) because semiconductor films deposited on the insulator substrate or the flexible film have been assumed in the discussion. Calculation results of excess carrier density distributions at 300 K are shown in Fig. (**3**), where Fig. (**3a**) shows those for the Si homo-junction and Fig. (**3b**) for the Ge homo-junction. The aspects of carrier density distributions have almost been the same regardless of materials between Fig. (**3a**) and Fig. (**3b**), although the carrier densities of Si homo-junction have been higher than those of Ge homo-junction. The electron density at the edge has been higher than the hole density due to the difference in the diffusion coefficient (Fig. **4**).

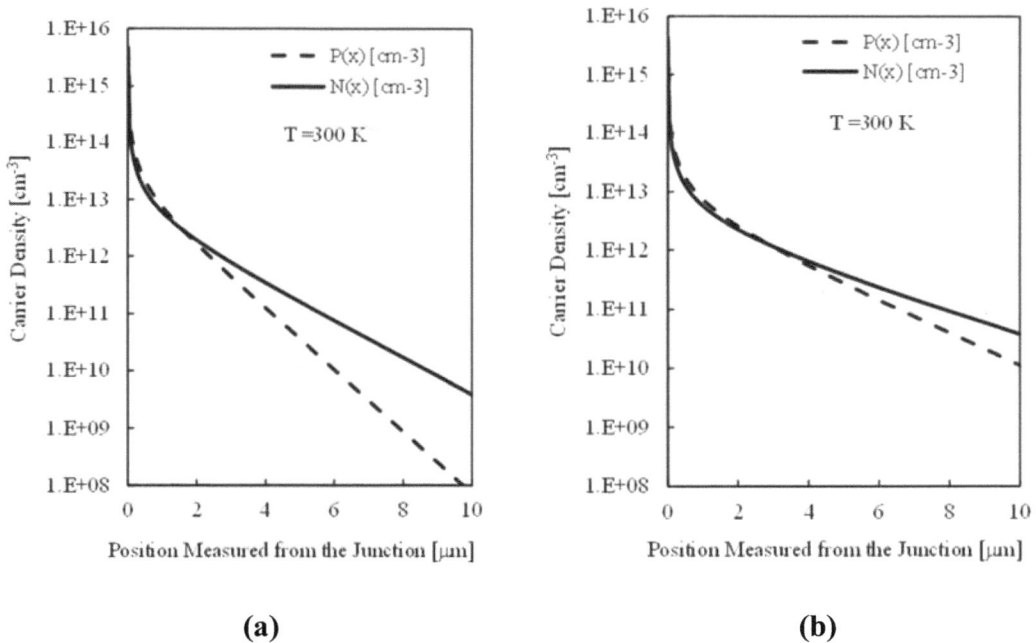

(a)   (b)

Fig. (3). Calculated local excess carrier density distributions from the metallurgical junction at 300 K. (a) Si film junction, (b) Ge film junction.

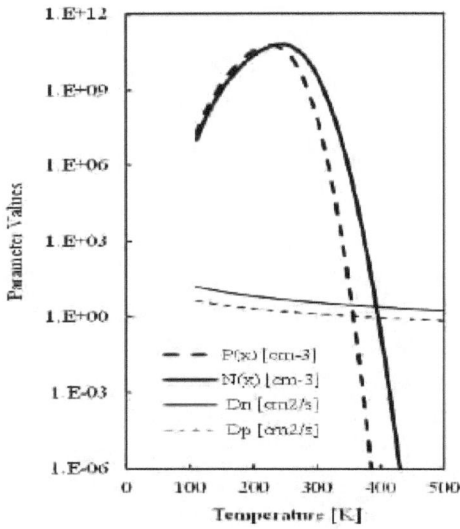
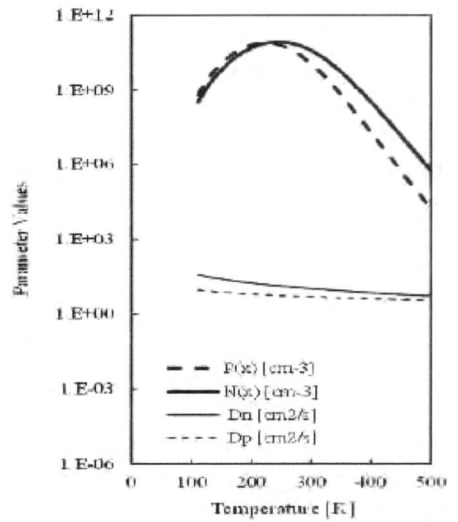

(a)   (b)

Fig. (4). Calculated excess carrier density and diffusion coefficient *versus* temperature. (a) Si film junction, (b) Ge film junction.

Calculated excess carrier densities at the film edge are shown as a function of temperature in Fig. (4), where Fig. (4a) shows those for the Si homo-junction and Fig. (4b) for the Ge homo-junction. Since the carrier density at the edge rules the maximal current level, this parameter is shown here. It is seen that Si films have much lower carrier density than Ge films at higher temperatures, which is due to the difference in the bandgap scale. The peak value of carrier density at the edge is obtained around 250 K. The decrease in carrier density at temperatures lower than 150K is due to the freeze-out effect. Fig. (4) suggests that, in solar batteries, a narrow bandgap is preferable to high performance because high carrier densities at the device edge guarantee a maximal current source potential.

Calculated Performance Parameters as a Function of Temperature in Homo-junction Devices

Calculation results of performance parameters [20] are shown in Fig. (5), where Fig. (5a) shows the parameters of Si film devices and Fig. (5b) shows those of Ge film devices. Fig. (5a) shows that the Si film device has a higher maximal power supply level than the Ge device at higher temperatures although the maximal current level of Ge film is higher than that of Si film at higher temperatures. The low output voltage of the Ge film device is due to the low impedance of the device. Since the efficiency of Si and Ge film devices is very low, these devices should be restricted to sensors implemented in local networks.

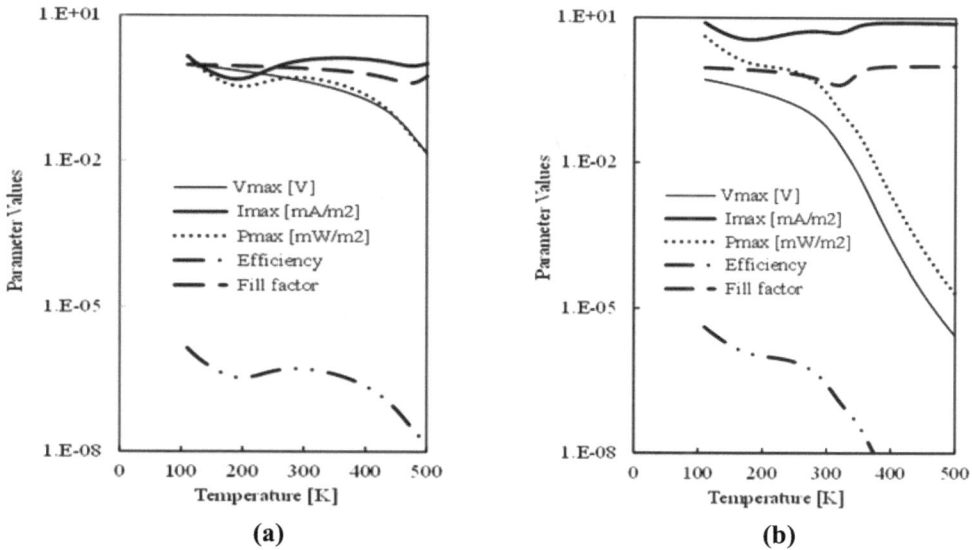

Fig. (5). Calculation results of various performance parameters *versus* temperature. (**a**) Si film junction, (**b**) Ge film junction.

Calculation Results of Carrier Diffusion from the Metallurgical Junction in Hetero-junction Devices

Calculation results of excess carrier density distributions at 300 K are shown in Fig. (**6**), where Fig. (**6a**) shows those for the n-ZnO/p-Si type-I hetero-junction and Fig. (**6b**) for the n-GaN/p-Si type-II hetero-junction. The aspects of carrier density distributions are quite different between Fig. (**6a**) and Fig. (**6b**); that is, the electron density in the n-ZnO film is very low because of the energy barrier at the metallurgical junction (Fig. **2a**). Excess hole density at the edge is lower than that of the Si homo-junction (Fig. **3a**) because fewer holes are generated in ZnO and GaN than Si due to their wider bandgap.

Calculated excess carrier densities at the film edge are shown as a function of temperature in Fig. (**7**), where Fig. (**7a**) shows those for the n-ZnO/p-Si hetero junction and Fig. (**7b**) for the n-GaN/p-Si heterojunction. It can be seen that the electron density of n-ZnO films is much lower than that of n-GaN films in the temperature range from 100 K to 500 K, which may be due to the energy barrier at the metallurgical junction in the n-ZnO/p-Si type-I heterojunction. The peak value of carrier density at the edge is obtained at around 250 K. This aspect is almost the same for Si, ZnO, and GaN films. The drastic decrease in carrier density at temperatures under 250K is due to the freeze-out effect in wide bandgap semiconductors.

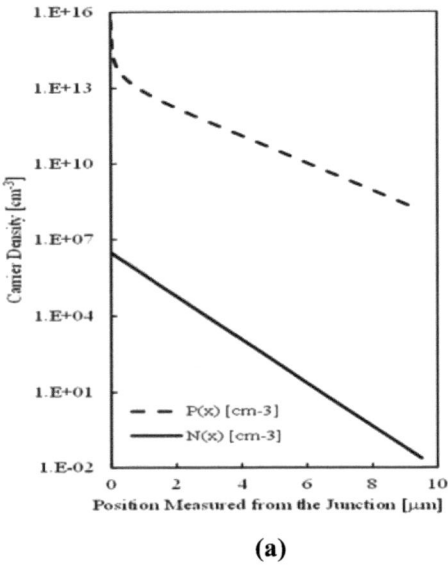

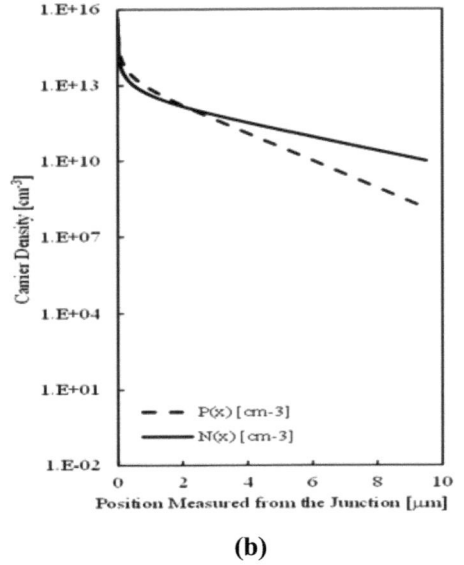

Fig. (6). Calculated local carrier distributions from the metallurgical junction. (**a**) n-ZnO/p-Si junction, (**b**) n-GaN/p-Si junction.

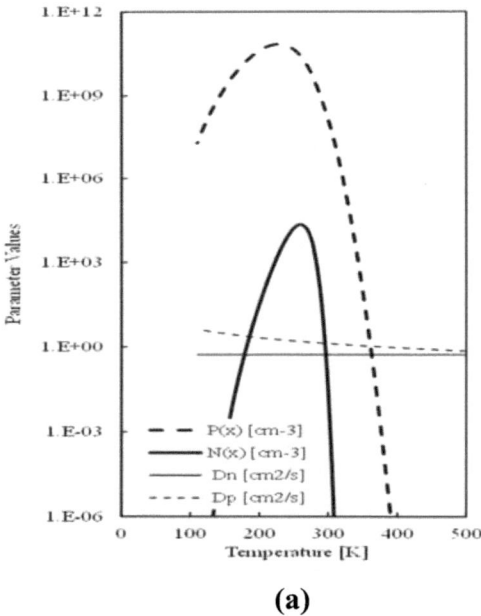

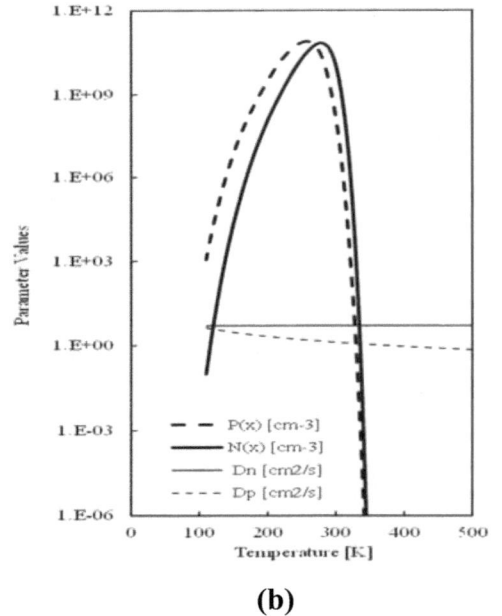

Fig. (7). Calculated excess carrier density and diffusion coefficient *versus* temperature. (**a**) n-ZnO/p-Si junction, (**b**) n-GaN/p-Si junction.

Performance Comparison of Homo-junction Devices and Hetero-junction Devices

Calculation results of various performance parameters of n-ZnO/p-Si and n-GaN/p-Si solar devices are shown as a function of temperature in Fig. (**8**); Fig. (**8a**) shows the parameter values of n-ZnO/p-Si device and Fig. (**8b**) shows those of n-GaN/p-Si device. Behaviors of parameter values shown in Fig. (**8a**) are roughly identical to those in Fig. (**5a**) because the p-type regions are composed of the same material (Si), although primary parameter values of n-ZnO/p-Si hetero-junction are lower than those in Si homo-junction device due to the lower electron density of the ZnO film region, as shown in Fig. (**6a**).

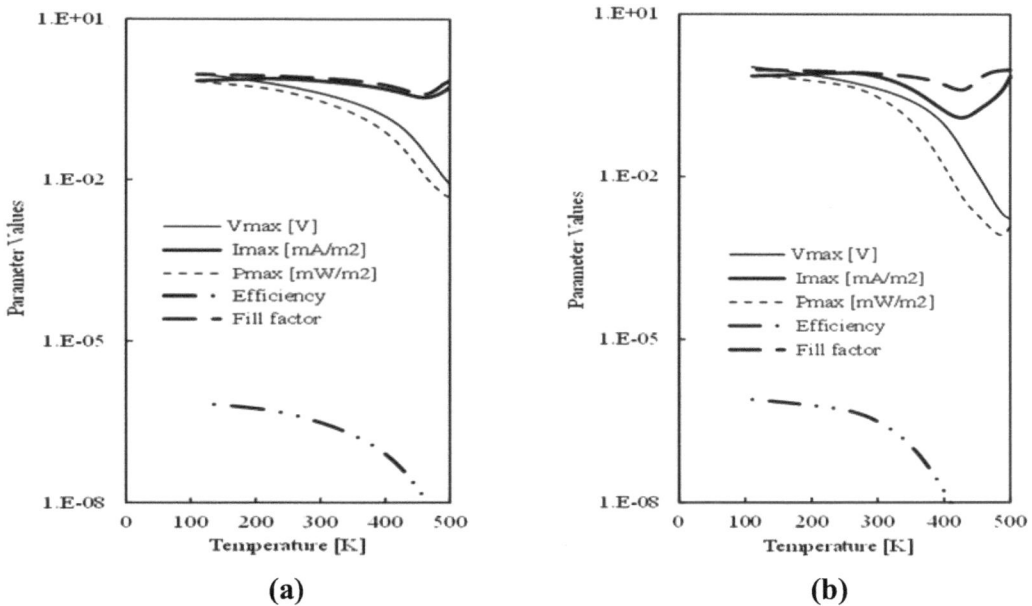

Fig. (8). Calculation results of various performance parameters *versus* temperature. (**a**) n-ZnO/p-Si junction, (**b**) n-GaN/p-Si junction.

On the other hand, primary parameter values in the n-GaN/p-Si device are almost the same as those in the n-ZnO/p-Si device despite having an electron density higher than that of the n-ZnO/p-Si device, which is a slightly unexpected result. It is considered that the electron density in the n-GaN film without any illumination is not so high and most electrons under illumination come from the p-Si region because of the large bandgap of GaN films. Consequently, the n-ZnO/p-Si and n-GaN/p-Si solar devices are not superior to Si homo-junction solar devices in terms of performance.

Performance Comparison of Various Lateral Pn-junction Solar Devices

Since various lateral thin-film pn-junction solar devices have already been investigated recently, their performances at room temperature are compared in Table **5**; it is to be noted that no low-temperature performance data, except those from this study, are shown because such experiments and theoretical consideration are missing from past studies. It is known that low-temperature environments yield better performance than room-temperature ones [38]. As shown in Table **5**, at room temperature, Si-based lateral homo-pn-junction solar devices, ZnO/Si hetero-pn-junction solar devices, and GaN/Si hetero-pn-junction solar devices have comparable performance to chalcogenide solar devices, except for efficiency. V_{max} value of Ge-based lateral homo-pn-junction solar devices is very low due to low internal resistance. The efficiency of the Ge-based homo-p--junction at 150 K becomes better than Si-based homo-pn-junction, ZnO/Si hetero-pn-junction, and GaN/Si hetero-pn-junction solar devices at room temperature.

Table 5. Photovoltaic metrics of lateral pn-junction solar devices (300K).

Materials	Wavelength (nm)	V_{max} (V)	Fill Factor	Efficiency	Refs.
WSe$_2$/MoS$_2$	White light source	0.22	0.39	0.2	[5]
WS$_2$/WSe$_2$	514	0.47	---	---	[47]
MoSe$_2$/WSe$_2$	532	0.9	0.2	0.08	[48]
WSe$_2$	White light source	0.85	0.5	0.5	[49]
WSe$_2$	522	0.72	---	---	[50]
bP	640	0.05	0.1	---	[51]
ZnO/Si	550	0.43 (300K) 0.85 (150 K)	0.80 (300 K) 0.93 (150 K)	3.0×10^{-7} (300 K) 6.3×10^{-7} (150 K)	This work
GaN/Si	550	0.43 (300 K) 0.92 (150 K)	0.80 (300 K) 0.93 (150 K)	3.0×10^{-7} (300 K) 6.9×10^{-7} (150 K)	This work
Si	550	0.43 (300 K) 0.84 (150 K)	0.81 (300 K) 0.93 (150 K)	5.2×10^{-7} (300 K) 5.5×10^{-7} (300 K)	This work
Ge	550	0.054 (300 K) 0.41 (150 K)	0.47 (300 K) 0.87 (150 K)	2.7×10^{-7} (300 K) 1.7×10^{-6} (150 K)	This work

The above simulation results suggest that Ge-based multi-stacked thin-film lateral pn junction solar devices formed on a transparent panel can be used in battery systems for low-energy devices in satellites. On the other hand, n-ZnO/p-Si (type-I) and n-GaN/p-Si (type-II) hetero-pn-junction solar devices formed on a transparent panel are not better than expected. In these applications, it is expected that high-efficiency peripheral circuits, like Schenkel circuits [39, 40], and storage devices will be required. Such lateral pn junction film devices [41 - 43] have already been investigated for medical applications [8] and others [44 - 46]. However, more investigation is needed from the viewpoint of the availability of various semiconductor materials.

Design Issue of Lateral Pn Junction Film Solar Battery

Calculation results of V_{max}, I_{max}, and photovoltaic efficiency at 300 K are shown as a function of each length of n-type and p-type regions in Fig. (9), where the performances of Si homo-junction, Ge homo-junction, ZnO/Si type-I hetero-junction, and GaN/Si type-II hetero-junction are compared. It can be seen in Fig. (9a) that V_{max} is insensitive to film length. The Ge homo-junction has very low V_{max} due to lower internal resistance because the Ge film has the largest excess carrier density. Fig. (9b) reveals that the I_{max} values fall as L is increased due to the increase in internal resistance. Ge film devices attain the highest I_{max} value. In Fig. (9c), it can be seen that the efficiency is degraded as L is increased, and that Si film devices offer perfectly adequate efficiency. It can be concluded that only at very short film lengths do ZnO/Si type-I hetero-junction and GaN/Si type-II hetero-junction film devices have better performance than the Si homo-junction equivalents.

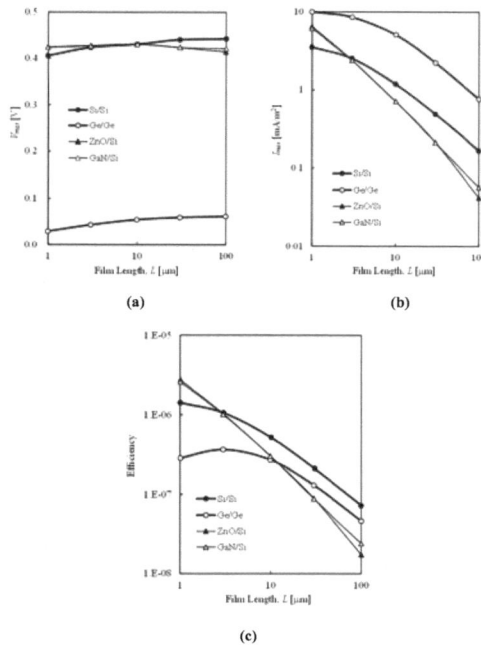

Fig. (9). Calculated performance parameters as a function of device length at 300 K. (a) V_{max} (V), (b) I_{max} (mA/m^2), (c) efficiency.

In summary, it is considered that Si-film-based lateral homo-pn-junction solar devices with film lengths shorter than 10 μm are promising for the IoT sensor net application from the viewpoints of performance and production cost. Though a small atomic energy battery was proposed for commercial use [47 - 53], it is anticipated that its cost will be very high, and that the environmental issue of waste disposal will remain difficult to solve.

CONCLUSION

This chapter has reconsidered the theoretical base for photovoltaic performance estimation of thin film lateral pn-junction solar devices and comprehensively discussed the performances of homo- and hetero-junction devices because it is anticipated that such lateral pn-junction solar devices are much more robust against local damage in comparison to the conventional vertical junction structure. This characteristic is needed in future various flexible electronic devices. Though the primary theoretical base has already been given, the physical conditions have not been deeply considered. Accordingly, the theoretical formulation has been reviewed to widen the consideration, and some additional theoretical concepts have been introduced for hetero-junction structures and the device potential has numerically been examined. Various calculations have been performed for Si- and Ge-based homo-junction devices and ZnO/Si (Type-I) and GaN/Si (Type-II) based hetero-junction devices from the viewpoint of future possible applications. It has been demonstrated that Si-film-based lateral homo-pn-junction solar devices with film lengths shorter than 10 μm are promising for the IoT sensor net application from the viewpoint of performance and production cost.

KEY POINTS

1. The practical theoretical base to consider the performances of various thin film lateral pn-junction solar devices is given.
2. The expected performance of Si- and Ge-based homo-junctions and ZnO/Si (Type-I) and GaN/Si (Type-II) based hetero-junctions is compared.
3. Low-temperature and high-temperature behaviors of solar devices are also addressed from the viewpoint of various applications.
4. It is demonstrated that highly-doped Si- and Ge-based homo-junction architectures are very promising for high-performance solar devices.

REFERENCES

[1]   Y. Omura, "Rotor", *Jpn. Pat. No. 6470993,* 2019.

[2]   Y. Omura, *Potential of Nano-scale Optical Rotor Based on a pn-Junction Wire* JSAP/IEEE Si Nanoelectron. Workshop (June, Kyoto), 2019. 6-28, pp. 87-88.
[http://dx.doi.org/10.23919/SNW.2019.8782946]

[3]   N. Inomata, Y. Yamanishi, and F. Arai, "Manipulation and observation of carbon nanotubes in water

under an optical microscope using a microfluidic chip", *IEEE Trans. Nanotechnol.,* vol. 8, no. 4, pp. 463-468, 2009.
[http://dx.doi.org/10.1109/TNANO.2008.2012346]

[4]   S. Masum Nawaz, M. Saha, N. Sepay, and A. Mallik, "Energy-from-waste: a triboelectric nanogenerator fabricated from waste polystyrene for energy harvesting and self-powered sensor", *Nano Energy,* vol. 104, no. 107902, 2022.

[5]   M-Y. Li, Y. Shi, C-C. Cheng, L-S. Lu, Y-C. Lin, H-L. Tang, M-L. Tsai, C-W. Chu, K-H. Wei, and H. Jr, He, W.-H. Chang, K. Suenaga, and L.-J. Li, "Epitaxial growth of a monolayer WSe2-MoS2 lateral pn junction with an atomically sharp interface", *Science,* vol. 349, pp. 524-528, 2015.
[http://dx.doi.org/10.1126/science.aab4097] [PMID: 26228146]

[6]   D. Muchahary, S. Maity, and S.K. Metya, "Modelling and analysis of temperature-dependent carrier lifetime and surface recombination velocity of Si–ZnO heterojunction thin film solar cell", *Micro & Nano Lett.,* vol. 14, no. 4, pp. 399-403, 2019.
[http://dx.doi.org/10.1049/mnl.2018.5147]

[7]   M. Alrefaee, U.P. Singh, and S.K. Das, "Fabrication of a-Si/ZnO heterojunction diode using a nonconventional sol-gel method", *J. Electron. Mater.,* vol. 52, no. 7, pp. 4369-4374, 2023.
[http://dx.doi.org/10.1007/s11664-023-10335-8]

[8]   1st USA Medical Battery Conf. 2023, Available from: https://www.medicalbatteryconference.com/

[9]   A.K. Dubey, V. Sugumaran, H.J.C. Peter, Ed., *Advanced IoT Sensors, Networks and Systems: Select Proceedings of SPIN 2022 (Lecture Notes in Electrical Engineering, 1027).* Springer, 2023.
[http://dx.doi.org/10.1007/978-981-99-1312-1]

[10]   A.D. Pathak, S. Saha, V.K. Bharti, M.M. Gaikwad, and C.S. Sharma, "A review on battery technology for space application", *J. Energy Storage,* vol. 61, p. 106792, 2023.
[http://dx.doi.org/10.1016/j.est.2023.106792]

[11]   G.K. Bhaumik, A.K. Nath, and S. Basu, "Laser annealing of zinc oxide thin film deposited by spray-CVD", *Mater. Sci. Eng. B,* vol. 52, no. 1, pp. 25-31, 1998.
[http://dx.doi.org/10.1016/S0921-5107(97)00272-9]

[12]   J.G. Lu, T. Kawaharamura, H. Nishinaka, Y. Kamada, T. Ohshima, and S. Fujita, "Zno-based thin films synthesized by atmospheric pressure mist chemical vapor deposition", *J. Cryst. Growth,* vol. 299, no. 1, pp. 1-10, 2007.
[http://dx.doi.org/10.1016/j.jcrysgro.2006.10.251]

[13]   D. Shinohara, and S. Fujita, "Heteroepitaxy of corundum-structured α-Ga 2 O 3 thin films on α-Al 2 O 3 substrates by ultrasonic mist chemical vapor deposition", *Jpn. J. Appl. Phys.,* vol. 47, no. 9R, pp. 7311-7313, 2008.
[http://dx.doi.org/10.1143/JJAP.47.7311]

[14]   T. Oshima, T. Nakazono, A. Mukai, and A. Ohtomo, "Epitaxial growth of γ-Ga2O3 films by mist chemical vapor deposition", *J. Cryst. Growth,* vol. 359, pp. 60-63, 2012.
[http://dx.doi.org/10.1016/j.jcrysgro.2012.08.025]

[15]   M. De Souza, O. Bulteel, D. Flandre, and M.A. Pavanello, "Temperature and silicon film thickness influence on the operation of lateral SOI PIN photodiodes for detection of short wavelengths", *Journal of Integrated Circuits and Systems,* vol. 6, no. 2, pp. 107-113, 2020.
[http://dx.doi.org/10.29292/jics.v6i2.346]

[16]   C. Novo, R. Giacomini, R. Doria, A. Afzalian, and D. Flandre, "Illuminated to dark ratio improvement in lateral SOI PIN photodiodes at high temperatures", *Semicond. Sci. Technol.,* vol. 29, no. 7, p. 075008, 2014.
[http://dx.doi.org/10.1088/0268-1242/29/7/075008]

[17]   E.J. Rodrigues, and M. De Souza, "Temperature, silicon thickness and intrinsic length influence on the operation of lateral SOI PIN photodiodes", *Journal of Integrated Circuits and Systems,* vol. 15, no. 2,

pp. 1-5, 2020.
[http://dx.doi.org/10.29292/jics.v15i2.158]

[18]   F.A. Silva, R.T. Doria, E. Simoen, and M.G.C. Andrade, "Lateral PIN Photodiode with germanium and silicon layer on SOI wafers", *J. Integr. Cir. Syst.,* vol. 18, pp. 1-8, 2023.
[http://dx.doi.org/10.29292/jics.v18i2.690]

[19]   Y. Omura, "Theoretical consideration on potential and scalability of optical rotors based on a pn-junction rod", *IFSA, Sensors & Transducers,* vol. 235, no. 7, pp. 1-8, 2019. [Special Issue].

[20]   Y. Omura, "Theoretical Estimation of power generation performance of nano-sheet planar lateral pn-junction under illumination", *Jordan J. Elect. Eng.,* vol. 9, no. 3, pp. 272-284.

[21]   Y. Omura, "Potential of Nano-sheet Lateral pn-Junction Flexible Solar Panel," AM-FPD 2023, (Kyoto,   2023),   Session   PVp,   P_14,   2023   Available   from: https://ieeexplore.ieee.org/document/10264980

[22]   N. Takahashi, S. Sato, Y. Omura, and T. Saitoh, "Measuring impact of light on the resistance of non-doped ZnO films", *ECS J. Solid State Sci. Technol.,* vol. 8, no. 1, pp. P57-P61, 2019.
[http://dx.doi.org/10.1149/2.0201901jss]

[23]   Y. Omura, and S. Sato, "Theoretical analysis of the impacts of light illumination on the transient current of sputter-deposited non-doped ZnO films", *AIP Adv.,* vol. 11, no. 1, p. 015030, 2021.
[http://dx.doi.org/10.1063/5.0036882]

[24]   K.M.A. Saron, M.R. Hashim, M. Ibrahim, M. Yahyaoui, and N.K. Allam, "Temperature-dependent transport properties of CVD-fabricated n-GaN nanorods/p-Si heterojunction devices", *RSC Advances,* vol. 10, no. 55, pp. 33526-33533, 2020.
[http://dx.doi.org/10.1039/D0RA05973K] [PMID: 35515063]

[25]   X. Liu, M. Wang, J. Wei, C.P. Wen, B. Xie, Y. Hao, X. Yang, and B. Shen, "GaN-on-Si quasi-vertical p-n diode with junction termination extension based on hydrogen plasma treatment and diffusion", *IEEE Trans. Electron Dev.,* vol. 70, no. 4, pp. 1636-1640, 2023.
[http://dx.doi.org/10.1109/TED.2023.3247366]

[26]   M. Benhaliliba, A. Tiburcio-Silver, A. Avila-Garcia, A. Tavira, Y.S. Ocak, M.S. Aida, and C.E. Benouis, "The sprayed ZnO films: nanostructures and physical parameters", *J. Semicond.,* vol. 36, no. 8, p. 083001, 2015.
[http://dx.doi.org/10.1088/1674-4926/36/8/083001]

[27]   V. Kabra, L. Aamir, and M.M. Malik, "Low cost, p-ZnO/n-Si, rectifying, nano heterojunction diode: Fabrication and electrical characterization", *Beilstein J. Nanotechnol.,* vol. 5, pp. 2216-2221, 2014.
[http://dx.doi.org/10.3762/bjnano.5.230] [PMID: 25551049]

[28]   Anwar, A. F. M., Wu, S., & Webster, R. T. "Temperature dependent transport properties in GaN, Al/sub x/Ga/sub 1-x/N, and In/sub x/Ga/sub 1-x/N semiconductors", *IEEE Transactions on Electron devices,* vol. 48, no. 3, pp. 567-572, 2001.

[29]   S.M. Sze, *Physics of Semiconductor Devices.* 2nd ed. J. Wiley & Sons, 1981.

[30]   S.O. Kasap, *Principles of Electronic Materials and Devices.* 3rd ed. McGraw Hill, 2006.

[31]   M.B. Prince, "Drift Mobilities in Semiconductors. I. Germanium", *Phys. Rev.,* vol. 92, no. 3, pp. 681-687, 1953.
[http://dx.doi.org/10.1103/PhysRev.92.681]

[32]   D. Muchahary, S. Maity, and S.K. Metya, "Modelling and analysis of temperature-dependent carrier lifetime and surface recombination velocity of Si–ZnO heterojunction thin film solar cell", *Micro & Nano Lett.,* vol. 14, no. 4, pp. 399-403, 2019.
[http://dx.doi.org/10.1049/mnl.2018.5147]

[33]   M. Ilegems, R. Dingle, and J. Appl. Phys, *Luminescence of Be- and Mg-doped GaN* vol. 44. , 1973, pp. 4234-4235.

[http://dx.doi.org/10.1063/1.1662930]

[34]   R.C. Rai, M. Guminiak, S. Wilser, B. Cai, and M.L. Nakarmi, "Elevated temperature dependence of energy band gap of ZnO thin films grown by e-beam deposition", *J. Appl. Phys.*, vol. 111, no. 7, p. 073511, 2012.
[http://dx.doi.org/10.1063/1.3699365]

[35]   H. Teisseyre, P. Perlin, T. Suski, I. Grzegory, S. Porowski, J. Jun, A. Pietraszko, and T.D. Moustakas, "Temperature dependence of the energy gap in GaN bulk single crystals and epitaxial layer", *J. Appl. Phys.*, vol. 76, no. 4, pp. 2429-2434, 1994.
[http://dx.doi.org/10.1063/1.357592]

[36]   Z.Z. Bandić, P.M. Bridger, E.C. Piquette, and T.C. McGill, "Minority carrier diffusion length and lifetime in GaN", *Appl. Phys. Lett.*, vol. 72, no. 24, pp. 3166-3168, 1998.
[http://dx.doi.org/10.1063/1.121581]

[37]   O. Lopatiuk-Tirpak, and L. Chernyak, "Studies of minority carrier transport in ZnO", *Superlattices Microstruct.*, vol. 42, no. 1-6, pp. 201-205, 2007.
[http://dx.doi.org/10.1016/j.spmi.2007.04.030]

[38]   S. Dubey, J.N. Sarvaiya, and B. Seshadri, "Temperature dependent photovoltaic (PV) efficiency and its effect on pv production in the world – A review", *Energy Procedia,* vol. 33, pp. 311-321, 2013.
[http://dx.doi.org/10.1016/j.egypro.2013.05.072]

[39]   Y. Omura, and Y. Iida, "Performance prospects of fully-depleted SOI MOSFET-based diodes applied to schenkel circuit for RF-ID chips", *Circuits and Systems,* vol. 4, no. 2, pp. 173-180, 2013.
[http://dx.doi.org/10.4236/cs.2013.42024]

[40]   Y. Omura, A. Mallik, and N. Matsuo, *MOS Devices for Low-Voltage and Low-Energy Applications.* 1st ed. IEEE Press – J. Wiley & Sons, 2017.

[41]   F.A. Chaves, and D. Jiménez, "Electrostatics of two-dimensional lateral junctions", *Nanotechnology,* vol. 29, no. 27, p. 275203, 2018.
[http://dx.doi.org/10.1088/1361-6528/aabeb2] [PMID: 29664417]

[42]   F.A. Chaves, P.C. Feijoo, and D. Jiménez, "2D pn junctions driven out-of-equilibrium", *Nanoscale Adv.,* vol. 2, no. 8, pp. 3252-3262, 2020.
[http://dx.doi.org/10.1039/D0NA00267D] [PMID: 36134281]

[43]   A. Fournol, J.E. Blond, A. Aliane, H. Kaya, J.E. Meilhan, and L. Dussopt, "Thermal sensing performances of thin-film lateral pin diodes at 80 K and 300 K", In: *The 52nd European Solid State Device Research Conference*, 2022, pp. 285-288.
[http://dx.doi.org/10.1109/ESSDERC55479.2022.9947136]

[44]   H. Pal, S. Singh, C. Guo, W. Guo, O. Badami, T. Pramanik, and B. Sarkar, "Lateral P–N junction photodiodes using lateral polarity structure GaN films: A theoretical perspective", *J. Electron. Mater.,* vol. 52, no. 3, pp. 2148-2157, 2023.
[http://dx.doi.org/10.1007/s11664-022-10166-z]

[45]   J. Jia, A. Takasaki, N. Oka, and Y. Shigesato, "Experimental observation on the fermi level shift in polycrystalline Al-doped ZnO films", *J. Appl. Phys.*, vol. 112, no. 1, p. 013718, 2012.
[http://dx.doi.org/10.1063/1.4733969]

[46]   M. Oshikiri, Y. Imanaka, F. Aryasetiawan, and G. Kido, "Comparison of the electron effective mass of the n-type ZnO in the wurtzite structure measured by cyclotron resonance and calculated from first principle theory", *Physica B,* vol. 298, no. 1-4, pp. 472-476, 2001.
[http://dx.doi.org/10.1016/S0921-4526(01)00365-9]

[47]   V. Bougrov, M.E. Levinshtein, S.L. Rumyantsev, and A. Zubrilov, *Properties of Advanced Semiconductor Materials GaN, AlN, InN, BN, SiC, SiGe* Eds. Levinshtein M.E., Rumyantsev S.L., Shur M.S.,(John Wiley & Sons, Inc., New York, 2001, pp. 1-30.

[48]   X. Duan, C. Wang, J.C. Shaw, R. Cheng, Y. Chen, H. Li, X. Wu, Y. Tang, Q. Zhang, A. Pan, J. Jiang,

R. Yu, Y. Huang, and X. Duan, "Lateral epitaxial growth of two-dimensional layered semiconductor heterojunctions", *Nat. Nanotechnol.,* vol. 9, no. 12, pp. 1024-1030, 2014.
[http://dx.doi.org/10.1038/nnano.2014.222] [PMID: 25262331]

[49]   P.K. Sahoo, S. Memaran, F.A. Nugera, Y. Xin, T. Díaz Márquez, Z. Lu, W. Zheng, N.D. Zhigadlo, D. Smirnov, L. Balicas, and H.R. Gutiérrez, "Bilayer lateral heterostructures of transition-metal dichalcogenides and their optoelectronic response", *ACS Nano,* vol. 13, no. 11, pp. 12372-12384, 2019.
[http://dx.doi.org/10.1021/acsnano.9b04957] [PMID: 31532628]

[50]   A. Pospischil, M.M. Furchi, and T. Mueller, "Solar-energy conversion and light emission in an atomic monolayer p–n diode", *Nat. Nanotechnol.,* vol. 9, no. 4, pp. 257-261, 2014.
[http://dx.doi.org/10.1038/nnano.2014.14] [PMID: 24608229]

[51]   B.W.H. Baugher, H.O.H. Churchill, Y. Yang, and P. Jarillo-Herrero, "Optoelectronic devices based on electrically tunable p–n diodes in a monolayer dichalcogenide", *Nat. Nanotechnol.,* vol. 9, no. 4, pp. 262-267, 2014.
[http://dx.doi.org/10.1038/nnano.2014.25] [PMID: 24608231]

[52]   M. Buscema, D.J. Groenendijk, G.A. Steele, H.S.J. van der Zant, and A. Castellanos-Gomez, "Photovoltaic effect in few-layer black phosphorus PN junctions defined by local electrostatic gating", *Nat. Commun.,* vol. 5, no. 1, p. 4651, 2014.
[http://dx.doi.org/10.1038/ncomms5651]

[53]   Available from: https://www.greencarcongress.com/2024/01/20240114-betavolt.html

APPENDIX

This appendix provides additional information discussed in Chapter 9.

The solution to the surface potential is given below:

$$\sigma_k = -\alpha_k / \lambda_k{}^2, \lambda_k{}^2 = 2\beta_k / t_{body}{}^2, \alpha_k = \left[\left(qN_k / \varepsilon_{si} - 2\beta_k / t_{body}{}^2 \right) \left(V_{gseff,k} \right) \right]$$

$$T = (\sigma_2 - \sigma_1) \cosh(\lambda_1 X_1), O = e^{\lambda_2 X_1} \left[\cosh(\lambda_1 X_1) - \left(\frac{\lambda_2}{\lambda_1} \right) \sinh(\lambda_1 X_1) \right]$$

$$P = e^{-\lambda_2 X_1} \left[\cosh(\lambda_1 X_1) + \left(\frac{\lambda_2}{\lambda_1} \right) \sinh(\lambda_1 X_1) \right], W = 0.5(\sigma_3 - \sigma_2) \left[Oe^{-\lambda_2 (X_1 + X_2)} + Pe^{\lambda_2 (X_1 + X_2)} \right]$$

$$F = 0.5 \left[Oe^{(\lambda_3 - \lambda_2)(X_1 + X_2)} \left(1 + \frac{\lambda_3}{\lambda_2} \right) + Pe^{(\lambda_3 + \lambda_2)(X_1 + X_2)} \left(1 - \frac{\lambda_3}{\lambda_2} \right) \right]$$

$$G = 0.5 \left[Oe^{(-\lambda_3 - \lambda_2)(X_1 + X_2)} \left(1 - \frac{\lambda_3}{\lambda_2} \right) + Pe^{(-\lambda_3 + \lambda_2)(X_1 + X_2)} \left(1 + \frac{\lambda_3}{\lambda_2} \right) \right]$$

$$A_1 = 0.5(\sigma_4 - \sigma_3) \left[Fe^{-\lambda_3 (X_1 + X_2 + X_3)} + Ge^{\lambda_3 (X_1 + X_2 + X_3)} \right]$$

$$A_2 = 0.5 \left[Fe^{(\lambda_4 - \lambda_3)(X_1 + X_2 + X_3)} \left(1 + \frac{\lambda_4}{\lambda_3} \right) + Ge^{(\lambda_4 + \lambda_3)(X_1 + X_2 + X_3)} \left(1 - \frac{\lambda_4}{\lambda_3} \right) \right]$$

$$A_3 = 0.5 \left[Fe^{(-\lambda_4 - \lambda_3)(X_1 + X_2 + X_3)} \left(1 - \frac{\lambda_4}{\lambda_3} \right) + Ge^{(-\lambda_4 + \lambda_3)(X_1 + X_2 + X_3)} \left(1 + \frac{\lambda_4}{\lambda_3} \right) \right]$$

Ashish Raman, Prabhat Singh, Naveen Kumar & Ravi Ranjan (Eds.)
All rights reserved-© 2025 Bentham Science Publishers

$$B_1 = 0.5\left(\sigma_5 - \sigma_4\right)\left[A_2 e^{-\lambda_4 \sum_{k=1}^{4} X_k} + A_3 e^{\lambda_4 \sum_{k=1}^{4} X_k} \right],$$

$$B_2 = 0.5\left[A_2 e^{(\lambda_5 - \lambda_4)\sum_{k=1}^{4} X_k}\left(1 + \frac{\lambda_5}{\lambda_4}\right) + A_3 e^{(\lambda_5 + \lambda_4)\sum_{k=1}^{4} X_k}\left(1 - \frac{\lambda_5}{\lambda_4}\right) \right]$$

$$B_3 = 0.5\left[A_2 e^{(-\lambda_5 - \lambda_4)\sum_{k=1}^{4} X_k}\left(1 - \frac{\lambda_5}{\lambda_4}\right) + A_3 e^{(-\lambda_5 + \lambda_4)\sum_{k=1}^{4} X_k}\left(1 + \frac{\lambda_5}{\lambda_4}\right) \right]$$

$$C_1 = 0.5\left(\sigma_6 - \sigma_5\right)\left[B_2 e^{-\lambda_5 \sum_{k=1}^{5} X_k} + B_3 e^{\lambda_5 \sum_{k=1}^{5} X_k} \right],$$

$$C_2 = 0.5\left[B_2 e^{(\lambda_6 - \lambda_5)\sum_{k=1}^{5} X_k}\left(1 + \frac{\lambda_6}{\lambda_5}\right) + B_3 e^{(\lambda_6 + \lambda_5)\sum_{k=1}^{5} X_k}\left(1 - \frac{\lambda_6}{\lambda_5}\right) \right]$$

$$C_3 = 0.5\left[B_2 e^{(-\lambda_6 - \lambda_5)\sum_{k=1}^{5} X_k}\left(1 - \frac{\lambda_6}{\lambda_5}\right) + B_3 e^{(-\lambda_6 + \lambda_5)\sum_{k=1}^{5} X_k}\left(1 + \frac{\lambda_6}{\lambda_5}\right) \right]$$

$$D_1 = 0.5\left(\sigma_7 - \sigma_6\right)\left[C_2 e^{-\lambda_6 \sum_{k=1}^{6} X_k} + C_3 e^{\lambda_6 \sum_{k=1}^{6} X_k} \right], D_2 = 0.5\left[C_2 e^{(\lambda_7 - \lambda_6)\sum_{k=1}^{6} X_k}\left(1 + \frac{\lambda_7}{\lambda_6}\right) + C_3 e^{(\lambda_7 + \lambda_6)\sum_{k=1}^{6} X_k}\left(1 - \frac{\lambda_7}{\lambda_6}\right) \right]$$

$$D_3 = 0.5\left[C_2 e^{(-\lambda_7 - \lambda_6)\sum_{k=1}^{6} X_k}\left(1 - \frac{\lambda_7}{\lambda_6}\right) + C_3 e^{(-\lambda_7 + \lambda_6)\sum_{k=1}^{6} X_k}\left(1 + \frac{\lambda_7}{\lambda_6}\right) \right]$$

$$E_1 = 0.5\left(\sigma_8 - \sigma_7\right)\left[D_2 e^{-\lambda_7 \sum_{k=1}^{7} X_k} + D_3 e^{\lambda_7 \sum_{k=1}^{7} X_k} \right], E_2 = 0.5\left[D_2 e^{(\lambda_8 - \lambda_7)\sum_{k=1}^{7} X_k}\left(1 + \frac{\lambda_8}{\lambda_7}\right) + D_3 e^{(\lambda_8 + \lambda_7)\sum_{k=1}^{7} X_k}\left(1 - \frac{\lambda_8}{\lambda_7}\right) \right]$$

$$E_3 = 0.5\left[D_2 e^{(-\lambda_8 - \lambda_7)\sum_{k=1}^{7} X_k}\left(1 - \frac{\lambda_8}{\lambda_7}\right) + D_3 e^{(-\lambda_8 + \lambda_7)\sum_{k=1}^{7} X_k}\left(1 + \frac{\lambda_8}{\lambda_7}\right) \right]$$

$$z_1 = e^{\lambda_7 \left(\sum_{l=1}^{8} X_l \right)} , z_2 = e^{-\lambda_7 \left(\sum_{l=1}^{8} X_l \right)} , z_3 = M + N + A_1 + B_1 + C_1 + D_1 + E_1$$

$$U_8 = \frac{\left(V_{bi,source} - \sigma_1 \right) z_2 - \left(V_{bi,drain} + V_{ds} - \sigma_8 \right) E_3 - \left(z_3 z_2 \right)}{z_2 E_2 - z_1 E_3};$$

$$V_8 = \frac{\left(V_{bi,source} - \sigma_1 \right) z_1 - \left(V_{bi,drain} + V_{ds} - \sigma_8 \right) E_2 - \left(z_1 z_3 \right)}{z_1 E_3 - z_2 E_2}$$

$$U_k = 0.5 \begin{bmatrix} U_{k+1} exp\left((\lambda_{k+1} - \lambda_k) \sum_{l=1}^{k} X_l \right)\left(1 + \left(\frac{\lambda_{l+1}}{\lambda_l} \right) \right) + \\ V_{k+1} exp\left((-\lambda_{l+1} - \lambda_l) \sum_{l=1}^{k} X_l \right)\left(1 - \left(\frac{\lambda_{l+1}}{\lambda_l} \right) \right) + \\ (\sigma_{l+1} - \sigma_l) exp\left(-\lambda_l \sum_{l=1}^{k} X_l \right) \end{bmatrix}, k = 1, 2, 3, \dots 7$$

$$V_k = 0.5 \begin{bmatrix} U_{k+1} exp\left((\lambda_{k+1} + \lambda_k) \sum_{l=1}^{k} X_l \right)\left(1 - \left(\frac{\lambda_{k+1}}{\lambda_k} \right) \right) + \\ V_{k+1} exp\left((-\lambda_{k+1} + \lambda_k) \sum_{l=1}^{k} X_l \right)\left(1 + \left(\frac{\lambda_{k+1}}{\lambda_k} \right) \right) + \\ (\sigma_{k+1} - \sigma_k) exp\left(\lambda_k \sum_{l=1}^{k} X_l \right) \end{bmatrix}, k = 1, 2, 3, \dots 7$$

APPENDIX A: SOLVING EQ. (15)

This appendix provides additional information discussed in Chapter 12.

After $\frac{\partial P(x,T)}{\partial x}$ is multiplied to both sides of eq. (15), the integration yields:

$$\int \frac{\partial P}{\partial x'} \frac{\partial^2 P}{\partial x'^2} dx' - A_P \int P \frac{\partial P}{\partial x'} \frac{\partial P}{\partial x'} dx' - C_P \int P \frac{\partial P}{\partial x'} dx' = 0, \qquad \textbf{(A1)}$$

$$\frac{1}{2}\left(\frac{\partial P}{\partial x'}\right)^2\bigg|_0^x - A_P\int_0^x P\frac{\partial P}{\partial x'}\frac{\partial P}{\partial x'}dx' - C_P\,P^2\bigg|_0^x = 0, \tag{A2}$$

Here, the second term of eq. (A2) is partially integrated as:

$$\int_0^x P\frac{\partial P}{\partial x'}\frac{\partial P}{\partial x'}dx' = \left(\frac{1}{2}P^2\frac{\partial P}{\partial x'}\right)\bigg|_0^x - \frac{1}{2}\int_0^x P^2\frac{\partial^2 P}{\partial x'^2}dx', \tag{A3}$$

In addition, from a rough estimation of parameter values, we can confirm that $A_p\dfrac{\partial P(x,T)}{\partial x} \gg C_P$. In this case, from eq. (15), we have:

$$\frac{\partial^2 P(x,T)}{\partial x^2} - A_p P(x,T)\frac{\partial P(x,T)}{\partial x} \cong 0, \tag{A4}$$

By inserting eq. (A4) into eq. (A3), we have:

$$\int_0^x P\frac{\partial P}{\partial x'}\frac{\partial P}{\partial x'}dx' = \left(\frac{1}{2}P^2\frac{\partial P}{\partial x'}\right)\bigg|_0^x - \frac{A_p}{8}\left(P^4\right)\bigg|_0^x, \tag{A5}$$

Equation (A5) is inserted into eq. (A2) as:

$$\frac{1}{2}\left(\frac{\partial P}{\partial x'}\right)^2\bigg|_0^x - A_p\left\{\left(\frac{1}{2}P^2\frac{\partial P}{\partial x'}\right)\bigg|_0^x - \frac{A_p}{8}\left(P^4\right)\bigg|_0^x\right\} - C_P\left(P^2\right)\bigg|_0^x = 0, \tag{A6}$$

This is rewritten to:

$$\left(\frac{\partial P(x,T)}{\partial x}\right)^2 - A_p P^2(x,T)\frac{\partial P(x,T)}{\partial x} + \frac{A_p^2}{4}P^4(x,T) - 2C_P P^2(x,T) + P_{00} = 0, \tag{A7}$$

Where:

$$P_{00} = -\left(\frac{\partial P(x,T)}{\partial x}\right)^2\bigg|_0 + A_p P^2(0,T)\frac{\partial P(x',T)}{\partial x'}\bigg|_0^x - \frac{A_p^2}{4}P^4(0,T) + 2C_P P^2(0,T), \tag{A8}$$

From eq. (A7), we have the following form for $\dfrac{\partial P(x,T)}{\partial x}$.

$$\frac{dP(x,T)}{dx} = \frac{1}{2}A_p P^2(x,T) \pm \sqrt{2C_P P^2(x,T) - P_{00}}, \tag{A9}$$

This yields the following integration form:

$$\int \frac{dP}{P^2(x,T) \pm 2\frac{\sqrt{2C_P}}{\alpha A_P}\sqrt{P^2(x,T) - P_{00}/2C_P}} = \int \frac{\alpha A_P}{2} dx, \tag{A10}$$

Equation (A10) cannot be solved analytically as it is, so possible solutions are considered in Appendix B.

APPENDIX B: POSSIBLE SOLUTIONS FOR EQ. (A10)

When we solve eq. (A10), the magnitude of P_{00} must be estimated. Eq. (A8) and Tables **1** and **2** yield the following estimation:

$$P_{00} \sim \left(\frac{P(0,T)}{L_p}\right)^2 \left\{1 - A_p L_p P(0,T) - \frac{1}{4}\left(A_p L_p P(0,T)\right)^2\right\} < 0 \tag{B1}$$

Where,

$$A_p L_p P(0,T) \sim 1 \qquad \text{for } P(0,T) = 1\times10^{17} \text{ cm}^{-3} \tag{B2}$$

$$A_p L_p P(0,T) \sim 100 \qquad \text{for } P(0,T) = 1\times10^{20} \text{ cm}^{-3} \tag{B3}$$

As a result, we have:

$$-\left(\frac{P(0,T)}{L_p}\right)^2 > P_{00} > -100\left(\frac{P(0,T)}{L_p}\right)^2 \tag{B4}$$

So, we consider the following two cases.

(i) $P(0,T) < 10^{17}\text{cm}^{-3}$

In this condition, the influence of the term of P_{00} in eq. (A10) is limited, and so it is discarded. Thus, we have:

$$\int \frac{dP}{P^2(x,T) \pm 2\frac{\sqrt{2C_P}}{A_P}\sqrt{P^2(x,T)}} = \int \frac{A_P}{2} dx, \tag{B5}$$

This integration is possible.

(ii) $P(0,T) \gg 10^{17}\text{cm}^{-3}$

Equation (B1) is reduced to:

$$P_{00} \sim -\frac{\alpha^2 A_P^2}{4} P^4(0,T) \tag{B6}$$

APPENDIX C: DERIVATION OF EQUATION (23)

The electron density, $(n_{p,Si}(x,T))$, of the conduction band of the Si film is expressed as:

$$n_{p,Si}(x,T) = \int_{E_{C,Si}}^{\infty} D_{os,Si}(E(x)) f_{FD}(T, E(x)) dE \tag{C1}$$

Where, $E_{C,Si}$ is the energy level of the conduction bottom of Si, $D_{os,Si}(E)$ is the density of states of the conduction band, and $f_{FD}(T, E)$ is the Fermi-Dirac function. Since the interface of the conduction band of the hetero-junction has the notch of Ec, as shown in Fig. (**2a**), the electron density $[n_{n,ZnO}(x, T)]$, which can be injected into the conduction band of the ZnO film, is expressed as:

$$n_{n,ZnO}(0,T) = \int_{E_{C,Si}+\Delta E_C}^{\infty} D_{os,Si}(E(0)) f_{FD}(T, E(0)) dE \tag{C2}$$

Because $\Delta E_c \gg k_B T$, the Fermi-Dirac function can be replaced, approximately, with the following exponential function:

$$n_{n,ZnO}(0,T) \cong \int_{E_{C,Si}+\Delta E_C}^{\infty} D_{os,Si}(E(0)) \exp\left(-\frac{E(0)-E_F(T)}{k_B T}\right) dE$$

$$\cong n_{p,Si}(0,T) \int_{\Delta E_C/k_B T}^{\infty} \sqrt{Z} \exp(-Z) dZ \tag{C3}$$

Equation (C-3) can be approximately calculated as:

$$n_{n,ZnO}(0,T) \cong n_{p,Si}(0,T) \sqrt{\frac{\Delta E_C}{k_B T}} \exp\left(-\frac{\Delta E_C}{k_B T}\right) \tag{C4}$$

APPENDIX D: FERMI-LEVEL DEFINITIONS OF MATERIALS

Since we have to take into account the ionization fraction of impurities at high-doping levels, the following expressions are introduced. For the Si and Ge films, we have:

$$n_{n0}(0,T) = \frac{N_D}{1 + \frac{1}{2}\exp\left(\frac{E_D - E_F(0)}{k_B T}\right)} \tag{D1}$$

$$p_{po}(0,T) = \frac{N_A}{1 + 4\exp\left(\dfrac{E_A - E_F(0)}{k_B T}\right)} \tag{D2}$$

$$E_D - E_F(0) = 2\ln(2) \cdot k_B T$$
$$-k_B T \ln\left[\sqrt{1 + 8\frac{N_D}{N_C}\exp\left(\frac{E_C - E_D}{k_B T}\right)} - 1\right] \tag{D3}$$

$$E_A - E_F(0) = -\ln(8) \cdot k_B T$$
$$+k_B T \ln\left[\sqrt{1 + 16\frac{N_A}{N_V}\exp\left(\frac{E_A - E_V}{k_B T}\right)} - 1\right] \tag{D4}$$

Where, *ND, NA, ED, EA, EF, NC,* and *NV* take the conventional meanings. For degenerate p-type Si and Ge films, the simulations use equations (D2) and (D4). On the other hand, when degenerate n-type ZnO and GaN films are assumed, we have [45]:

$$E_F(x) - E_C = \frac{\hbar^2}{2m^*_{C,ZnO}}\left[3\pi^2\left(n_{n,ZnO}(x,T) + n_{p,Si}(0,T)\right)\right]^{2/3} \tag{D5}$$

for n-ZnO/p-Si junction devices or:

$$E_F(x) - E_C = \frac{\hbar^2}{2m^*_{C,GaN}}\left[3\pi^2\left(n_{n,GaN}(x,T) + n_{p,Si}(0,T)\right)\right]^{2/3} \tag{D6}$$

for n-GaN/p-Si junction devices , where $m_{c,ZnO}{}^*$ is the electron effective mass of the conduction band of ZnO film [46] and $m_{c,GaN}{}^*$ is the electron effective mass of the conduction band of GaN film [47].

SUBJECT INDEX

A

Advanced 201, 217, 288, 291
  driver assistance systems (ADAS) 201, 217
  mobile phone system (AMPS) 288, 291
Anti-reflective coatings 114
Applications 4, 63, 65, 67, 68, 112, 113, 115,131, 134, 136, 275, 338, 354
  medical 275, 354
  quantum computing 4
  radiation-sensitive 275
  sensor 63, 65, 67, 68, 131, 134, 136
  sensor-network 338
  solar cell 112, 113, 115
Arsenic 46, 55, 56, 61
  atoms 55, 61
  atoms interlace 46
  trichloride 56
Atomic 9, 10, 116, 233, 235, 240, 241
  force microscopy (AFM) 9, 240, 241
  layer deposition (ALD) 10, 116, 233, 235
Audio 311, 312, 317, 326, 327, 330, 331, 333
  interfaces 330
  processing 311, 312, 317, 326, 327, 330, 333
  signals 317, 330
  technology 331

B

Ball milling, high-energy 45, 54
Bandgap energy 54, 57, 84, 89, 95, 96, 137
Binary logic processes 28
Binding energy 242
Biomedical 10, 11, 220
  nanotechnology 11
  sensing in semiconductor technology 220
  sensors 10
Biomolecular 266, 277
  binding 266
  detection 277

Biosensors 247, 263, 264, 265, 266, 267, 268, 269, 270, 274, 276, 277, 278
  traditional MOSFET-based 267
  wearable 247
Building-integrated photovoltaics (BIPV) 93

C

Carbon 5, 7, 228, 229
  -based nanomaterials 5, 228
  nanotube nano-electro-mechanical systems 7
  quantum dots (CQDs) 228, 229
Cascaded integrator 321
  -feedback (CIFB) 321
  -feedforward (CIFF) 321
Catastrophic breakdown 97
CFD simulations 17
Charge 114, 120, 122, 136, 218, 247
  -coupled Device (CCDs) 218, 247
  storage 120
  storage devices 122
  transfer kinetics 136
  transportation 114
Chemical 44, 75, 219
  sensing 219
  solution deposition 75
  stresses 44
Chemical vapor 6, 7, 40, 42, 45, 47, 50, 53, 56, 61, 65, 66, 67, 68, 71, 74, 233
  deposition (CVD) 6, 7, 40, 42, 45, 47, 50, 53, 65, 66, 67, 233
  transport (CVT) 56, 61, 68, 71, 74
Code division multiple access (CDMA) 289, 291
Coefficient of thermal expansion (CTE) 37
Communication 28, 199, 201, 246, 302, 303, 304, 305, 306, 327, 328
  networks 302, 304
  systems 28, 199, 201, 303, 305, 306, 328
  wireless 201, 246, 327, 328

Communication technologies 289, 290, 293, 304
  mobile 289
Complementary metal oxide semiconductor (CMOS) 30, 264
Computational fluid dynamics (CFD) 17
Computed tomography (CT) 331
Computer-aided design (CAD) 18
Conditions 248, 297
  chronic 248
  monitoring microclimatic 248
  soil 297
Conduction band energy 88
Conductivity 67, 69, 72, 73, 80, 120, 124, 125, 131, 134, 217, 219
  electronic 80
Configuration 120, 232
  electrode 232
  energy resource 120
Consumer 32, 94, 98, 100, 216, 217, 248, 251, 322, 330
  audio devices 330
  electronics 32, 94, 98, 100, 216, 217, 248, 251, 322
  electronics and internet 248
Contaminants, chemical 233
Contemporary electronic gadgets 90
Control, environmental monitoring and pollution 248
COVID-19 pandemic 292
CVD, thermal 233, 234
CVT synthesis 71, 74

D

Deposition 39, 45, 59, 66, 71, 94, 115, 128, 233, 236, 275
  chemical bath 71
  methods 66
  radiation's energy 275
  thin-film 45
  vapor 128
Deposition techniques 14, 42, 61, 63, 233, 235
  thin-film 42
Design multipurpose devices 303
Devices 7, 11, 19, 20, 28, 32, 39, 81, 93, 94, 110, 111, 120, 121, 125, 126, 138, 174, 183, 199, 204, 218, 231, 266, 267, 269, 271, 273, 275, 276, 277, 291, 297, 304, 305, 322, 329, 330, 338, 340

battery-powered 322, 338
biomedical 32, 218
biosensing 276
biosensor 266, 267, 271
communication 305
electrical 93
energy-efficient 19, 111, 330
flexible 20
microfluidic 231
mobile 329
photoelectrochemical 121
thermoelectric 110, 111, 125, 126, 138
wireless 28
Dielectric 55, 217
  constant, aluminum arsenide's 55
  layer, hygroscopic 217
Digital signal processing (DSP) 28, 316
Disease(s) 221, 248
  cardiovascular 248
  detection 221
DNA 12, 230, 231, 263, 267, 268, 269, 272, 273, 274, 281
  biomolecules 268, 269, 272, 273, 274
  detection 263, 267, 268, 281
  hybridization 230, 231
  nanorobots 12
  sequencing 12
Dynamic random-access memory (DRAM) 174, 184, 185, 186

E

Economical production methods 114
Effective number of bits (ENOB) 316
Electrical 33, 37, 42, 70, 73, 82, 87, 91, 110, 111, 115, 121, 125, 127, 128, 129, 130, 135, 139, 227, 228, 229, 230
  conductivity 33, 37, 42, 70, 73, 127, 128, 129, 130, 227, 228, 229, 230
  energy 82, 87, 91, 110, 111, 115, 121, 125, 130, 135, 139
Electrocardiogram 329
Electrocatalysis and energy conversion 139
Electrocatalysts 135
Electrochemical 43, 123, 124, 125, 135, 220
  double-layer capacitors (EDLCs) 123, 124
  methods 43
  processes 125
  reactions 135, 220
Electrodes 65, 67, 219

semiconductor-based 219
transparent 65, 67
Electrolytes 120
Electron 6, 8, 44, 55, 57, 88, 193, 198, 236,
    239, 240, 340, 341
  beam lithography (EBL) 6, 8, 236
  energy loss spectroscopy (EELS) 240
  mobility 44, 55, 57, 88, 193, 198, 340, 341
  tomography 239
Electronic(s) 7, 9, 17, 19, 20, 28, 30, 31, 32,
    44, 46, 60, 62, 67, 75, 80, 82, 85, 88, 89,
    90, 91, 99, 133, 193, 200, 201, 276, 306,
    338
  applications 46, 200
  automobile 28
  automotive 201
  contemporary 193
  designing TFET-based 276
  devices 7, 9, 17, 19, 20, 28, 30, 31, 32, 60,
    62, 80, 82, 90, 91
  industry 75, 85, 99, 338
  properties 44, 60, 67, 82, 88, 89
  wearable 20, 133, 306
Electronic systems 28, 95, 96, 121, 138, 215,
    311, 333
  contemporary 311
Emerging photovoltaic technology 116
Energy 21, 23, 121, 124, 133, 239
  -dispersive X-ray Spectroscopy 239
  storage 21, 23, 121, 124, 133
Energy storage 5, 13, 21, 110, 112, 120, 121,
    122, 125, 139
  capabilities 110, 139
  devices 5, 13, 21, 112, 121, 122, 125
  technologies 120, 121
Energy systems 28, 32, 120, 122, 193, 194,
    201
  renewable 28, 32, 122, 193, 194, 201
  sustainable renewable 120
Environment, ultra-high vacuum 61
Environmental 9, 135, 225, 250, 329
  conditions 135, 225, 250, 329
  stressors 9, 250
Evaporation, thermal 66, 233, 235

F

Fabricating pressure sensors 42
Fabrication 2, 7, 23, 94, 116, 128, 202, 213,
    238, 304

methods 7, 23, 128, 202
processes 2, 94, 116, 213, 238, 304
Fabrication techniques 2, 4, 6, 10, 13, 22, 213,
    220, 231, 232, 246, 250, 251
  innovative 22, 250
FDMA technology 288
Field-effect transistors (FETs) 6, 39, 65, 173,
    174, 177, 186, 206, 231, 274, 275
Finite 17, 327
  element analysis (FEA) 17
  impulse response (FIR) 327
Fluorescence-based tests 268
Flux growth techniques 64
Fuel cells 121, 135
  solar 121
Function, electron density distribution 346

G

Gallium nitride vaporizes 58
Gas(s) 50, 98, 137, 215, 216, 217
  concentration 215, 216, 217
  electron 98
  nitrogen-containing 50
  oxygen 137
Gas sensors 131, 214, 226, 248
  catalytic 226

H

Heat energy 73, 125
HEMT's threshold voltage 199
High 46, 65, 127, 306
  -frequency electronic devices 65, 306
  -power electronic devices 46
  -pressure techniques 127
High-resolution 236, 239, 241, 247
  images 239, 241
  imaging techniques 247
  interface patterning methods 236
High-speed 62, 85, 305
  devices for telecommunication 85
  electronic devices 62, 305

I

Imaging techniques 12, 238
Infinite impulse response (IIR) 327
Integral nonlinearity (INL) 314

Integrated circuits, energy-efficient Photonic 20
Integration techniques 247
Ionization process 276

J

Junctionless tunnel FET (JLTFET) 275

K

Key performance indicators (KPIs) 286, 287, 288

L

Landscape, sustainable energy 130
Large bandgap, relatively 48
LEDs and photovoltaic cells 59, 69
Light 42, 134, 215
    activation techniques 134
    sensors 42, 215
Lighting 95, 297
    automotive 95
    intelligent street 297
Localized surface plasmon resonance (LSPR) 226, 237

M

Machine-type communications (MTC) 287
Magnetic resonance imaging (MRIs) 28, 331
Maldonado technique 163
Memory technology 299
Memristive devices 11
MEMS 32, 94, 338
    applications 338
    technology 32, 94
Metal 47, 50, 53, 55, 56, 58, 60, 63, 68, 73, 133, 173, 174, 214, 219
    -organic chemical vapour deposition (MOCVD) 47, 50, 53, 55, 56, 58, 60, 63, 68, 73
    oxide semiconductor field-effect transistors (MOSFETs) 173, 174
    oxide semiconductors (MOS) 133, 214, 219
Metalorganic chemical vapour deposition 55
Micro electro mechanical system 338
Microelectromechanical systems 32, 93, 246

Microfluidic systems 267
Microwave 98, 201
    applications 98
Mobile communications 289, 290, 293, 294, 295
    wireless 289
    networks 290, 294
Molecular beam epitaxy (MBE) 6, 7, 47, 50, 55, 61, 63, 68, 73, 74, 233
Monitoring devices 13
Moore's law 30

N

Nanoelectromechanical systems (NEMS) 5, 12
Nanofluidic 12, 17
    devices 12, 17
    fabrication techniques 12
Nanoscale semiconductor 233, 235
    devices 235
    sensing devices 233
Nanoscale 6, 20, 231, 232, 233, 234, 238, 241, 242, 243, 244, 245, 246, 251, 252
    semiconductor sensors 231, 232, 238, 242, 243, 244, 245, 246, 251, 252
    sensor devices 243
    sensor fabrication 234, 241
    thin-film deposition 233
    transistors 6
    waveguides and optical resonators 20
Nanostructured electrode materials 21
Nature, electron wave 4
Noise 267, 319, 321, 325, 326, 332
    energy 319
    shaping techniques 326, 332
    thermal 267, 325
    transfer function 321

O

Omega method 169
Organic 112, 114, 115, 116
    photovoltaics (OPVs) 112, 115
    solar cells (OSCs) 112, 114, 115, 116
Oxygen evolution reaction (OER) 110, 135, 139

P

Photoelectrochemical hydrogen evolution
  reaction (PECHER) 137
Photoelectrolysis processes 137
Photolithography 6, 8, 70, 94, 235, 236, 251
  traditional 236
Photomask fabrication techniques 236
Photonics 1, 3, 10, 19, 20, 22, 34, 39, 218,
  230, 300
  devices 230
  integrated 300
  resonance 218
Photovoltaic 91, 93, 110, 116, 119, 218
  effect 91, 93, 218
  solar cell 91, 218
  technologies 110, 116, 119
Photovoltaic systems 31, 93
  silicon-based 93
Physical vapor 40, 43, 45, 47, 50, 59, 66, 233
  deposition (PVD) 40, 43, 45, 47, 50, 66,
    233
  transport (PVT) 59
Plasma-enhanced CVD (PECVD) 233, 235
Pollutants, environmental 225, 229
Polymerase chain reaction (PCR) 268
Power 116, 117, 118, 319, 320
  conversion efficiency (PCE) 116, 117, 118
  spectral density (PSD) 319, 320
Production 8, 9, 14, 40, 42, 84, 95, 115
  industrial 40
  solar energy 115
Pulsed laser deposition (PLD) 40, 66, 233
PVD techniques 40, 43

Q

Quantum 11, 300, 303, 306
  dot LEDs (QLEDs) 306
  key distribution (QKD) 300, 303
  sensors 11

R

Radio frequency (RF) 28, 97, 201, 305, 327
Reaction, carbothermal reduction 50
Reliability degradation pathways 14

S

Scanning 9, 236, 238
  electron microscopy 238
  probe lithography (SPL) 236
  transmission electron microscopy (STEM)
    9
Semiconductor 4, 30, 42, 69, 72, 83, 85, 86,
  90, 215, 216, 218, 219, 220, 221, 222,
  223, 225, 238, 240, 244, 246, 247, 248,
  249, 250, 251
  device applications 42
  quantum dots 4, 220
  sensing technologies 216, 246
  sensors 215, 216, 221, 222, 223, 225, 238,
    240, 244, 246, 247, 248, 249, 250, 251
  technology 30, 69, 72, 83, 85, 86, 90, 218,
    219, 220, 221
Semiconductor properties 41, 51, 90, 91, 92
  exceptional 91
  remarkable 41
Sensing 19, 213, 214, 215, 216, 225, 233, 238,
  240, 241, 242, 244, 246, 247, 249, 251,
  252, 263
  devices 213, 233, 238, 240, 241, 242, 244,
    252
  mechanisms 249, 263
  technologies 19, 213, 214, 215, 216, 225,
    246, 247, 251, 252
Sensor 135, 300
  fabrication 135
  fusion 300
  fusion chips 300
Sensor devices 243, 246, 248
  multifunctional 246
Signal 81, 225
  processing techniques 225
  transmission 81
Software-defined Radio (SDR) 328
Solar cells, silicon-based 31, 113
Solid-state reaction methods 45, 69
Source 112, 135, 136
  sustainable energy 112, 136
  traditional energy 135
Surface 14, 226, 227
  modification techniques 14, 227
  plasmon resonance (SPR) 226
Surgery, remote 297
Sustainable energy ecosystems 138

Systems 7, 9, 32, 91, 126, 131, 133, 218, 248, 250, 298, 330
  automotive 91, 218
  cooling 9, 126
  digital audio processing 330
  digital audio recording 330
  humidity control 250
  immunological 133
  industrial monitoring 131
  infrastructure management 248
  nanoscale electromechanical 7
  scalable quantum computing 32
  traffic control 298

T

Techniques, photolithography 236
Technological advancements 10, 21, 23, 99, 101, 102
Technology 3, 28, 100, 101, 112, 113, 126, 130, 135, 138, 225, 264, 265, 281, 287, 290, 292, 297, 301, 338
  biosensing 264, 265, 281
  contemporary 28
  flexible electronics 338
  mobile 287, 290, 292
  renewable energy 3, 135
  semiconductor fabrication 225
  silicon-based 100, 101
  silicon solar cell 113
  sustainable 112, 138
  thermoelectric 130
  transistor 301
  wearable 126, 130, 297
Telecommunications, mobile 288
Thermal 37, 44, 45, 65, 67, 70, 73, 113, 125, 126, 162, 168, 193, 208, 246
  conductance technique 162
  decomposition 45
  energy, wasted 126
  measurement methods 168
  stability 44, 65, 67, 70, 73, 113, 125, 193, 208
  stress 37, 246
Thermoelectric applications 72
Thin-film 93, 238
  deposition processes 93
  morphology 238
Traditional lithographic 6, 8
  methods 8

techniques 6
Transistors 42, 82, 84, 91
  bipolar 84
  crafting 91
  microwave 82
  thin-film 42
Transition 5, 10, 82, 86, 87, 88, 90, 97, 102, 120, 137, 138, 173, 213, 229, 305
  electronic 86, 87, 88
  metal dichalcogenides 5, 10, 82, 102, 137, 229, 305
  sustainable 120
Transmission electron microscopy 239
Transparent conductive oxide (TCO) 65

U

Ultra-reliable low latency communication (URLLC) 294

V

Virtual reality (VR) 287, 294, 298
Voltage, electric 125

X

X-ray photoelectron spectroscopy (XPS) 241, 242